# Musculoskeletal Injections Manual

Baris Kocaoglu · Lior Laver
Laura de Girolamo
Riccardo Compagnoni

*Editors*

# Musculoskeletal Injections Manual

## Basics, Techniques and Injectable Agents

*Editors*
Baris Kocaoglu
Acibadem Mehmet Ali
Aydinlar University
Faculty of Medicine
Department of Orthopedics
and Traumatology
Istanbul, Türkiye

Laura de Girolamo
Orthopaedic Biotechnology Laboratory
Ospedale Galeazzi Sant'Ambrogio
Milano, Italy

Lior Laver
Hilll Yaffe Medical Center (HYMC)
Technion University Hospital
Hadera, Israel

Riccardo Compagnoni
Clinica Ortopedica, 1°
Istituto Ortopedico Gaetano Pini
Milano, Italy

ISBN 978-3-031-52605-3        ISBN 978-3-031-52603-9   (eBook)
https://doi.org/10.1007/978-3-031-52603-9

# Preface

In recent years, the field of injection therapy for the treatment of pathologies in joints has witnessed remarkable advancements, particularly in combination with orthobiologics. These innovations have transformed the way we diagnose and treat various musculoskeletal conditions, providing patients with new hope and improved outcomes. It is with great pleasure that ESSKA U45 Committee and ORBIT introduce this comprehensive book, which serves as an essential guide to these front-line techniques and therapies specially for young orthopedics and sports medicine specialists.

*Musculoskeletal Injections Manual: Basics, Techniques and Injectable Agents* is a testament to the collective expertise and dedication of the appreciated authors who have contributed their knowledge and experiences to this remarkable resource. As a multidisciplinary field, injection therapies rely on the collaboration of specialists from various fields, including orthopedics, sports medicine, and basic science. This book brings together the insights of these experts, providing a comprehensive and confident reference for especially young surgeons.

The book encompasses a wide range of topics, covering the fundamental principles and techniques of musculoskeletal injections, orthobiologics, and their usage. From basic concepts to advanced procedures, the surgeons will find detailed discussions on joint injections and regenerative therapies. The integration of orthobiologics, such as platelet-rich plasma (PRP), cell-based, and blood-derived products, adds another dimension to the field, offering innovative approaches to tissue healing and regeneration.

What sets this book apart is its emphasis on evidence-based practice. Each chapter combines the authors' clinical experience with the latest scientific research, ensuring that readers have access to the most up-to-date information and treatment recommendations. Additionally, the book includes practical tips, step-by-step procedural guidelines, and illustrative figures, enabling clinicians to apply these techniques with confidence and exactness.

Whether you are young surgeon who is seeking to improve your skills or a novice exploring the world of musculoskeletal medicine, this book will prove to be an invaluable resource. It serves as a comprehensive reference for sports medicine doctors, surgeons, physical therapists, and other healthcare professionals involved in the management of musculoskeletal conditions. By incorporating the latest advancements and emerging therapies, this book enables clinicians to provide optimal care to their patients, improving their quality of life and functional outcomes.

As ESSKA U45 Committee and ORBIT, we would like to express our deepest appreciation to the authors for their dedication to advancing the field of musculoskeletal medicine and for sharing their expertise through this remarkable publication. We are confident that *Musculoskeletal Injections Manual: Basics, Techniques and Injectable Agents* will become an essential publication for young surgeons who are motivated to deliver the best possible care in the scene of sports traumatology.

Istanbul, Turkey<br>Hadera, Israel<br>Milano, Italy<br>Milano, Italy
Baris Kocaoglu<br>Lior Laver<br>Laura de Girolamo<br>Riccardo Compagnoni

# Contents

### Part III   Ortho-biologic Agents for Injections



### Part IV   Injections of Anatomical Regions and Diseases

# Part I

# Basics of Musculoskeletal Injections

# Philosophy of Musculoskeletal Injections

Behiç Çelik and Gökhan Meriç

## 1.1 Introduction

While providing orthopedic healthcare services, various invasive interventions are applied to patients according to their clinical status. Injection, which is an invasive procedure, is one of the most commonly used medical procedures in the world. Many other invasive procedures also require injections. In addition, injections are also applied to healthy individuals for the purpose of maintaining health and protection apart from its therapeutic purpose. According to the 2002 data of the World Health Organization (WHO), it is estimated that 16 billion injections are made annually in developing countries, 95% of which are therapeutic in purpose [1]. Providing health care without injections through needles is merely possible. And here is how to paint this as a theme; the needle represents the ability to heal through pain. As Brokensha portraits in 1999, the needle, like the hollow teeth of the serpent that wraps around Aesculapius' staff, penetrates and reinforces doctors' authority [2].

The philosophy of musculoskeletal injections in medicine is based on the idea of delivering therapeutic agents, such as medications or local anesthetics, directly into the affected area of the musculoskeletal system to relieve pain, inflammation, and other symptoms. Musculoskeletal injections can be used to treat a variety of conditions, including arthritis, tendonitis, bursitis, and other inflammatory or degenerative diseases like osteoarthritis. The common use of injections, their use for preventive, therapeutic, or recreational purposes, and their varied routes of administration make one forget that the person credited with inventing the method over a century years ago was merely looking to treat the agony of neuralgia. The goal is to provide targeted relief to the affected area while minimizing systemic side effects that can occur with oral medications.

Modern medicine followed the footsteps of nature, just as in any other field which has technological aspects that have to stay up to date. Sharp objects which were necessary for hunting and piercing through thing were always necessary in the ancient times. Snake bites and arrows loaded with poison were fine examples. Even though many references were present in nature, it still took several centuries for the syringe and the needle, which are the current must-have devices for modern medicine, to be developed. Before any medical subdivision even occurred, invasive procedures such as dissections, injections, etc. were all part of the general medical practice in countries that were forerunners of science and research. It is no coincidence that these developments took place in Europe, where medicine took its modern turn via experimentations arising from its ancient roots.

B. Çelik · G. Meriç (✉)
Department of Orthopedics and Traumatology,
Yeditepe University Medical Faculty,
Istanbul, Turkey

## 1.2 Philosophy of Musculoskeletal Injections

Physicians have experimented with a multitude of strategies to get medications into the body over the decades. Rubbing of ointments and oils onto the skin, one of the earliest procedures, was described in the Old Testament, Homer's Odyssey, and by Aristotle and was employed by practically all ancient physicians. However, severe adverse effects and questionable results prompted the search for alternate treatments. According to our knowledge, the earliest intravenous injection trials were conducted in 1642 in eastern Germany and in 1656 in England by Christopher Wren [3].

The philosophy of injections, just as any other invention that ever took place in the history of mankind, got his origin idea from nature. In 1628, William Harvey presented fresh knowledge about human blood circulation, which spurred these studies. According to our knowledge, the earliest intravenous injection trials were conducted in 1642 in eastern Germany and in 1656 in England by Christopher Wren. Subcutaneous injection was also one of the new methods; predecessors of this application either used a vaccination-lancet to introduce the medication beneath the epidermis or they'd remove the epidermis before introducing the medication to the stripped epidermal layer. Lafargue, Lembert, and Lesieur reported these procedures starting early in the 1800s, and they remained in use until the discovery of subcutaneous injection in the second half of the century.

The syringe for subcutaneous injection was invented by Alexander Wood of Edinburgh and Charles-Gabriel Pravaz of Lyon. In Lyon in 1853, the French surgeon C. Pravaz devised a miniature syringe, the piston of which could be moved by a screw, allowing precise dosage. Using a sharp needle with a trocar, he eliminated the need for a dissection. Pravaz used his syringe to obliterate artery aneurysms with ferric chloride injections. His device sparked the development of calibrated syringes made of glass or metal coupled with glass [4]. The gradual steps introduced by Wood, Pravaz, and Luer resulted in increased utilization, safety, and accuracy. As a result, the Luer syringe was designed for aseptic heating and a sharp needle that could easily pierce the skin. Pasteur, Chamberland, and Koch created autoclave sterilization after regulating aseptic conditions. Another invention, Limousin's ampoule, introduced in 1886, provided a secure technique for storing sterile injectates, followed by the advent of multi-dose containers. Nowadays, with the emergence of transdermal drug delivery via micron-scale needles and monitored drug delivery, the evolution of the syringe and its needle goes on [5].

Considering all the historical processes and developments that took place, just as there is a need for improvement, management and keeping things up-to-date in any practice in medical science, the practice of injection depends on the purpose, the right indication, the dose, the material and sterility, the right patient selection, and the right agent. It is a clear fact that research and inventions under appropriate conditions must be done, keeping things up to date, and while doing that, all the following must be taken into account: the right place and timing, the right method and circumstance of application, the right follow-up, ideal doctor and patient communication, the patients' expectation, and all that might affect the outcome in terms of improvement of the patients' condition and well-being. An understanding of the determinants of current injection practices in the sociocultural-economic context is necessary in order to plan relevant and effective interventions.

From a doctor's perspective, we use injections with various application methods in cases that do not respond to conservative treatment, with the logic of transitioning to a higher-level treatment. At the same time, various injection techniques can be used in advanced cases where surgery is not considered, in patients who do not want surgery but need an outpatient invasive procedure that does not require surgery. From a patient's perspective, there are several reasons why injections may be necessary. Injections are often required when oral medications or alternative treatments are not as effective or suitable for a particular condition. Injections allow medications to be delivered directly into the body, ensuring quicker and more potent effects.

Musculoskeletal injections are an important diagnostic and therapeutic technique for orthopedist. Some medical conditions require ongoing treatment and management. While regular injections may be inconvenient, they can significantly improve the patient's quality of life and overall health outcomes (Fig. 1.1). Injections are sometimes used as part of diagnostic procedures to help identify or evaluate certain medical conditions. It can help with pain management and overall function [6]. The choice of injection technique and medication depends on the specific condition being treated, as well as the patient's individual needs and preferences. Overall, the philosophy of musculoskeletal injections is to provide safe, effective, and minimally invasive treatments for musculoskeletal conditions, with the goal of improving patients' quality of life and functional outcomes while minimizing the use of more invasive interventions, such as surgery.

Injections play a crucial role in sports medicine as they can provide targeted and rapid relief

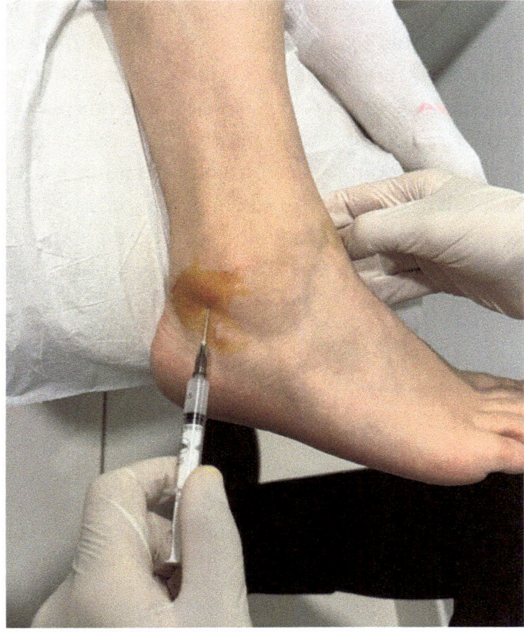

**Fig. 1.1** Injection for peroneal tendon synovitis

for pain and inflammation, facilitate healing, and enhance performance. Here are some ways in which injections are important in sports medicine. Injections can be used to provide rapid pain relief to athletes who have sustained injuries or are experiencing chronic pain. Platelet-rich plasma (PRP) injections, for instance, contain growth factors that can stimulate the repair and regeneration of damaged tissues. Injections of hyaluronic acid can also help improve the health of damaged joint cartilage by lubricating and cushioning the joint [7]. In some cases, injections can be used to enhance athletic performance. For example, erythropoietin (EPO) injections can increase the production of red blood cells, which can improve an athlete's endurance by delivering more oxygen to the muscles [8].

Musculoskeletal pain is a prevalent problem that many primary care physicians, orthopedic surgeons, and pain specialists identify and try to treat on a daily basis. Treating pain with a multimodal strategy is critical to provide patients with safe and effective results. In addition to this common ground of application, each specialty has its own particular aspects. When we take a look at the top of the specialties which use musculoskeletal injections in their practice routine, we see orthopedics, physical therapy and rehabilitation, algology, and anesthesia. It is possible to extend the list depending on the application of some regimes of therapy. Physicians should comprehend the target anatomy and injection indications. Injections can be used to both identify the source of discomfort and reduce pain to allow for a more thorough evaluation. Aspiration may be used in injections to aid in diagnostics, provide pain relief, and restore joint motion. It is critical to explain activity restrictions, postinjection pain expectations, and the planned treatment course, and as always, just before any other medical intervention, informed consent should always be acquired prior to the procedure [9, 10]. Injections could be performed blind or through guiding. When compared to landmark guiding, ultrasound guidance has been shown in numerous trials to be more accurate and effective [6].

Musculoskeletal injections also differ according to the specific agent that is used and its specific mechanism of action. Anesthetic drugs are mostly used in injections to reduce pain and help to diagnose the medical condition [10]. Corticosteroids are effective immune suppressants and pain relievers. Intra-articular corticosteroid injections might reduce pain temporarily, especially when used to treat osteoarthritis [11]. Hyaluronic acid (HA) is a naturally occurring polysaccharide chain that the synovium secretes into the joint area. In the mid- to long-term treatment of knee osteoarthritis, higher molecular weight formulations seem to be more helpful and efficient [12, 13]. Trigger point injections are a treatment option for myofascial trigger points, particularly in symptomatic individuals. Chronic or episodic headaches, temporomandibular joint pain, back pain, restricted range of motion due to trigger points, and groin pain are common concurrent symptoms. Trigger point injections can produce meaningful results and should be considered as a therapy option in the right circumstances [14]. Dry needling is a procedure that uses a small needle which pierces the epidermis, subcutaneous tissues, and muscle in order to mechanically disrupt tissue without the use of an anesthetic. The physiological mechanism underlying the effects of dry needling is still unknown. Dry needling, on the other hand, has been hypothesized to elicit both local and central neural responses to restore homeostasis at the site, resulting in a reduction in both peripheral and central sensitivity to pain [15]. Prolotherapy, also known as regenerative injection therapy, is a medical procedure used to treat chronic musculoskeletal pain and promote tissue healing. It involves the injection of a substance, typically a proliferant solution, into damaged or weakened ligaments, tendons, or joints to stimulate the body's natural healing response [16].

It is important to note that the philosophy of injections is a broad topic that encompasses various ethical, practical, and metaphysical dimensions. Philosophers, medical professionals, and ethicists engage in ongoing discussions to deepen our understanding of these issues and inform ethical decision-making in medical practice. From a philosophical standpoint, some before mentioned key aspects of this topic are the thera-

peutic purpose in which the underlying principle is the promotion of health and the alleviation of suffering; secondly, the autonomy and informed consent, which ensures that individuals understand the potential benefits, risks, and alternatives before agreeing to undergo an injection; and thirdly, and maybe most importantly, beneficence and non-maleficence which emphasizes the duty to promote benefit and avoid harm at all costs.

## 1.3 Conclusion

In conclusion, musculoskeletal injections can provide a safe and effective treatment option for a variety of conditions affecting the bones, joints, and muscles. Whether it be corticosteroids for inflammation, hyaluronic acid for joint lubrication, or platelet-rich plasma for tissue regeneration, there are a variety of injection therapies that can help alleviate pain and improve mobility. However, it is important to note that these injections should only be administered by qualified healthcare professionals and that patients should be thoroughly evaluated beforehand to ensure that they are suitable candidates for the procedure. Overall, with careful consideration and appropriate use, musculoskeletal injections can be a valuable tool in the management of musculoskeletal conditions. Overall, our chapter emphasizes the critical role that injections play in modern medicine and underscores the importance of ongoing research and education in this area. By continually striving for safer, more effective injection techniques, we can improve patient outcomes and advance the field of health care as a whole.

## References

1. WHO. Injection safety. World Health Organization revised April 2002. DSÖ Fact Sheet No:231. http://www.who.int/injection_safety/toolbox/en/InjectionFactSheet2002.pdf
2. Brokensha G. The hollow needle: inappropriate injection in practice. Aust Prescr. 1999;22:145–7.
3. Kotwal A. Innovation, diffusion and safety of a medical technology: a review of the literature on injection

practices. Soc Sci Med. 2005;60(5):1133–47. https://doi.org/10.1016/j.socscimed.2004.06.044.

4. Die FH, der Injektionen G. Bilder aus der Geschichte der Hals-Nasen-Ohren-Heilkunde, dargestellt an Instrumenten aus der Sammlung im Deutschen Medizinhistorischen Museum in Ingolstadt [History of injections. Pictures from the history of otorhinolaryngology highlighted by exhibits of the German History of Medicine Museum in Ingolstadt]. Laryngorhinootologie. 2000;79(4):239–46. German. https://doi.org/10.1055/s-2000-8797.

5. Norn S, Kruse PR, Kruse E. Traek af injektionens historie [on the history of injection]. Dan Medicinhist Arbog. 2006;34:104–13. Danish

6. McNabb JW. A practical guide to joint and soft tissue injections. 3rd ed. Philadelphia (PA): Wolters Kluwer Health; 2014. p. 383.

7. Everts P, Onishi K, Jayaram P, Lana JF, Mautner K. Platelet-rich plasma: new performance understandings and therapeutic considerations in 2020. Int J Mol Sci. 2020;21(20):7794. https://doi.org/10.3390/ijms21207794. PMID: 33096812; PMCID: PMC7589810

8. Scott J, Phillips GC. Erythropoietin in sports: a new look at an old problem. Curr Sports Med Rep. 2005;4(4):224–6. https://doi.org/10.1007/s11932-005-0040-6.

9. Rai K, Sylvester J. Musculoskeletal injections. Curr Sports Med Rep. 2020;19(6):191–3. https://doi.org/10.1249/JSR.0000000000000715.

10. O'Connor F, Lutrzykowksi CJ, Barkdull T. ACSM's sports medicine: a comprehensive review. Philadelphia (PA): Lippincott, Williams, and Wilkins; 2012. p. 859.

11. Stone S, Malanga GA, Capella T. Corticosteroids: review of the history, the effectiveness, and adverse effects in the treatment of joint pain. Pain Physician. 2021;24(S1):S233–46.

12. Trojian TH, Concoff AL, Joy SM, et al. AMSSM scientific statement concerning viscosupplementation injections for knee osteoarthritis: importance for individual patient outcomes. Br J Sports Med. 2016;50:84–92.

13. Weiss BD, Smith MA, Rew KT, et al. Musculoskeletal therapies. FP Essent. 2018;470:21–6.

14. Hammi C, Schroeder JD, Yeung B. Trigger point injection. 2022 Nov 25. In: StatPearls [internet]. Treasure Island (FL): StatPearls Publishing; 2023.

15. Gattie E, Cleland JA, Snodgrass S. The effectiveness of trigger point dry needling for musculoskeletal conditions by physical therapists: a systematic review and meta-analysis. J Orthop Sports Phys Ther. 2017;47(3):133–49. https://doi.org/10.2519/jospt.2017.7095. Epub 2017 Feb 3

16. Bae G, Kim S, Lee S, Lee WY, Lim Y. Prolotherapy for the patients with chronic musculoskeletal pain: systematic review and meta-analysis. Anesth Pain Med (Seoul). 2021;16(1):81–95. https://doi.org/10.17085/apm.20078. Epub 2020 Dec 16. PMID: 33348947; PMCID: PMC7861898

# The Evidence-Based Medicine for Injection Therapy



Marko Ostojić

## 2.1 Introduction

As a therapeutic modality, injections have been widely used in treating musculoskeletal disorders, usually in an outpatient clinic setting. This treatment modality is defined as a method whereby therapeutical agents are delivered to the selected location with the use of a syringe and hypodermic needle. It is common practice that injection therapy, in terms of treatment sequence, finds a place between the conservative first-line treatment (physical therapy, lifestyle modification, pain management, etc.) and the surgical intervention. The type and timing of injections depend on many factors: from the pathology that is being treated and the patient's functional demands to the *modus operandi* of the attending physician. Regarding the latter, some physicians are prone to deliver conservative treatment, while others favor the operative approach. Some physicians are quick to adopt new emerging techniques, yet others are skeptical and require hard clinical evidence to be convinced. A plethora of different factors affect the type of treatment that patients will receive. During medical training, doctors gain a lot of anecdotal knowledge in this field, mostly concerning corticosteroids and their beneficial effects. Approaches to injection therapy as a field of treatment have been known for a lack of standardization. A more evidence-based approach to this field is necessary to oppose some well-established dogmas with the final goal of delivering the best available care to patients. For future studies, it is of outmost importance to define the outcome measures leading to clinical improvement, with a special focus on "responders" and "nonresponders." The remission of symptoms and the deceleration of disease progression are not the sole objectives of an optimal injectable. A scientific research perspective would pronounce the importance of developing an injectable agent with disease-modifying properties. In other words, an agent that reverses disease progression, by breaking the vicious circle of tissue deterioration and leads to tissue healing [1, 2]. Since orthobiologic agents show a pronounced healing potential, it is not a surprise that some people see it as the holy grail of nonsurgical treatment.

Let us briefly go through the available options in an evidence-based and critical manner. First, it is of paramount importance to make a clear distinction between **intra-articular** and **extra-articular** injections.

M. Ostojić (✉)
Department of Orthopaedics and Traumatology,
University Hospital Mostar,
Mostar, Bosnia and Herzegovina

## 2.2 Intra-Articular Injections

Intra-articular injections are used for the treatment of osteoarthritis and other degenerative diseases on one hand and for fresh injuries where we strive for faster tissue healing and pain management on the other. A joint is an enclosed organ possessing its equilibrium. The synovial membrane of a joint acts as a barrier that holds in any injected substance for a defined period. Any kind of arthritis or fresh injury affecting a part of the joint causes an inflammatory reaction that leads to a vicious circle that leads to further tissue damage [3, 4]. This induces the release of inflammatory factors in the synovial fluid that promote further synovial inflammation [5, 6]. There is a (probably oversimplified) rule that any intra-articular injection that is given in a conventional medicine setting leads to the improvement of clinical symptoms in osteoarthritic joints. For example, in hip osteoarthritis, the injection of saline into an osteoarthritic hip joint resulted in symptomatic relief that was comparable to some of the commercially most often used treatments [7]. Many studies evaluating the treatment of osteoarthritis use saline as a placebo, which is a methodologically questionable procedure. Saline, the isotonic solution of sodium chloride and water, has the effect of diluting the synovial fluid and "washing out" inflammatory factors. It does not have a long-term effect, but it does bring on the temporary improvement in clinical function and pain [8]. A true placebo for this kind of trials should be a "sham procedure" in which the joint capsule is not penetrated and where the injectable placebo agent is given subcutaneously [9]. Corticosteroids are naturally produced in the cortex of adrenal glands. Synthetic variants have been used for over 70 years, and they still are one of the most widely used anti-inflammatory medicines. In orthopedics, long-releasing (depo) products are routinely used in the form of injections, both intra- and extra-articularly. In combination with local analgesics, they are a potent injectable agent. This combination has a quick effect and brings on short-term pain relief. This translates to patients walking out of their physician's office satisfied. Studies have indeed shown that their short-term effect is beneficial concerning pain relief, but, unfortunately, their long-term positive effect has not been observed [10, 11]. As mentioned before, corticosteroids are often given with local anesthetics, which are proven to be chondrotoxic, causing the apoptosis of the chondrocytes [12]. Corticosteroids also have a time- and dose-dependent negative effect on hyaline cartilage, being beneficial at low doses and detrimental at high doses [13]. In combination with local analgesics, the chondrotoxicity of corticosteroids is more pronounced due to the synergistic effect these two injectables possess [14, 15]. Hyaluronic acid, a large molecule that binds water and that is naturally produced by the synovial membrane, has anti-inflammatory and shock absorption effects and acts as a lubricant during joint motion, with possible disease-modifying effects [16]. Theoretically, it has all the necessary qualities to be a highly successful intra-articular injectable agent. Indeed, meta-analyses have shown the superiority of hyaluronic acid as compared to corticosteroids. However, HA falls behind when compared to orthobiologic agents [17, 18]. Orthobiologic therapies that have gained popularity in the last two decades, including platelet-rich plasma (PRP) and cellular-based products (*e.g.*, stem cells), show the most promising results in the middle- and the long-term results in the treatment of early osteoarthritis. The main difference to the previously mentioned agents is that, due to their biological origin, they are autotransplants that possess more disease-modifying effects. Meta-analyses have demonstrated the superiority of orthobiologic agents to other injectables. However, some controversies exist due to the diversity of the products and a lack of high-quality randomized control trials [17]. PRP is the most widely used blood-derived product, and therefore, details on other blood-derived products are not going to be discussed (e.g., alpha-2 macroglobulin, autologous conditioned serum). Let's try to clarify how PRP works practically. After most injuries, bleeding occurs, and blood is the first source of healing agents. With PRP, the fraction of venous blood from which we expect the strongest healing potential (serum with growth factors and platelets) is extracted and applied to the desired site. Platelets comprise several growth factors, which play crucial roles in tissue repair and regeneration mechanisms [2]. To put it in a nutshell, a fresh injury is mimicked to start the healing cascade. The literature supporting the use of PRP is growing, with meta-analyses and ESSKA "ORBIT Delphi consensus" supporting its use for knee osteoarthritis [18]. The injectable agents, which the orthopedic community is putting most hopes in recently, are cellular-based products, which show promising long-term effects. Mesenchymal stem cells (MSCs) are multipotent cells that are usually derived from fat tissue or bone marrow and can be used as autotransplant or homotransplant. They can be given immediately after the explantation process, in a minimal manipulation manner, or they can also be expanded in vitro for application at a later date. New insights show that most of their therapeutic potency is due to their paracrine effect. Hence it is suggested that the MSC abbreviation should stand for medicinal signaling cells [19]. In vivo studies have shown that there is a scarce chance of these cells surviving for a longer period of time after intra-articular injection and, afterward, transforming to the desired missing chondrocyte cells in the diseased joint. More hope is put in their exosome, which is expected to have a healing potential, through the previously mentioned paracrine effect. Due to the heterogeneity of MSCs products, the results, although promising, have some confronting conclusions, with high-quality randomized controlled studies still needed to support long-lasting effects of this therapeutic agent [20].

## 2.3 Extra-Articular Injections

Extra-articular injections are most commonly used in inflammatory conditions, degenerative conditions (e.g., tendinopathies), and acute injuries. Some of the chronic overuse injuries have characteristics of chronic inflammation and degenerative conditions, so they are treated accordingly. In the case of inflammatory conditions, it is crucial to seek benefit of anti-inflammatory potential of the therapeutic agent

that has the least adverse effects on the surrounding tissues. It is known that corticosteroids can cause weakening of the tendon tissue. For example, using corticosteroids is effective in the treatment of tendovaginitis, but potentially negative side effects can occur, iatrogenic tendon rupture being of the most disastrous one [21]. Hyaluronic acid can be used as an extra-articular injection but preferably delivered inside the space that is encircled by a tendon sheath. This space partially resembles the intra-articular *milieu* and has hyaluronic acid produced naturally in small quantities. Again, orthobiologic agents, particularly PRP, show the most promising results with their anti-inflammatory properties that oppose extra-articular inflammation. For acute injuries, like muscle and tendon rupture, corticosteroid and hyaluronic acid use does not have a biological rationale, and PRP is advised to be first option as injection therapy [22]. The other therapy modality that is commonly used for extra-articular tissue disorders is prolotherapy. It is a modality that uses irritative agent (like hyperosmotic dextrose or saline) causing a local inflammatory response that leads to fibrosis and potential healing of the diseased tissue. It is mostly used in tendinopathies, where it shows successful results [23]. The use of botulinum toxin is common in neuromuscular disorders, where spastic muscles are relaxed by intramuscular application. Botulinum toxin blocks the neuromuscular junction thus paralyzing the muscle. It has shown clinical improvement in patients with the spastic type of cerebral palsy.

The most important factor in getting a successful outcome in this type of treatment is careful selection of the adequate injectable agents for a certain patient. All of the beforementioned have their advantages and disadvantages. It is therefore crucial to choose an agent that brings on an improvement and minimizes iatrogenic harm. The *primum nil nocere* rule is of headmost importance in this field, where paramedical factors often influence the physician's decision-making where the desire for a lucrative practice can cloud her/his judgment. Also, we have to outline that placebo effect could be encountered, but we cannot, in any instance, rely on it.

In the hands of a trained professional, musculoskeletal injections have a low percentage of complications [24]. The learning curve is not trivial, but it is far quicker than that of surgeries. Ultrasound has been used to improve precision, even in the case of joints that are easily approachable and positioned close to the skin [25]. For deeper joints, like the hip joint and the smaller joints of the spine, anatomical landmark orientation with fluoroscopic control is advised [26, 27]. It is important to stress that when it comes to precision in extra-articular applications, the use of imaging guidance is necessary. Without skillful orientation in surface anatomy and the usage of imaging guidance, there is a high chance of missing the desired spot [28]. When the correct spot is missed, not only will the therapeutic agent have no effect, but it can also produce iatrogenic damage. In an enclosed space, the volume injected can physically damage the surrounding tissue. The best example of this is the intratendinous application of the drug, which produces a tear and a resulting defect in a tendon [23]. Also, the same agent can have a beneficial effect on one tissue with a specific condition and a detrimental on another tissue. An example for this is missed corticosteroid injection, reducing the quality of tendon tissue that it is in contact with.

Different injection agents require a different number of injection applications for the successful treatment. In clinical practice, anecdotally it was said that 6 weeks should be the minimum time between applications [29]. A recommendation from the American College of Rheumatology states that corticosteroid injections should not be given more than every 3 months per joint and at a maximum of four per year. The ORBIT consensus for PRP states that the interval between PRP injections should be 1–4 weeks and that multiple injections can be done. For hyaluronic acid and cellular-based products, there is, till date, neither strong scientific evidence nor consensus on the frequency of injections. Optimal time and frequency of injections are still debatable and require further research.

## 2.4  Conclusion

In conclusion, musculoskeletal injections are safe and affordable and require less logistics than operative treatment. By its widely used definition, evidence-based medicine is the conscientious, explicit, and judicious use of current best evidence in making decisions about the care of individual patients. Physicians should put their effort into providing adequate care to their patients by giving them the best possible treatment, in this instance the best possible injection, based on updated, critically reviewed literature and their clinical expertise.

## References


1. DePhillipo NN, Aman ZS, Dekker TJ, Moatshe G, Chahla J, LaPrade RF. Preventative and disease-modifying investigations for osteoarthritis management are significantly under-represented in the clinical trial pipeline: a 2020 review. Arthroscopy. 2021;37(8):2627–39.
2. Oo WM, Yu SP, Daniel MS, Hunter DJ. Disease-modifying drugs in osteoarthritis: current understanding and future therapeutics. Expert Opin Emerg Drugs. 2018;23(4):331–47.
3. Berenbaum F. Osteoarthritis as an inflammatory disease (osteoarthritis is not osteoarthrosis!). Osteoarthr Cartil. 2013;21(1):16–21.
4. Hunter DJ, Bierma-Zeinstra S. Osteoarthritis. Lancet. 2019;393(10182):1745–59.
5. Felson DT, Niu J, Neogi T, Goggins J, Nevitt MC, Roemer F, et al. Synovitis and the risk of knee osteoarthritis: the MOST study. Osteoarthr Cartil. 2016;24(3):458–64.
6. Sokolove J, Lepus CM. Role of inflammation in the pathogenesis of osteoarthritis: latest findings and interpretations. Ther Adv Musculoskelet Dis. 2013;5(2):77–94.
7. Gazendam A, Ekhtiari S, Bozzo A, Phillips M, Bhandari M. Intra-articular saline injection is as effective as corticosteroids, platelet-rich plasma and hyaluronic acid for hip osteoarthritis pain: a systematic review and network meta-analysis of randomised controlled trials. Br J Sports Med. 2021;55(5):256–61.
8. Saltzman BM, Leroux T, Meyer MA, Basques BA, Chahal J, Bach BR Jr, et al. The therapeutic effect of intra-articular normal saline injections for knee osteoarthritis: a meta-analysis of evidence level 1 studies. Am J Sports Med. 2017;45(11):2647–53.

9. Bradley JD, Heilman DK, Katz BP, Gsell P, Wallick JE, Brandt KD. Tidal irrigation as treatment for knee osteoarthritis: a sham-controlled, randomized, double-blinded evaluation. Arthritis Rheum. 2002;46(1):100–8.
10. Jüni P, Hari R, Rutjes AW, Fischer R, Silletta MG, Reichenbach S, et al. Intra-articular corticosteroid for knee osteoarthritis. Cochrane Database Syst Rev. 2015;2015(10):Cd005328.
11. He WW, Kuang MJ, Zhao J, Sun L, Lu B, Wang Y, et al. Efficacy and safety of intraarticular hyaluronic acid and corticosteroid for knee osteoarthritis: a meta-analysis. Int J Surg. 2017;39:95–103.
12. Jayaram P, Kennedy DJ, Yeh P, Dragoo J. Chondrotoxic effects of local anesthetics on human knee articular cartilage: a systematic review. PM R. 2019;11(4):379–400.
13. Wernecke C, Braun HJ, Dragoo JL. The effect of intra-articular corticosteroids on articular cartilage: a systematic review. Orthop J Sports Med. 2015;3(5):2325967115581163.
14. Braun HJ, Wilcox-Fogel N, Kim HJ, Pouliot MA, Harris AH, Dragoo JL. The effect of local anesthetic and corticosteroid combinations on chondrocyte viability. Knee Surg Sports Traumatol Arthrosc. 2012;20(9):1689–95.
15. Seshadri V, Coyle CH, Chu CR. Lidocaine potentiates the chondrotoxicity of methylprednisolone. Arthroscopy. 2009;25(4):337–47.
16. Altman RD, Manjoo A, Fierlinger A, Niazi F, Nicholls M. The mechanism of action for hyaluronic acid treatment in the osteoarthritic knee: a systematic review. BMC Musculoskelet Disord. 2015;16:321.
17. Singh H, Knapik DM, Polce EM, Eikani CK, Bjornstad AH, Gursoy S, et al. Relative efficacy of intra-articular injections in the treatment of knee osteoarthritis: a systematic review and network meta-analysis. Am J Sports Med. 2022;50(11):3140–8.
18. Zhao D, Pan JK, Yang WY, Han YH, Zeng LF, Liang GH, et al. Intra-articular injections of platelet-rich plasma, adipose mesenchymal stem cells, and bone marrow mesenchymal stem cells associated with better outcomes than hyaluronic acid and saline in knee osteoarthritis: a systematic review and network meta-analysis. Arthroscopy. 2021;37(7):2298–314.e10.
19. Yu H, Huang Y, Yang L. Research progress in the use of mesenchymal stem cells and their derived exosomes in the treatment of osteoarthritis. Ageing Res Rev. 2022;80:101684.
20. Song Y, Zhang J, Xu H, Lin Z, Chang H, Liu W, et al. Mesenchymal stem cells in knee osteoarthritis treatment: a systematic review and meta-analysis. J Orthop Translat. 2020;24:121–30.
21. Coombes BK, Bisset L, Vicenzino B. Efficacy and safety of corticosteroid injections and other injections for management of tendinopathy: a systematic review of randomised controlled trials. Lancet. 2010;376(9754):1751–67.
22. Setayesh K, Villarreal A, Gottschalk A, Tokish JM, Choate WS. Treatment of muscle injuries with platelet-rich plasma: a review of the literature. Curr Rev Musculoskelet Med. 2018;11(4):635–42.
23. Aicale R, Bisaccia RD, Oliviero A, Oliva F, Maffulli N. Current pharmacological approaches to the treatment of tendinopathy. Expert Opin Pharmacother. 2020;21(12):1467–77.
24. Nichols AW. Complications associated with the use of corticosteroids in the treatment of athletic injuries. Clin J Sport Med. 2005;15(5):370–5.
25. Fang WH, Chen XT, Vangsness CT Jr. Ultrasound-guided knee injections are more accurate than blind injections: a systematic review of randomized controlled trials. Arthrosc Sports Med Rehabil. 2021;3(4):e1177–e87.
26. Masoud MA, Said HG. Intra-articular hip injection using anatomic surface landmarks. Arthrosc Tech. 2013;2(2):e147–9.
27. Peh W. Image-guided facet joint injection. Biomed Imaging Interv J. 2011;7(1):e4.
28. Kumar Sahu A, Rath P, Aggarwal B. Ultrasound-guided injections in musculo-skeletal system—an overview. J Clin Orthop Trauma. 2019;10(4):669–73.
29. Douglas RJ. Corticosteroid injection into the osteoarthritic knee: drug selection, dose, and injection frequency. Int J Clin Pract. 2012;66(7):699–704.

# Contraindications and Potential Side Effects of Injections

**3**

Riccardo Compagnoni, Rossella Ravaglia, and Pietro Randelli

## 3.1 Introduction

Musculoskeletal injections are considered relatively safe procedures, if performed following the indications and avoiding cases in which there are absolute contraindications. Contraindication of intra-articular injections is a known hypersensitivity to any component of the injection, skin cellulitis, or broken skin over the needle entry site. The risk of infection is increased in these cases and in cases of joints with septic effusion of a bursa or a periarticular structure. Intra-articular injections cannot be performed in patients with an unstable coagulopathy for the risk of hemarthrosis.

Intra-articular or osteochondral fractures are a contraindication to a corticosteroid injection. The presence of joint prostheses represents a relative contraindication. Other relative contraindication would be the lack of response to prior injections and more than three prior injections in the last year to a weight-bearing joint [1].

If delivered properly and with correct indications, injections can be very rewarding for both the patient and the clinicians. Nonetheless, literature reports some possible complications.

It is possible to classify these complications in three groups (Table 3.1): intra-articular effects, surrounding tissue effects, and systemic effects.


R. Compagnoni (✉)
Department of Biomedical, Surgical and Dental Sciences, Università degli Studi di Milano. Via della Commenda, Milan, Italy
e-mail: riccardo.compagnoni@unimi.it

R. Ravaglia
1° Clinica Ortopedica, ASST Centro Specialistico Ortopedico Traumatologico Gaetano Pini-CTO, Milan, Italy
e-mail: rossella.ravaglia@unimi.it

P. Randelli
1° Clinica Ortopedica, ASST Centro Specialistico Ortopedico Traumatologico Gaetano Pini-CTO, Milan, Italy

Laboratory of Applied Biomechanics, Department of Biomedical Sciences for Health, Università degli studi di Milano, Milan, Italy
e-mail: pietro.randelli@unimi.it


**Table 3.1** Types of different complications after injections

| Intra-articular effect | Surrounding tissue effects | Systemic effects |
| --- | --- | --- |
| Flare of pain | Pericapsular calcification | Cushing syndrome |
| Infection | Tendon rupture | Hyperglycemia |
| Pseudoseptic arthritis | Skin alteration | Facial flushing |
| Steroid arthropathy | | Anaphylaxis |
| | | Neuropsychiatric changes |

© The Author(s), under exclusive license to Springer Nature Switzerland AG 2024
B. Kocaoglu et al. (eds.), *Musculoskeletal Injections Manual*,
https://doi.org/10.1007/978-3-031-52603-9_3

## 3.2 Intra-Articular Effects

Regarding the first group, the most common complication of intra-articular injection is a postejection flare of pain which has a 2–10% chance of occurrence. In this setting, the patient develops a flare of pain in the immediate 6- to 12-h period after an injection. The etiology is not sure, considering the timing of the reaction is thought to be secondary to a local reaction to microcrystalline steroid suspension. This complication is generally self-limiting and may be treated with activity modification, ice, and a short course of a nonsteroidal anti-inflammatory drug (NSAID). When, despite these precautions, pain persists after 36 h, patients should be evaluated for a septic joint. Infection is the most feared and dangerous local complication. Iatrogenic septic arthritis has been reported in patients following hyaluronic acid, steroids injections, and even ozone [2]. The risk of this complication has been estimated at 0.005% and 0.0002% for joint injections [3]. Risk factors include preexisting joint diseases such as rheumatoid arthritis, alcoholism, diabetes, cutaneous ulcers, previous steroid injections, previous knee surgery intravenous drug abuse, and immunosuppression. The likelihood of infection increases if these risk factors are all associated with a poor intra-articular injection technique [2, 4].

*Staphylococcus aureus* is the most common pathogen (51.4%), including methicillin-resistant *Staphylococcus aureus* (MRSA). It is followed by gram-negative enteric bacillus such as *E. coli* (5.7%), which is more prevalent in elderly patients, immunocompromised patients, and patients with intravascular devices and urinary catheters. *Streptococcus pyogenes* is less common (4.7%) [4].

The most common acute clinical aspects are poor articular functionality, local erythema, and swelling, all associated with laboratory abnormalities (elevated ESR and CRP) [5].

Knee is the joint involved in about 50% of cases of infection after injection [2]. Potentially any joint after injection is prone to develop an infection, but hip, shoulder, and elbow infection are less common than knee [2]. The diagnosis of septic arthritis continues to be a challenging task for clinicians. Accurate diagnosis is crucial; in fact, delayed diagnosis may lead to irreversible joint damage, while overdiagnosis may cause patients to undergo unnecessary medical and surgical treatments. A positive culture from the affected joint is the gold standard diagnostic test [6].

Cultures often take days to produce results; for this reason, alternative laboratory data are necessary for clinicians to initiate prompt treatment. Common blood tests are ESR, CRP, and CBC. In addition, the synovial fluid cell count and percentage of synovial fluid polymorphonuclear cells from the joint aspiration are perhaps the most critical determinant of native septic arthritis [6]. A synovial fluid cell count of 50,000 cells/mm$^3$ or higher is typically concerning for septic arthritis, while lower values are more consistent with a crystalline or inflammatory arthropathy [7]. However, the literature to support this cutoff is rather limited, yet this value is treated in a dogmatic manner.

Although synovial cultures are the gold standard, they are estimated to be only 75–95% sensitive [8]. It is thought that if synovial cultures are used in isolation, then some instances of septic arthritis may be missed.

In 1976, Newman defined four criteria to diagnose septic arthritis. According to his study, only one of four needs to be present to make a diagnosis:

1. An organism isolated from the affected joint.
2. An organism isolated from elsewhere with a clinically swollen, painful joint.
3. No organism isolated, but histologic or radiologic evidence of infection.
4. Turbid fluid aspirated from the joint in a patient that has previously received antibiotics [9].

This definition remains popular today and is often used. Septic arthritis following joint injections is much feared due to its various complications described in literature: mortality appears to be approximately 11% in monoarticular septic arthritis and a permanent loss of joint function nearly 40% [10]. Different treatment options in managing acute septic arthritis have been proposed, ranging from oral/e.v. antibiotic therapy, arthroscopic/open articular washing/debridement, antibiotic cemented spacers, and joint replacements/arthrodesis to amputations, with obvious costs for the health system and often poor patient's outcomes [2].

Pseudo septic arthritis is a rare complication of hyaluronic acid injections that can be difficult to differentiate from septic arthritis. In most cases, pseudo sepsis develops when patients are sensitized to the agent, following the second or third injection. Nevertheless, more rarely, the symptoms may develop after the first injection. Usually, it has been reported within 2 and 48 h after the injection [11]. Even after completing appropriate investigations, it can be difficult to differentiate between pseudo septic arthritis and septic arthritis. Investigations for pseudo septic arthritis generally include blood work and joint aspiration to rule out septic arthritis. Patients can present with elevated leucocyte counts, C-reactive protein (CRP), erythrocyte sedimentation rates (ESRs), and joint aspirates with elevated synovial leucocyte counts. Most notably, patients present with negative culture results, and their symptoms improve without antibiotics or operative intervention [12].

The pathophysiology responsible for pseudo septic arthritis has not yet been fully explored. Multiple theories have been suggested with minimal evidence, including inaccurate extra-articular hyaluronic acid injection [13]. The currently most accredited theory suggests a proinflammatory process, likely in keeping with a type IV cell-mediated hypersensitivity reaction with secondary activation of the complement pathway [14]. The sensitization theory suggests that new antigenic material (a second injection) triggered a new inflammatory cascade mediated by complement C5a and C5b-9, in addition to a chronic macrophagic reaction [15]. This concept explains why pseudo septic arthritis occurred more significantly in patients receiving more than a single course of injection. Hence, no parameters to suggest pseudo septic arthritis can be concluded, and a high suspicion for septic arthritis given its

severity should be maintained [11]. Another possible complication is steroid arthropathy.. Avascular necrosis of the femoral or humeral heads may follow systemic steroids of varying dose and duration. After intra-articular administration, there may be a rapid progression of pre-existing degenerative joint disease or frank aseptic necrosis [16].

## 3.3 Surrounding Tissue Effects

The second group of injections complications involves effects on surrounding tissue, such as pericapsular calcification, tendon rupture, and skin alteration. Pericapsular calcification occurs in more than 40% of cases. The accumulation of insoluble steroid acts as a foreign body and induces a chronic granulomatous inflammatory process, with subsequent dystrophic calcification [17]. Tendon rupture is another important complication described following steroid injection. It can have significant consequences, particularly in an active or athletic cohort. It is difficult to imply causality between injection and tendon rupture despite the number and variety of reported cases. For example, the previous condition of the tendon that requires the injection can be implicated. The incidence of tendon rupture increases with the number of injections undergone by the patient [18]. This may reflect that worsening native tendon abnormality, the accumulative effect of multiple injections, or a combination of these factors may be involved in the etiopathology of tendon rupture [19]. Injection technique must be considered, given the potentially deleterious mechanical effects of intratendinous needle placement. There is some evidence that the choice of corticosteroid may affect the incidence of tendon rupture following injection: triamcinolone acetonide is associated with a higher incidence of rupture than betamethasone, methylprednisolone acetate, or hydrocortisone [19]. The postinjection tears most frequently encountered in the literature regards patellar, quadriceps, Achilles, and biceps tendons. Tears of the common extensor origin at the lateral epicondyle, the supraspinatus, tibialis anterior, biceps femoris, triceps, and the small flexor tendons of the hand are less commonly described [18].

Skin alteration may occur when local steroid is applied too close to the surface of the skin. Alteration may include depigmentation or hyperpigmentation. These changes may be irreversible. Furthermore, intramuscular injection of corticosteroids may lead to atrophy of the skin and subcutaneous tissue at the site of administration. This occurs because of the dose-dependent inhibitory action of corticosteroids on the proliferation of fibroblasts and the accelerated decomposition of collagen. Clinically, the disappearance of the subcutaneous adipose tissue is observed without evidence of an inflammatory type. Generally, the subcutaneous atrophy is reversible within a year, but in some cases, it is necessary to resort to the injection of a "fat bolus," which often causes cysts and calcifications associated with necrosis.

## 3.4 Systemic Effects

The third group includes systemic complications, developing above all after corticosteroid injections. The systemic side effects associated with both intra- and extra-articular corticosteroid injections are not completely understood. The systemic absorption of corticosteroid occurring following a local injection is variable. In most cases, particularly where there is accurate intra-articular administration, systemic absorption of locally injected musculoskeletal corticosteroid is minimal [20]. Less soluble corticosteroid formulations, such as triamcinolone and methylprednisolone acetate, appear to deliver a greater and more prolonged suppression of endogenous serum cortisol than more soluble corticosteroid preparations such as betamethasone [21]. Patients who are concomitantly receiving certain medications, in particular cytochrome P450 3A4 inhibitors such as ritonavir, appear to be at risk of suffering very severe and prolonged systemic side effects attributable to corticosteroids following even a single intra-articular injection, including Cushing's syndrome and adrenal suppression.

Cytochrome P450 3A4 inhibitors appear to significantly reduce the clearance of administered corticosteroids from the patient's system, while they not affect absorption [20]. Exogenous Cushing syndrome can occur from locally injected glucocorticoids [20]. Cushing's syndrome derives from an excessive production of cortisol. It is associated with several physiological changes, including facial and trunk obesity, stretch marks, hypertension, weakness, facial hair growth in females, and osteoporosis. Hyperglycemia after intra-articular injection has been demonstrated in patients with diabetes. It may consist in short-term difficulties with glycemic control. More rarely, an increase in serum glucose can also occur in patients without diabetes. Usually, this situation has no real consequences and resolves between 24 h and 1 week.

Facial flushing is a much more frequent complication, developing in 15% of cases. It is described mostly in women, and it occurs secondary to a histamine-mediated response [22]. It almost always resolves within a short time. As with any situation in which a medication is administered, anaphylaxis must be considered; however, this is extraordinarily rare in the setting of corticosteroid injection. This may occur more commonly secondary to polyethylene glycol (macrogol) which acts as a solvent for particulate corticosteroids. Injected glucocorticoids can lead to temporary mania and psychosis in some patients. There are multiple case reports, but no known incidence or dose correlation. Most of these cases involve patients with preexisting psychiatric conditions, but many do not [20].

## 3.5 Conclusion

Musculoskeletal injections can be an effective treatment option, but they are not without risks. To ensure safety, the doctor must carefully review the patient's medical history and current medications, including any recent glucocorticoid injections in other areas. During the process of obtaining informed consent and making decisions, it is important to clearly educate all patients receiving injections about the possible local and systemic side effects. Patients should be informed that these effects can vary among individuals.

# References


1. Stephens MB, Beutler AI, O'Connor FG. Musculoskeletal injections: a review of the evidence. Am Fam Physician. 2008;78(8):971–6.
2. Larghi MM, Grassi M, Placenza E, Faugno L, Cerveri P, Manzotti A. Septic arthritis following joint injections: a 17 years retrospective study in an Academic General Hospital. Acta Biomed. 2021;92(6):e2021308.
3. Geirsson ÁJ, Statkevicius S, Víkingsson A. Septic arthritis in Iceland 1990–2002: increasing incidence due to iatrogenic infections. Ann Rheum Dis. 2008;67(5):638–43.
4. Mohamed M, Patel S, Plavnik K, Liu E, Casey K, Hossain MA. Retrospective analysis of septic arthritis caused by intra-articular viscosupplementation and steroid injections in a single outpatient Center. J Clin Med Res. 2019;11(7):480–3.
5. García-Arias M, Balsa A, Mola EM. Septic arthritis. Best Pract Res Clin Rheumatol. 2011;25(3):407–21.
6. Rasmussen L, Bell J, Kumar A, et al. A retrospective review of native septic arthritis in patients: can we diagnose based on laboratory values? Cureus. 2020;12:6.
7. Mathews CJ, Coakley G. Septic arthritis: current diagnostic and therapeutic algorithm. Curr Opin Rheumatol. 2008;20(4):457–62.
8. Li SF, Cassidy C, Chang C, Gharib S, Torres J. Diagnostic utility of laboratory tests in septic arthritis. Emerg Med J. 2007;24(2):75–7.
9. Newman JH. Review of septic arthritis throughout the antibiotic era. Ann Rheum Dis. 1976;35(3):198–205.
10. Mathews CJ, Weston VC, Jones A, Field M, Coakley G. Bacterial septic arthritis in adults. Lancet. 2010;375(9717):846–55.
11. Sedrak P, Hache P, Horner NS, Ayeni OR, Adili A, Khan M. Differential characteristics and management of pseudoseptic arthritis following hyaluronic acid injection is a rare complication: a systematic review. J ISAKOS. 2021;6(2):94–101.
12. Leopold SS, Warme WJ, Pettis PD, Shott S. Increased frequency of acute local reaction to intra-articular hylan GF-20 (Synvisc) in patients receiving more than one course of treatment. J Bone Joint Surg. 2002;84(9):1619–23.
13. Adams ME, Lussier AJ, Peyron JG. A risk-benefit assessment of injections of hyaluronan and its derivatives in the treatment of osteoarthritis of the knee. Drug Saf. 2000;23(2):115–30.
14. Marino AA, Waddell DD, Kolomytkin OV, Pruett S, Sadasivan KK, Albright JA. Assessment of immunologic mechanisms for flare reactions to Synvisc®. Clin Orthop Relat Res. 2006;442:187–94.
15. Dragomir CL, Scott JL, Perino G, Adler R, Fealy S, Goldring MB. Acute inflammation with induction of anaphylatoxin C5a and terminal complement complex C5b-9 associated with multiple intra-articular injections of hylan G-F 20: a case report. Osteoarthr Cartil. 2012;20(7):791–5.
16. Miller WT, Restifo RA. Steroid Arthropathy1. 1966;86(4):652–657. https://doi.org/10.1148/86.4.652.
17. Conti RJ, Shinder M. Soft tissue calcifications induced by local corticosteroid injection. J Foot Surg. 1991;30(1):34–7.
18. Hynes JP, Kavanagh EC. Complications in image-guided musculoskeletal injections. Skelet Radiol. 2022;51(11):2097–104.
19. Nichols AW. Complications associated with the use of corticosteroids in the treatment of athletic injuries. Clin J Sport Med. 2005;15(5):370–5.
20. Stout A, Friedly J, Standaert CJ. Systemic absorption and side effects of locally injected glucocorticoids. PM&R. 2019;11(4):409–19.
21. Horani MH, Silverberg AB. Secondary Cushing's syndrome after a single epidural injection of a corticosteroid. Endocr Pract. 2005;11(6):408–10.
22. Common Injections in Sports Medicine: General Principles and Specific Techniques I Musculoskeletal Key.

# Informing Patients 4


Daniel Pérez-Prieto, Ana Soria, Marta Torruella,
and Narcís Pérez de Puig


## 4.1 Possible Complications of Injections

Infiltrations and musculoskeletal injections are common procedures in the treatment of various musculoskeletal conditions such as arthritis, tendinopathy, or osteoarthritis. These procedures involve the injection of medications into the affected soft tissues and joints to reduce inflammation and pain and improve function. Musculoskeletal infiltrations and injections are generally safe and effective, but like any medical procedure, they carry risks and potential complications [1, 2]. Therefore, it is essential for patients to be informed about the risks and benefits of the treatment before the intervention [3].

Among the different types of complications, it should be distinguished between adverse events at the local level and systemic complications that may arise from the process. It is also important to have in mind that adverse reactions can appear even after several weeks after the injection has been performed.

Some of the most common complications are explained hereafter.

- Infection [4, 5]: Injections into the skin or joints can lead to bacterial or fungal infections. Symptoms of an infection may include redness, swelling, pain, and fever. The incidence of infection after an injection varies depending on the injection site and the technique used for the injection. It is crucial to perform the injection under sterile conditions (see correspondent chapter). The presence of a particular germ may vary depending on the environment, patient population, and other factors. Additionally, the risk of infection may also be influenced by the aseptic technique used during the injection and individual patient characteristics, such as immunity and the presence of comorbidities. It is always recommended to follow appropriate aseptic and sterilization guidelines to prevent infections associated with musculoskeletal injections.


D. Pérez-Prieto (✉)
Orthopedic Surgery Department, Hospital del Mar, Barcelona, Spain

Department of Traumatology and Orthopaedic Surgery, Hospital del Mar, Barcelona, Spain

IcatKnee – Hospital Dexeus, Barcelona, Spain

Universitat Autònoma de Barcelona (UAB), Barcelona, Spain
e-mail: dperezprieto@hospitaldelmar.cat

A. Soria · M. Torruella
Orthopedic Surgery Department, Hospital del Mar, Barcelona, Spain
e-mail: ana.soria.blazquez@psmar.cat;
marta.torruella.grive@psmar.cat

N. P. de Puig
Legal Department, Hospital del Mar, Barcelona, Spain
e-mail: nperez@psmar.cat

- Pain and discomfort [1]: After an injection, patients may experience pain or discomfort at the injection site, which can last for several days. In some cases, the pain can be intense and may require analgesics.
- Nerve injury [1, 2, 4]: Nerve injuries are a potentially serious complication of musculoskeletal injections. Although relatively rare, they can occur due to various factors such as incorrect injection technique, improper needle placement, lack of image guidance, injection in an area near a nerve, or the presence of anatomical abnormalities. Peripheral nerves, such as the median nerve in carpal tunnel syndrome or the radial nerve in elbow region injections, can be affected during the injection. This can result in symptoms such as pain, numbness, weakness, or loss of motor function in the area innervated by the affected nerve.
- Allergic reactions [4]: These reactions can occur due to an immune response of the body to one or more components of the injection, such as the drug, injection vehicle, or preservatives used in the medication. Some signs and symptoms of an allergic reaction include the following:
  - Hives. Appearance of red, elevated wheals on the skin, which can cause intense itching.
  - Sensation of itching on the skin or throughout the body.
  - Edema. Localized or generalized swelling of the skin or mucous membranes.
  - Redness of the skin at the injection site or in other areas of the body.
  - Difficulty breathing, chest tightness, or wheezing. In type 1 allergy, a more severe allergic reaction can trigger an anaphylactic reaction, which is a potentially life-threatening medical emergency.
  - To reduce the risk of allergic reactions, it is necessary to inquire about patient allergies prior to administering any medication or performing a musculoskeletal injection. Additionally, if an allergy is suspected, specific allergy tests can be conducted to identify the responsible allergens.
- Bleeding [1]: Bleeding after musculoskeletal injections is a potential complication, especially if damage to blood vessels occurs during the injection. Some common causes of bleeding include incorrect injection technique, inappropriate needle gauge, or the use of anticoagulant medications that increase the risk of bleeding. Bleeding can manifest in different ways, depending on the severity of the bleeding and the location of the injury. Some signs and symptoms of bleeding after a musculoskeletal injection may include:
  - Bleeding at the injection site. There may be excessive bleeding at the site where the injection was performed. The bleeding may be visible externally or may form a hematoma beneath the skin.
  - Hematomas may appear in the injection area, which are accumulations of blood beneath the skin. Hematomas can vary in size and may cause pain and tenderness in the affected area.
  - In general, bleeding may be controlled by compression and cryotherapy. In exceptional cases, if a big vessel is damaged, a pseudo-aneurism may be formed, and surgery can be required to correct it.
- Muscular tissue injury [2]: In rare cases, the injection can directly damage the muscular tissue. This can cause localized pain, muscle weakness, and difficulty in moving the affected muscle.
- Articular cartilage and ligament injury. This is a rare but possible complication. This injury can occur if the needle used during the injection comes into contact with the articular cartilage or injures the ligaments in the joint. It is also important to note the potential concern that repeated corticosteroid infiltrations may increase the risk of tendon or ligament rupture in some patients [6].
- Effects of corticosteroids [4]: Corticosteroids have anti-inflammatory properties and can help reduce inflammation and pain in joints and soft tissues. However, they can also weaken connective tissue, such as tendons and ligaments, when administered at high doses or

for prolonged periods. Moreover, they can produce hypopigmentation of the skin in the site of infiltration and fat tissue atrophy that in severe cases may produce important deformities in the site of infiltration.

Several factors can increase the aforementioned risks (specially tendon or ligament rupture after corticosteroid infiltrations) such as repeated injections in the same location, high doses of corticosteroids, systemic diseases like diabetes, advanced age, preexisting tendon or ligament weakness, and intense physical activity after the infiltration. The tendons most commonly associated with rupture after corticosteroid infiltrations are the Achilles tendon, the patellar tendon, and the rotator cuff tendons [2].

## 4.2 Obtaining Consent from the Patient

It is important to note that the incidence of complications after a musculoskeletal injection may vary depending on the patient, the treated condition, and the technique used for the injection. Therefore, it is crucial for patients to be informed about the risks and benefits of the treatment by obtaining their informed consent [7, 8].

Informed consent is a process by which a patient agrees to receive medical treatment after being provided with information about the risks and benefits of the treatment. It is important for physicians to obtain informed consent from patients before performing a musculoskeletal infiltration or injection. The process of obtaining informed consent should include the following information [9]:

- Procedure description: The physician should explain in detail the procedure of the musculoskeletal infiltration or injection, including how it will be carried out and what can be expected after the procedure.
- Risks and benefits: The physician should discuss the risks and benefits of the procedure, including the possible complications mentioned above.
- Alternatives: The physician should explain any alternatives to the musculoskeletal infiltration or injection procedure.
- Questions: The patient should have the opportunity to ask questions about the procedure and the associated risks and benefits.

Whether the informed consent for infiltrations should be oral or written is still of controversy [10, 11]. There is not a uniform regulation among different countries regarding how patients should express his/her informed consent to a certain treatment, nor a uniform practice. For instance, the European Alliance of Associations for Rheumatology (EULAR) recommends that the patient must be fully informed of the nature of the procedure, the injectable, and potential benefits and risks; informed consent should be obtained and documented according to local habits [9]. In a recent survey, only 10% of UK upper extremity surgeon used written informed consent [11].

It is generally accepted that any serious or frequently occurring risks of any procedure must be explained to the patient, and the most frequent complications should be also discussed. Then, written informed consent could be recommended to prove that this process of explanation, advice, and patient's acceptance has been duly performed by physicians. There is a clear obligation to obtain written informed consent in certain invasive procedures, such as surgical interventions or invasive diagnostic or therapeutic procedures (Spanish law 41/2002, 14th November). Even in those cases, in which written informed consent is mandatory, physicians must notice that it is not a contract but the result of a permanent dialogue with patients for the decision-making that should be noted in the medical records at least [12]. Therefore, although the aforementioned different regulations, a general recommendation would be to explain risks, benefits, and alternatives to the patient and to write down the patient acceptance in the medical records.

# References


1. Hynes JP, Kavanagh EC. Complications in image-guided musculoskeletal injections. Skelet Radiol. 2022;51(11):2097–104.
2. Nichols AW. Complications associated with the use of corticosteroids in the treatment of athletic injuries. Clin J Sport Med. 2005;15(5):370–5.
3. Stephens MB, Beutler AI, O'Connor FG. Musculoskeletal injections: a review of the evidence. Am Fam Physician. 2008;78(8):971–6.
4. Brinks A, Koes BW, Volkers ACW, Verhaar JAN, Bierma-Zeinstra SMA. Adverse effects of extra-articular corticosteroid injections: a systematic review. BMC Musculoskelet Disord. 2010;11:206.
5. Baums MH, Aquilina J, Pérez-Prieto D, Sleiman O, Geropoulos G, Totlis T. Risk analysis of peri-prosthetic knee joint infection (PJI) in total knee arthroplasty after preoperative corticosteroid injection: a systematic review : a study performed by the Early-Osteoarthritis group of ESSKA-European Knee Associates section. Arch Orthop Trauma Surg. 2022;
6. Donovan RL, Edwards TA, Judge A, Blom AW, Kunutsor SK, Whitehouse MR. Effects of recurrent intra-articular corticosteroid injections for osteoarthritis at 3 months and beyond: a systematic review and meta-analysis in comparison to other injectables. Osteoarthr Cartil. 2022;30(12):1658–69.
7. Aly MNS. Intra-articular drug delivery: a fast growing approach. Recent Pat Drug Deliv Formul. 2008;2(3):231–7.
8. Cole BJ, Schumacher HR. Injectable corticosteroids in modern practice. J Am Acad Orthop Surg. 2005;13(1):37–46.
9. Uson J, Rodriguez-García SC, Castellanos-Moreira R, O'Neill TW, Doherty M, Boesen M, et al. EULAR recommendations for intra-articular therapies. Ann Rheum Dis. 2021;80(10):1299–305.
10. Beitzel K, Allen D, Apostolakos J, Russell RP, McCarthy MB, Gallo GJ, et al. US definitions, current use, and FDA stance on use of platelet-rich plasma in sports medicine. J Knee Surg. 2015;28(1):29–34.
11. Lim CS, Miles J, Peckham TJ. Current practice of obtaining informed consent for local steroid injection among the shoulder and elbow surgeons in United Kingdom. Scott Med J. 2010;55(3):32–4.
12. Broggi MA. ¿Consentimiento informado o desinformado? El peligro de la medicina defensiva. Med Clin (Barc). 1999;112:95–6.

F. De Filippo and Maristella F. Saccomanno

## 5.1 Introduction

Joint injections are a useful diagnostic and therapeutic skill in a surgeon portfolio. Although it is a simple procedure, it requires a correct training. Diagnostic indications include the aspiration of fluid for cytologic or microbiological analysis as well as to provide pain relief and range of motion recovery in swollen joints. Therapeutic indications include the delivery of several biologic or non-biologic agents such as local anesthetics, corticosteroids, hyaluronic acid, growth factors, and stem cells. Side effects are few, but one of the worse is surely the development of a postinjection infection [1]. Moreover, joint and soft-tissue injections can be performed with or without imaging guidance. Imaging guidance, such as ultrasound, is meant to increase accuracy [2], and it is very helpful when the articular space cannot be easily palpated, such in case of hip injections. Advantages are mainly related to the fact that anatomic structure (capsule, vessels, nerve) can be easily visualized, so the needle trajectory can be fine-tuned in accordance.

In any case, three main rules must be fulfilled before performing an injection:

- Good knowledge of the anatomy of the area to be injected, in order to avoid neuromuscular bundles as well as skin and subcutaneous fat injections.
- Aseptic technique.
- The indications and contraindications of the agent you are about to inject should be carefully considered. The procedure should always be performed when acute or chronic symptoms are present. Proper timing can help minimize complications that are directly associated with the type of agent being used, as well as potential interactions with other therapies.

## 5.2 Injection Equipment

Injections or aspiration techniques require aseptic conditions to minimize the risk of infection [2].

The following items are mandatory:

- Alcohol and or povidone-iodine (Betadine) solution
- Sterile and nonsterile gloves
- Sterile drapes
- 21- to 23-gauge needle

F. De Filippo
Department of Medical and Surgical Specialties, Radiological Sciences, and Public Health, University of Brescia, Brescia, Italy

M. F. Saccomanno (✉)
Department of Medical and Surgical Specialties, Radiological Sciences, and Public Health, University of Brescia, Brescia, Italy

Department of Bone and Joint Surgery, Spedali Civili, Brescia, Italy

- 3–60 mL syringe for aspirations
- 1–10 mL syringe for injections
- Sterile gauzes
- Injectable agent
- Laboratory tubes for culture (in case of aspiration)
- Adhesive bandage or other adhesive dressing

An optimal setting includes a professionally clean, quiet, private, well-lit room, with the patient in a comfortable position. Careful patient positioning before the procedure facilitates safe and efficient access to the joint. For shoulder, elbow, and wrist injections, the patient can be seated. In contrast, for hip, knee, or ankle injections, the patient should be placed in the supine position. This helps prevent or mitigate the effects of a vasovagal or syncopal episode.

Once all the necessary equipment has been gathered, the operative field can be prepared, followed by the preparation of the injection site. Begin by wearing non-sterile gloves and placing a sterile drape. Then, put on sterile gloves, and arrange the needles, gauzes, syringe, and adhesive bandage on the sterile drape (Fig. 5.1).

If ultrasound guidance is utilized, a preprocedural scan is performed to plan the needle trajectory, locate the target site, and identify nearby neurovascular structures and tendons to avoid. Sterile gloves can now be worn. An assistant will provide alcohol or povidone-iodine solution on sterile gauzes to adequately prepare the injection site.

## 5.3 Preparation of Injection Site

A good exposure of the entire area is required. Disinfection of the target area is an important step. The goal is to minimize risk of infection at the site. It is recommended to start from the entry point and then to gradually include a bigger area with circular movements (Fig. 5.2). This step will allow the physician to have access to the entire joint (Fig. 5.3). In this way, it will be easier to find the entry point. Soft tissue spot or bony landmarks must be palpated. The entry point can be marked with an impression from a needle cap. At this point, the injection can be safely performed. An important tip is to always perform aspiration before injecting to prevent intravascular injection. The injection should flow smoothly and should not cause discomfort to the patient. Most pain experienced during the procedure is due to tissue stretching, and it can be minimized by injecting slowly. If there is

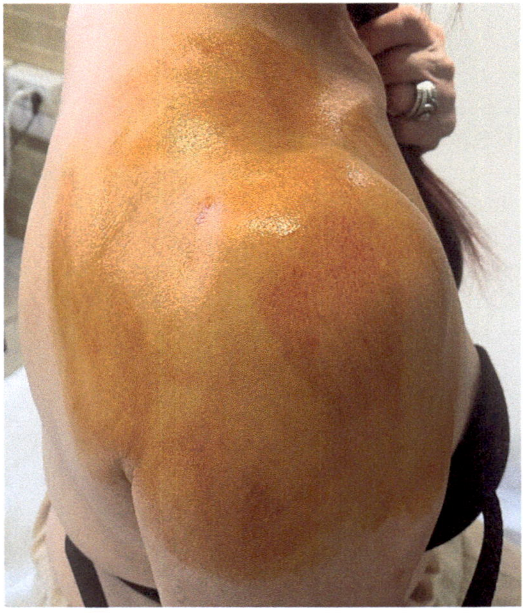

**Fig. 5.2** Right shoulder: it is recommended to disinfect a large area

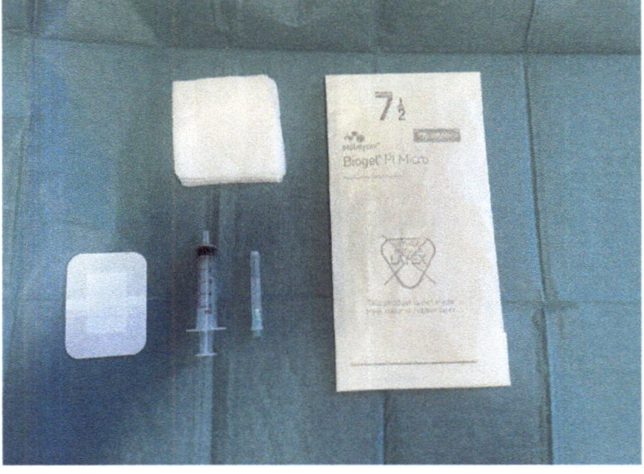

**Fig. 5.1** Sterile field: gloves, needles, gauzes, syringe, and adhesive bandage

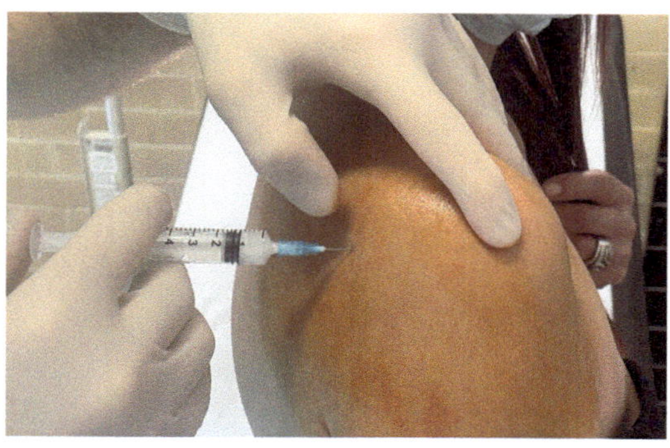

**Fig. 5.3** The physician can palpate several landmarks to ensure the entry point, even without imaging guidance

significant resistance encountered while injecting, it is advisable to readjust the needle trajectory.

## References

1. Zacay G, Heymann AD. Intra-articular and soft-tissue corticosteroid injections and risk of infections: population-based self-controlled-risk-interval design. Pharmacoepidemiol Drug Saf. 2023;32:718–25. https://doi.org/10.1002/pds.5602.
2. Uson J, Rodriguez-García SC, Castellanos-Moreira R, et al. EULAR recommendations for intra-articular therapies. Ann Rheum Dis. 2021;80:1299–305. https://doi.org/10.1136/annrheumdis-2021-220266.

# Things to Take into Consideration in Injection and Aspiration

**6**

Thorkell Snaebjörnsson

## 6.1 Indication for Treatment

When a healthcare provider gives an injection or does an aspiration, the underlying disease or condition can vary considerably. In recent years, injections have become a more widespread practice for degenerative spinal diseases with a range of different elective spinal injections available [1]. Another branch of healthcare providers treats patients with a great variety of injections with agents like neurotoxins and fillers [2]. In this chapter, the guidelines from WHO regarding injections and related procedures [3] have been taken into consideration. Within orthopedics, there is a long tradition of intra-articular injections in the knee and shoulder joints, either for therapeutic [4] or diagnostic purposes [5]. Other indications for treatment include aspiration from cysts or abscesses for diagnostic purposes in case of infection or treatment in case of drainage. In this chapter, our focus lays on traditional superficial injections as well as intra-articular injections and aspirations of joints [6], bursa, [7] or abscess's.


T. Snaebjörnsson (✉)
Department of Orthopaedics, Institute of Clinical Sciences, Sahlgrenska Academy, Gothenburg University, Gothenburg, Sweden

Department of Orthopaedics, Sahlgrenska University Hospital, Mölndal, Sweden
e-mail: thorkell.snabjornsson@vgregion.se


## 6.2 Settings

The first thing to address when preparing for either aspiration or injection is the indication for treatment as well as the appropriate environment for the chosen treatment form. Many patients are sensitive to vasovagal fainting under these circumstances and are therefore at risk of falling and injuring themselves during the procedure. It is therefore important to inform the patient about the planned procedure and possible side effects. After the information is given, it is time to make sure that the patient is in a stable position with appropriate support, preferentially lying on his back. It is equally important that the healthcare provider is well prepared and has good access to the area of interest with a satisfactory environment and sources to perform the task ahead.

## 6.3 Medical History

Every patient should be evaluated according to prior health problems, risk of infection, medications before treatment, and accessibility to the anatomical area of interest. If patients are on anticoagulation treatment, the healthcare provider must be sure that the benefit of the treatment given is bigger than the possible risk of bleeding after the operation. It is worth noting that according to Kotecha et al. [8], it is fairly uncommon for a patient on anticoagulant therapy to have bleeding complications after aspiration or injection. Prior history of allergic reactions to vaccinations or injections should be thoroughly evaluated before choosing to perform the injection.

Another supplementary option to conventional injection or aspiration is the ultrasound-guided approach to be able to penetrate the area of interest more accurately. This form of treatment can be especially valuable when treating patients that are overweight, the anatomical area is challenging, it is difficult to palpate manually, or even where it is difficult for the patients to position themselves to expose the area of interest. Ultrasound-guided aspiration is even a valuable treatment option to measure the amount of fluid in suspected arthroplasty infections and therefore provides a guide to the clinical treatment of measuring the amount of fluid and deciding if aspiration is a feasible option. The aspiration can then be performed subsequently with greater accuracy compared with conventional methods. According to Hoeber et al. [9], who performed a review and meta-analysis on the accuracy of hip injection, the accuracy of the injection is significantly better when performed ultrasound guided when compared with landmark-guided injections.

## 6.4	Anatomical Landmarks

In order to be able to provide the best possible treatment for the patient, the healthcare provider must know the local anatomical landmarks such as osseous points, palpate, and mark accordingly. The marking of the area then serves as an indicator for the draping or the sterile preparation. The physician marks the site of injection point; this is often done by skin marking pen, and unfortunately, this marking often disappears when the skin is cleaned. It is therefore recommended to make an indentation in the skin, using an instrument like the cap of a pen and pressing the skin for several seconds, leaving a mark that does not disappear when the skin is prepared with antiseptics [10].

## 6.5	Minimize Risk of Infection

It is important to emphasize the careful preparation of the injection site with a focus on any signs of redness, swelling, scratch marks, wounds, or other actual skin changes. This is especially important in elective procedures where it is recommended to postpone the procedure to lower the risk of local infection or an infectious abscess if any signs of damage to the skin barrier arise. In acute settings, the clinical decision of aspiration or injection must outweigh the risk of adverse effects and every patient evaluated separately. The decision to proceed with aspiration or injection in acute settings must be taken separately from the elective settings, bearing in mind that one does not always have the luxury to postpone the treatment without consequences to the patient.

## 6.6	The Procedure

Firstly, clean the skin before using local anesthesia. Then wait for 10–15 min, and make sure that the skin is numbed before you proceed. When penetrating the skin, the needle should be injected perpendicularly through the skin before adjusting to the point of interest. As previously mentioned, a sterile method is recommended for all procedures, but it is even more important when penetrating joints to lower the risk of infections (Table 6.1). The local anatomy of the injection site must be accurately known to avoid neuromuscular structures (damages) in the region. As an example, there is a risk of skin necrosis if corticosteroids are injected into the subcutaneous fat tissue of the skin. This is especially important when using steroid injections on muscle tendons to avoid skin necrosis. When performing an injection, negative aspiration must be performed to avoid intravenous injection (Table 6.1).

After the negative aspiration, the injection is then performed steadily and without causing any pain to the patient. Patients can experience pain caused because of tissue expansion; this can be

**Table 6.1** Preparation for the healthcare provider

- An aseptic technique is the golden rule using sterile equipment.
- Prepare the patient using gloves for you own protection.
- It is not mandatory to use sterile gloves when using no touch technique.
- Alcohol wiper or other disinfectant (chlorhexidine/povidone-iodine).
- Sterile drape.
- Choose appropriate size of needles depending on your task.
  Local anesthesia requires approximately 25–30 gauge needle.
  Injection requires a needle of approximately 22–25 gauge.
  Aspiration requires a needle of approximately 18–20 gauge.
- Choose an injector of adequate size for either aspiration or injection.
- Choose suitable local anesthetic agent, corticosteroid, or other agents.
- Prepare laboratory tubes for assembling or for storing aspirate.
- Prepare a gauze or wound dressing for compression.
- Control hemostasis, especially important when performing both aspiration and injection with the same needle.
- Apply wound dressing.

**Table 6.2** Preparation of the patient

- Safely position the patient so the patient and the muscles can relax.
- Identify the surface anatomy and the relevant landmarks.
- Any marking to identify point of entry should be made before sterilization.
- Clean the overlying skin with alcohol swab or other disinfectant.
- Local anesthetic can be applied (always for children, occasionally for adults).
- If ultrasound is required, use sterile gel on the ultra sound probe.
- Prepare wound dressing if necessary.

managed by lowering the speed of the injection. The injection should then flow easily, if there is any obstruction or strong resistance, care must be taken to make sure the needle is within the joint. The feeling of injection resistance may include the placement of a needle within the muscle, tendon, bone, or cartilage. If sharp pain is felt during the injection, the reason often is that the injection is going into the synovia; it is therefore recommended to temporarily stop the injection and adjust the tip of the needle before resuming the injection (Table 6.2). According to a systematic review performed by Hermans et al. [11], the accuracy of knee injections varied from 91% using the superolateral approach to 67% accuracy using the anterolateral approach of the knee joint. Bearing in mind the frequency of knee injections, it is therefore of great importance to follow guidelines and train your methods to increase the rate of successful procedure, even for more difficult or smaller joints. After the procedure is performed, care should be taken to

apply pressure with a gauze or other wound dressing and control hemostasis before the patient can be allowed to return home.

## 6.7  Conclusion

Injections and aspirations are valuable methods for patient treatment and are currently used within many branches of the healthcare system. As a general rule, the healthcare provider must thoroughly evaluate the indication for treatment and, together with the patient, decide that the advantage of the treatment is greater than the possible disadvantage. Patient's medical history should always be evaluated, with special emphasis on medications with anticoagulation effects or previous allergic reactions. Preparation for treatment includes marking the area, sterile environment, and anatomical knowledge of the healthcare provider. A valuable treatment option in difficult cases like suspicious arthroplasty infection is an ultrasound-guided aspiration. To let the patient relax, it is important that the skin is numb and the needle is perpendicular to the skin during insertion. When an injection is performed, a negative aspiration is necessary to avoid intravenous injection. If sharp pain is felt during the intra-articular injection, it is wise to adjust the needle, since sharp pain is often found if the needle is located in the synovia. When intra-articular injection is given, there should not be any significant resistance. Before applying wound dressing, satisfactory hemostasis should be achieved.

# References


1. Gimarc DC, Stratchko LM, Ho CK. Spinal injections. Semin Musculoskelet Radiol. 2021;25(6):756–68.
2. Alam M, Tung R. Injection technique in neurotoxins and fillers: indications, products, and outcomes. J Am Acad Dermatol. 2018;79(3):423–35.
3. WHO Guidelines Approved by the Guidelines Review Committee. WHO Best Practices for Injections and Related Procedures Toolkit. Geneva: World Health Organization. Copyright © 2010, World Health Organization.; 2010.
4. Bennell KL, Paterson KL, Metcalf BR, Duong V, Eyles J, Kasza J, et al. Effect of intra-articular platelet-rich plasma vs placebo injection on pain and medial tibial cartilage volume in patients with knee osteoarthritis: the RESTORE randomized clinical trial. JAMA. 2021;326(20):2021–30.
5. Hecker A, Jungwirth-Weinberger A, Bauer MR, Tondelli T, Uçkay I, Wieser K. The accuracy of joint aspiration for the diagnosis of shoulder infections. J Shoulder Elb Surg. 2020;29(3):516–20.
6. Courtney P, Doherty M. Joint aspiration and injection. Best Pract Res Clin Rheumatol. 2005;19(3):345–69.
7. O'Shea NE, Tadi P. Olecranon Bursa aspiration. StatPearls. Treasure Island (FL): StatPearls Publishing. Copyright © 2023, StatPearls Publishing LLC.; 2023.
8. Kotecha J, Gration B, Hunt BJ, Goodman AL, Malaiya R. The safety of continued oral anticoagulation therapy in joint injections and aspirations: a qualitative review of the current evidence. J Clin Rheumatol. 2022;28(4):223–8.
9. Hoeber S, Aly AR, Ashworth N, Rajasekaran S. Ultrasound-guided hip joint injections are more accurate than landmark-guided injections: a systematic review and meta-analysis. Br J Sports Med. 2016;50(7):392–6.
10. Chalmers PN, Ellman MB, Chahal J, Verma NN. Injection therapy in the Management of Musculoskeletal Injuries of the knee. Oper Tech Sports Med. 2012;20(2):172–84.
11. Hermans J, Bierma-Zeinstra SM, Bos PK, Verhaar JA, Reijman M. The most accurate approach for intra-articular needle placement in the knee joint: a systematic review. Semin Arthritis Rheum. 2011;41(2):106–15.

Thorkell Snaebjörnsson

## 7.1 Introduction

With the modern standard of health care for patients, it is of great value for every healthcare provider to be able to offer accurate and up-to-date patient-specific aftercare after injection treatment. Specific details are described in subsequent chapters.

## 7.2 Immediate Aftercare

In the immediate aftercare after injection, it is important to place the patient under observation for approximately 30 min. Although the risk of type 1 hypersensitivity reactions (urticaria, anaphylaxis, angioedema) is much lower than, for example, in vaccinations, the risk of these adverse effects is to be found especially when injecting other agents than corticosteroids [1]. Other delayed immunological reactions have mostly been described concerning vaccinations and may include type IV hypersensitivity, including large local lesions or other autoimmune-mediated skin reactions [2]. The prior medical history of

T. Snaebjörnsson (✉)
Department of Orthopaedics, Institute of Clinical Sciences, Sahlgrenska Academy, Gothenburg University, Gothenburg, Sweden

Department of Orthopaedics, Sahlgrenska University Hospital, Mölndal, Sweden
e-mail: thorkell.snabjornsson@vgregion.se

**Table 7.1** Information about aftercare for patients

- Remain at the clinic until feeling well after the procedure
- Who to contact in case of emergency
- When to remove the dressing
- When to shower or take a bath
- Any physical restrictions
- Need to keep journal of symptoms

patients is therefore of great value, and special care should be taken when treating patients who have previously had reactions to injections or aspirations (Table 7.1). In clinical settings, intra-articular injections of either platelet-rich plasma or hyaluronic acid are not known to be a source of great discomfort [3], according to a recent study by Park et al.

## 7.3 Dressing

It is important to slowly remove the needle and use cotton wipes or other wound dressing to maintain pressure for 10–15 s before you can observe if there is any bleeding or swelling in the area after the procedure. If there is an actual bleeding or swelling in the area, it is recommended to apply pressure for 5–10 min either with a cotton swab or a compression and even elevate the area when possible. Further prolonged compression to stop bleeding is occasionally required. Appropriate wound dressing is then subsequently applied after hemostasis is secured.

If no bleeding or swelling is to be found, a simple band aid is sufficient to close cover the incision. If bleeding occurs in the dressing, the wound should be examined again the following day and new dressing applied if needed. If the patient wants to take a shower after the procedure, the dressing must be waterproof, and the usage of soaps or lotions should be avoided; likewise the usage of bath or swimming pool the following 24 hours after procedure should be avoided.

## 7.4 Pressure and Cooling

Cold and compression therapy is often used in acute trauma or injury. Similar methods can be applied when performing injections or aspirations. The rationale is that the cold suppresses the metabolic rate in the local soft tissues [4]. This decrease in metabolism reduces then the tissue damage caused by hypoxia [5]. Hypothermia significantly lowers microcirculation and induces vasoconstriction with the cooling effect lasting up to 60 minutes [6]. The cold therapy is dependent on time length since the local area must be cooled down for several minutes to achieve the hypoxic state. Similar effects on edema and blood microcirculation can be achieved with local pressure, but the maximum effect is accomplished when both methods are used simultaneously. In clinical settings, this treatment is often given by applying compression socks or elastic bandages with cooling units or custom-made orthosis or braces with hypothermal function. Treatment is applied for 5–10 min initially, and then the length of treatment can vary depending on symptoms.

## 7.5 Mobilization and Return to Activity

After intra-articular injections, which are most often combined with a short-acting local anesthetic, it is recommended to cycle the joint at least ten times or 1–2 min to achieve diffusion within the joint space. It is then important to register any changes in symptoms, since the local anesthetic agent should affect the nerve ends in the joint and provide symptom relief if the injection was correctly performed, noting injection accuracy and indication for the procedure. There are no scientifically proven guidelines about activity after injections. It is of value to ask the patient to keep a journal of symptoms and activities the following days, especially if the procedure is extra-articular or for a diagnostic purpose. It is a common practice to ask the patients to rest or only do moderate activities approximately 24 hours after an intra-articular injection given the fact there is an increased amount of fluid in the joint that can cause pain and soreness in activities. This is especially important for load-bearing joints, but even for other joints, excessive training is not recommended the following days after treatment. Return to physically demanding work should be delayed until the day after an injection, while patients only undergoing aspiration should now need to have the same restrictions. Operating a vehicle cannot be recommended directly after injection, especially if the area is still numb after local anesthesia. This information must be available before treatment in order for patients to be able to react accordingly.

## 7.6 Postinjection Pain: General Complications

Many patients experience discomfort while receiving treatment like injection or aspiration. The majority of these symptoms are vasovagal and can be treated with a professional approach and adequate patient information before treatment. These symptoms are most often self-limiting and are frequently described the following 24 hours after injection. In the case of adverse effects after platelet-rich plasma or hyaluronic acid injections, symptoms are often nonspecific [7] and include headache, dizziness, nausea, gastric complaints, stiffness, or syncope. If the local anesthetic agent is given in combination with other ingredients, extra care should be taken to give patients appropriate pain medication when the effect of the local anesthetic fades away.

## 7.7  Postinjection Pain: Local Complications

Local swelling and discomfort during either aspiration or injection is a common nonspecific complication. It can depend on the nerve endings in the skin, bleeding, or volume expansion. There is also another important factor in the deeper tissues because of the increased volume of fluid that is injected in the joint or the soft tissues, causing expansion of volume, swelling, and pain because of local pressure increase on the adjacent tissues. In smaller joints, the margin of error is smaller with a higher risk of complication or local pain. This is mostly because of the relatively large volume of fluid injected in a small cavity, compared with injections in larger joints that seldom give similar symptoms. The pressure in the small joint is therefore temporarily increased with subsequent discomfort, causing pressure on adjacent tissues and in some cases even pain following extra-articular injection in sensitive areas. Patients experiencing pain, especially during injections to small joints, should be given pain medication if they experience pain during the treatment.

## 7.8  Elective Surgery after Injection

In many cases, the injections given to patients are aimed to complement rehabilitation or to treat pain and discomfort until the time comes to do more definitive surgery. While this is a common clinical practice, care should be taken not to do any injection shortly before elective surgery, since studies have shown that the risk of infection can be increased following steroid injection to the knee (Table 7.2).

**Table 7.2** Information about aftercare for healthcare providers

- Inform restrictions after the procedure
- When to remove dressing
- Schedule a follow-up
- Avoid doing arthroscopy for 2 weeks after injection
- Avoid doing total joint arthroplasty 3 months after injection
- Provide contact information in case of adverse effects

Studies have shown evidence of an increased risk of adverse effects following surgery if the time interval between injection and surgery is too short. In a recent article by Lee et al. [8], the risk of infection following cortisone injection in the knee was increased if an arthroscopy of the knee was performed within 2 weeks of the injection. When patients are set to do a total joint arthroplasty, there is a current consensus of avoiding intra-articular injections 3 months before the operation to lower the risk of adverse effects [9]. With this evidence in mind, it is wise to either avoid injections under the aforementioned time frame before elective surgery or postpone elective surgeries if patients have recently undergone intra-articular injection in order to lower the risk of unnecessary adverse effects.

## 7.9  Conclusion

Injections offer a valuable treatment option for many patients. Standardized settings with clinical guidelines provide safety for both patients and healthcare providers. Information and treatment for allergic reactions should be available at all times. Applying pressure with or without the aid of hypothermal treatment after injection or aspiration is recommended in all cases. Mobilization with the cycling of a joint is advised directly after injection for the diffusion of the agent injected. If treatment is performed for a diagnostic value, the patients should keep a diary of symptoms in the following hours or days, depending on the agent given. Patients should be advised to only do activities that require moderate physical effort the following 24 h after treatment. Many patients experience nonspecific symptoms like headache or nausea during or shortly after either aspiration or injections. Information regarding pain after injection should be provided by the healthcare provider before treatment. Intra-articular injection results in the expansion of volume in the joint cavity and therefore even swelling and pain because of local pressure increase on the adjacent tissues. Elective surgery should not be planned directly after injection. The appropriate time frame between injection and arthroscopy is 2 weeks, while a total joint arthroplasty should not be performed until 3 months from injection.

## References


1. Peng S, Liang Y, Xiao W, Liu Y, Yu M, Liu L. Anaphylaxis induced by intra-articular injection of chitosan: a case report and literature review. Clin Case Rep. 2022;10(12):e6596.
2. Gambichler T, Boms S, Susok L, Dickel H, Finis C, Abu Rached N, et al. Cutaneous findings following COVID-19 vaccination: review of world literature and own experience. J Eur Acad Dermatol Venereol. 2022;36(2):172–80.
3. Park Y-B, Kim J-H, Ha C-W, Lee D-H. Clinical efficacy of platelet-rich plasma injection and its association with growth factors in the treatment of mild to moderate knee osteoarthritis: a randomized double-blind controlled clinical trial as compared with hyaluronic acid. Am J Sports Med. 2021;49(2):487–96.
4. Block JE. Cold and compression in the management of musculoskeletal injuries and orthopedic operative procedures: a narrative review. Open Access J Sports Med. 2010;1:105–13.
5. Wright JG, Araki CT, Belkin M, Hobson RW 2nd. Postischemic hypothermia diminishes skeletal muscle reperfusion edema. J Surg Res. 1989;47(5):389–96.
6. Yanagisawa O, Homma T, Okuwaki T, Shimao D, Takahashi H. Effects of cooling on human skin and skeletal muscle. Eur J Appl Physiol. 2007;100(6):737–45.
7. Shen L, Yuan T, Chen S, Xie X, Zhang C. The temporal effect of platelet-rich plasma on pain and physical function in the treatment of knee osteoarthritis: systematic review and meta-analysis of randomized controlled trials. J Orthop Surg Res. 2017;12(1):16.
8. Lee W, Bhattacharjee S, Lee MJ, Ho SW, Athiviraham A, Shi LL. A safe interval between preoperative intra-articular corticosteroid injections and subsequent knee arthroscopy. J Knee Surg. 2022;35(1):47–53.
9. Tang A, Almetwali O, Zak SG, Bernstein JA, Schwarzkopf R, Aggarwal VK. Do preoperative intra-articular corticosteroid and hyaluronic acid injections affect time to total joint arthroplasty? J Clin Orthop Trauma. 2021;16:49–57.

# Part II

# Non Biologic Agents for Injections

# Corticosteroids and Local Anesthetics

**8**

Matthieu Ollivier and Ahmed Mabrouk

## 8.1 Background

Corticosteroids and local anesthetics are two classes of drugs that revolutionized the medical practice. Since corticosteroids were discovered in the 1940s, they have been utilized worldwide and were proven to be effective in a wide array of medical conditions including inflammatory conditions, allergic reactions, and autoimmune disorders [1, 2]. Conversely, local anesthetics were first introduced in the nineteenth century and have since become a cornerstone of modern anesthesia [3–5].

### 8.1.1 Corticosteroids (CS)

CS are synthetic equivalents to the natural steroid hormones produced by the adrenal cortex and include glucocorticoids and mineralocorticoids. These synthetic agents vary in their mineralocorticoid/glucocorticoids attributes. Glucocorticoids have immunosuppressive, anti-inflammatory, and vasoconstrictive effects, besides their chief role in metabolism, whereas mineralocorticoids are involved in electrolytes and water balance [2].

The injectable corticosteroids are mainly glucocorticoids, and they exert their anti-inflammatory effects by decreasing local infiltration of inflammatory cells and mediators such as cytokines [6, 7].

There are three main types of injectable corticosteroids: hydrocortisone, triamcinolone, and methylprednisolone. Each of these corticosteroids has unique properties that make them effective for certain conditions. Hydrocortisone is often used for acute inflammation, triamcinolone for longer-term inflammation, and methylprednisolone for severe inflammation [8]. The original corticosteroid is hydrocortisone acetate. However, this was replaced by longer-acting alternatives such as methylprednisolone acetate, triamcinolone acetonide, and triamcinolone hexacetonide. Triamcinolone hexacetonide has been proven to be the most effective with clinical effects lasting up to several months [9].

The injectable corticosteroids differ in their crystal and chemical structure, solubility, and duration of action [6]. An injectable compound with low solubility has a tendency to stay longer at the injection site and with low systemic absorption when compared to a highly soluble compound.


M. Ollivier (✉)
Institut du Mouvement et de L'appareil Locomoteur
Marseille, Marseille, France

A. Mabrouk
Leeds Teaching Hospitals, Leeds, UK

### 8.1.2 Local Anesthetics (LA)

LA blocks transduction and transmission of nociception [10]. This means that LA block nerve impulses in a specific area of the body, thus preventing pain signals from being transmitted to the brain. LA can be administered through injection or topical application and are available in a variety of formulations and strengths [11]. There are two types of local anesthetics: esters and amides. Esters, such as procaine and cocaine, are typically short-acting and may cause allergic reactions. Amides, such as lidocaine and bupivacaine, are longer-acting and less likely to cause allergic reactions [12].

The first local anesthetic was cocaine, which was isolated from the leaves of the coca plant in the 1850s. However, cocaine's addictive properties and potential for toxicity led to the development of newer, safer local anesthetics, such as procaine and lidocaine [4]. Among the most commonly used local anesthetics in combination with steroids are the lidocaine, bupivacaine hydrochloride, and ropivacaine [13, 14]. A few merits have been reported for their combined use with steroids, such as increasing the volume of the injectate which allows better dissemination of steroids throughout the injected tissue. Additionally, they could provide a degree of immediate symptom relief and reduction in patient's discomfort due to the rapid onset of action and the relative durable anesthetic criteria [13, 14].

### 8.2 An Overview of the Indications

Corticosteroid injections in combination with local anesthetics are commonly used in the management of a variety of musculoskeletal conditions. Corticosteroids are injected into articular, periarticular, or soft tissue structures for pain relief, inflammation control, and enhancing mobility. Additionally, corticosteroids can be used as a definitive treatment in cases such as De-Quervain tenosynovitis and trochanteric bursitis [15]. Corticosteroids can supplement physical rehabilitation due to their analgesic effects as in cases of rotator cuff syndrome and lateral epicondylitis [15].

Corticosteroids can serve diagnostic purposes and can be utilized for postoperative pain control [16]. The addition of a local anesthetic can also help to confirm the diagnosis of musculoskeletal conditions by providing temporary pain relief. Chou et al. [17] reported that the addition of a local anesthetic to a corticosteroid injection provided greater pain relief compared to a corticosteroid injection alone in patients with knee osteoarthritis. Intra-articular steroid injections make pain relief in rheumatoid arthritis and osteoarthritis. Corticosteroid injections have been shown to be effective with improvements in pain and function in patients with osteoarthritis, tendinitis, and bursitis. Jüni et al. [18], in a systematic review, found that corticosteroid injections were effective in reducing pain and improving function in patients with knee osteoarthritis. Similarly, a systematic review by Coombes et al. [19] found that corticosteroid injections were effective from an analgesic and functional perspectives in patients with rotator cuff tendinitis.

### 8.3 Contraindications

Despite the benefits of injectable corticosteroids and local anesthetics, their use is contraindicated in some patients. However, these contraindications could be absolute or relative [20]. Local infection, bacteremia, and sepsis are absolute contraindications for corticosteroid injections as CS can suppress the immune system and exacerbate the infection. An increased risk of intraoperative and postoperative infection has been reported if hip joint arthroplasty is performed within 3 months of the steroid injection to the hip [21]. Additionally, due to the suppressive effect of corticosteroids on bone healing, intra-articular fractures are another contraindication. Similarly, due to the risk of subchondral osteonecrosis and weakening of other articular structures, CS are contraindicated in joint instability. Also, patients with a history of allergy to corticosteroids or local anesthetics should not receive injections of these agents [20].

More relative contraindications include patients with bleeding disorders, or those on anticoagulant therapy should also be cautious, as these agents can increase the risk of bleeding. Other contraindications include pregnancy, breastfeeding, and certain medical conditions, such as diabetes, hypertension, and congestive heart failure. The use of corticosteroids can increase blood glucose levels and blood pressure, while the use of local anesthetics can affect fetal heart rate and cause neonatal depression [22].

Certain risk factors for adverse events of local anesthetics should be noted. Individual risk factors should also be taken into consideration such as medical comorbidities with higher pulmonary, cardiac, and nervous susceptibilities, other concomitant medications, the site of LA injection, e.g., vessel rich tissue. In addition to care towards specific local anaesthetics and large total dose of LA [23, 24].

## 8.4    Complications

With the increased usage of corticosteroids and local anesthetics, major adverse events and complications are being more reported. Despite rare, physicians who are performing these injections should be aware of these complications as some of them can be potentially dramatic. Complications can vary from trivial ones such as skin depigmentation or a more serious complications such as septic arthritis [20]. Furthermore, concerns have been raised about the possibility of chondrotoxicity following intra-articular corticosteroids [25–27]. In a systematic review, Wernecke et al. [28] suggested that the effect of steroids on the articular cartilage is time and dose dependent. So, low doses over a short period of time would have beneficial effects, whereas large doses over prolonged periods of time could have devastating effects [28]. Similarly, LAs can result in a wide range of complications that vary from either minor symptoms and signs to more severe

systemic toxicity which can result in a disability or pose a threat to life [29].

## 8.5    Applications of Corticosteroids and Local Anesthetic in the Knee Joint

The knee joint is one of the most frequently joints to receive corticosteroids with or without local anesthetic injections [26], in scenarios such as early osteoarthritis of the knee which is a common condition that can cause pain and disability. Intra-articular corticosteroid injections have been shown to reduce pain and improve function in patients with early knee osteoarthritis [30]. In 2012, the American College of Rheumatology (ACR) guidelines weakly recommended the use of IA CS in patients with OA, whose cases were refractory to basic treatment [31]. In 2014, the Osteoarthritis Research Society International (OARSI) guidelines for the nonsurgical management of knee osteoarthritis also recommended intra-articular corticosteroid injections as a treatment option irrespective of the osteoarthritis subtype or patients' comorbidities [32].

In a 2015 update of a 2006 Cochrane review [33], including a total of 27 trial (14 new) [18], with all studies included were either RCTs or quasi-RCTs, and a control group receiving sham or no intervention. And a median of 76 participants (range 16–205) were randomized across all trials. A median prednisolone equivalent dose of 50 mg was reported across all trials, and the median number of CS injections was one.

Based on the metanalysis, at the end of 4–6 weeks of treatment, Jüni et al. [18] reported more effective pain reduction and function when compared to other control interventions. The difference in pain scores, between corticosteroids and other interventions, was 1 cm on a 10 cm VAS scale, which corresponds to 17% improvement and when compared to sham injection, corticosteroids had 13% more functional improvement. However, IA CS had no effect on quality of life or joint space narrowing when compared to control interventions [18]. The

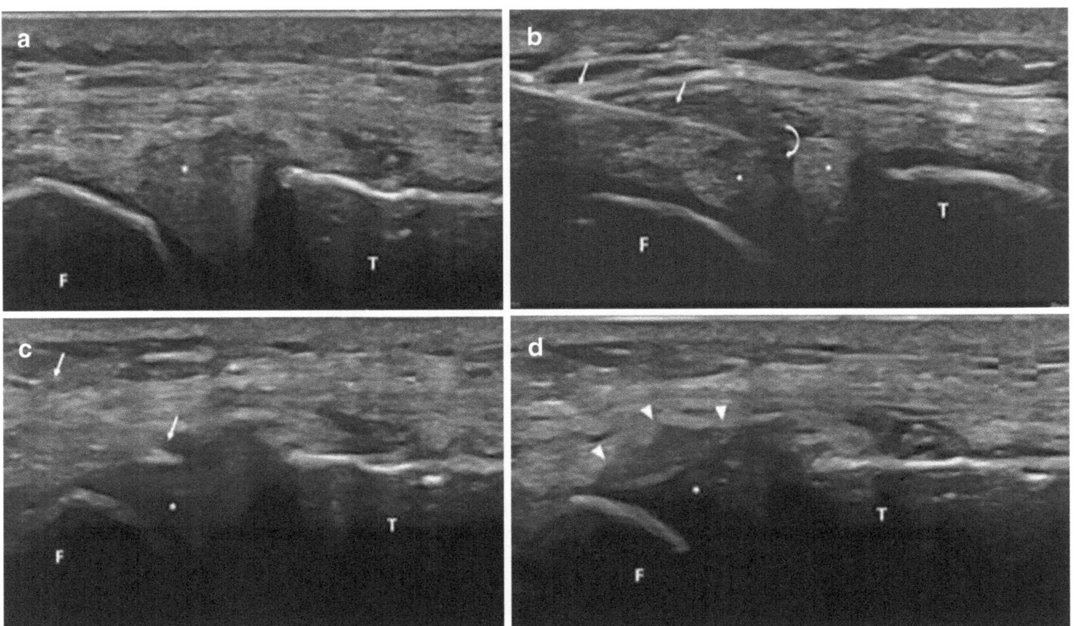

**Fig. 8.1** Peri-meniscal injection of a degenerative meniscus tear, the tear is visualized under ultrasonography (**a**), the needle approaches the lesion under direct control (**b**), the steroid injection can be seen (**c**), once the needle is removed, the steroid liquid surrounds the meniscus tissues (**d**)

American Academy of Orthopedic Surgeons (AAOS), in their recent guidelines for non-arthroplasty management of knee OA, advised that intra-articular corticosteroids may offer short-term relief for patients with symptomatic knee OA, with reported moderate strength recommendations [34].

Another application in knee joint for IA CS is meniscal injuries. Meniscal damage can set off an inflammatory cascade in the knee with local synovitis, effusion, and subsequent chondrosis [25]. Intra-articular corticosteroids can neutralize the inflammatory mediators and reverse the inflammatory cascade, hence can also be effective in treating meniscus or cartilage issues. The injection itself does not alter the meniscus; however, pain reduction or absence allows recovery of some biological and mechanical stability [35–37].

In patients with degenerative tears of the posterior horn of the medial meniscus, intra-articular steroid injections have been shown to result in short- and medium-term pain relief in the majority of the patients (81.7%) [38]. And more recently, peri-meniscal corticosteroid injections (Fig. 8.1) were demonstrated to result in significant symptom relief at 6 weeks in patients with meniscal pain [39].

## 8.6 Conclusion

Injectable corticosteroids and local anesthetics can be effective in the management of a variety of musculoskeletal conditions. Multiple knee joint disorders are frequently addressed with corticosteroids with or without local anesthetics injection. The use of corticosteroids should be carefully considered and monitored by a healthcare professional. Patients should also be informed of the potential risks and benefits of these medications.

## References

1. Coutinho AE, Chapman KE. The anti-inflammatory and immunosuppressive effects of glucocorticoids, recent developments and mechanistic insights. Mol Cell Endocrinol. 2011;335(1):2–13.
2. Liu D, Ahmet A, Ward L, Krishnamoorthy P, Mandelcorn ED, Leigh R, et al. A practical guide to the monitoring and management of the complications of systemic corticosteroid therapy. Allergy Asthma Clin Immunol. 2013;9(1):30.
3. Brown TCK. History of pediatric regional anesthesia. Paediatr Anaesth. 2012;22(1):3–9.
4. Grinspoon L, Bakalar JB. Coca and cocaine as medicines: an historical review. J Ethnopharmacol. 1981;3(2–3):149–59.
5. Ring ME. The history of local anesthesia. J Calif Dent Assoc. 2007;35(4):275–82.
6. Cole BJ, Schumacher HR. Injectable corticosteroids in modern practice. J Am Acad Orthop Surg. 2005;13(1):37–46.
7. Rhen T, Cidlowski JA. Antiinflammatory action of glucocorticoids—new mechanisms for old drugs. N Engl J Med. 2005;353(16):1711–23.
8. Corticosteroids. In: LiverTox: Clinical and Research Information on Drug-Induced Liver Injury [Internet]. Bethesda (MD): National Institute of Diabetes and Digestive and Kidney Diseases; 2012 [cited 2023 Apr 10]. Available from: http://www.ncbi.nlm.nih.gov/books/NBK548400/
9. Ostergaard M, Halberg P. Intra-articular corticosteroids in arthritic disease: a guide to treatment. BioDrugs. 1998;9(2):95–103.
10. Barletta M, Reed R. Local Anesthetics: pharmacology and special preparations. Vet Clin North Am Small Anim Pract. 2019;49(6):1109–25.
11. Nestor CC, Ng C, Sepulveda P, Irwin MG. Pharmacological and clinical implications of local anaesthetic mixtures: a narrative review. Anaesthesia. 2022;77(3):339–50.
12. Ruetsch YA, Böni T, Borgeat A. From cocaine to ropivacaine: the history of local anesthetic drugs. Curr Top Med Chem. 2001;1(3):175–82.
13. Rastogi AK, Davis KW, Ross A, Rosas HG. Fundamentals of joint injection. AJR Am J Roentgenol. 2016;207(3):484–94.
14. Seshadri V, Coyle CH, Chu CR. Lidocaine potentiates the chondrotoxicity of methylprednisolone. Arthrosc J Arthrosc Relat Surg. 2009;25(4):337–47.
15. Stephens MB, Beutler AI, O'Connor FG. Musculoskeletal injections: a review of the evidence. Am Fam Physician. 2008;78(8):971–6.
16. Wittich CM, Ficalora RD, Mason TG, Beckman TJ. Musculoskeletal injection. Mayo Clin Proc. 2009;84(9):831–6. quiz 837
17. Chou R, McDonagh MS, Nakamoto E, Griffin J. Analgesics for osteoarthritis: an update of the 2006 comparative effectiveness review [internet]. Rockville (MD): Agency for Healthcare Research and Quality (US). 2011 [cited 2023 Apr 10]. (AHRQ Comparative Effectiveness Reviews). Available from: http://www.ncbi.nlm.nih.gov/books/NBK65646/
18. Jüni P, Hari R, Rutjes AWS, Fischer R, Silletta MG, Reichenbach S, et al. Intra-articular corticosteroid for knee osteoarthritis. Cochrane Database Syst Rev. 2015;2015(10):CD005328.
19. Coombes BK, Bisset L, Vicenzino B. Efficacy and safety of corticosteroid injections and other injections for management of tendinopathy: a systematic review of randomised controlled trials. Lancet Lond Engl. 2010;376(9754):1751–67.
20. Šimurina T, Mraović B, Župčić M, Graf Župčić S, Vulin M. Local anesthetics and steroids: contraindications and complications—clinical update. Acta Clin Croat. 2019;58(Suppl 1):53–61.
21. Schairer WW, Nwachukwu BU, Mayman DJ, Lyman S, Jerabek SA. Preoperative hip injections increase the rate of periprosthetic infection after Total hip arthroplasty. J Arthroplast. 2016;31(9 Suppl):166–9.e1
22. MacMahon PJ, Eustace SJ, Kavanagh EC. Injectable corticosteroid and local anesthetic preparations: a review for radiologists. Radiology. 2009;252(3):647–61.
23. Neal JM, Barrington MJ, Fettiplace MR, Gitman M, Memtsoudis SG, Mörwald EE, et al. The third American Society of Regional Anesthesia and Pain Medicine practice advisory on local Anesthetic systemic toxicity: executive summary 2017. Reg Anesth Pain Med. 2018;43(2):113–23.
24. Sekimoto K, Tobe M, Saito S. Local anesthetic toxicity: acute and chronic management. Acute Med Surg. 2017;4(2):152–60.
25. Edd SN, Giori NJ, Andriacchi TP. The role of inflammation in the initiation of osteoarthritis after meniscal damage. J Biomech. 2015;48(8):1420–6.
26. Ayhan E, Kesmezacar H, Akgun I. Intraarticular injections (corticosteroid, hyaluronic acid, platelet rich plasma) for the knee osteoarthritis. World J Orthop. 2014;5(3):351–61.
27. McAlindon TE, LaValley MP, Harvey WF, Price LL, Driban JB, Zhang M, et al. Effect of intra-articular triamcinolone vs saline on knee cartilage volume and pain in patients with knee osteoarthritis: a randomized clinical trial. JAMA. 2017;317(19):1967–75.
28. Wernecke C, Braun HJ, Dragoo JL. The effect of intra-articular corticosteroids on articular carti-

lage: a systematic review. Orthop J Sports Med. 2015;3(5):2325967115581163.

29. Gitman M, Barrington MJ. Local Anesthetic systemic toxicity: a review of recent case reports and registries. Reg Anesth Pain Med. 2018;43(2):124–30.

30. Rutjes AWS, Jüni P, da Costa BR, Trelle S, Nüesch E, Reichenbach S. Viscosupplementation for osteoarthritis of the knee: a systematic review and meta-analysis. Ann Intern Med. 2012;157(3):180–91.

31. Hochberg MC, Altman RD, April KT, Benkhalti M, Guyatt G, McGowan J, et al. American College of Rheumatology 2012 recommendations for the use of nonpharmacologic and pharmacologic therapies in osteoarthritis of the hand, hip, and knee. Arthritis Care Res. 2012;64(4):465–74.

32. McAlindon TE, Bannuru RR, Sullivan MC, Arden NK, Berenbaum F, Bierma-Zeinstra SM, et al. OARSI guidelines for the non-surgical management of knee osteoarthritis. Osteoarthr Cartil. 2014;22(3):363–88.

33. Bellamy N, Campbell J, Robinson V, Gee T, Bourne R, Wells G. Intraarticular corticosteroid for treatment of osteoarthritis of the knee. Cochrane Database Syst Rev. 2006;2:CD005328.

34. Brophy RH, Fillingham YA. AAOS clinical practice guideline summary: management of osteoarthritis of the knee (nonarthroplasty), third edition. J Am Acad Orthop Surg. 2022;30(9):e721–9.

35. Adams JG, McAlindon T, Dimasi M, Carey J, Eustace S. Contribution of meniscal extrusion and cartilage loss to joint space narrowing in osteoarthritis. Clin Radiol. 1999;54(8):502–6.

36. Berthiaume MJ, Raynauld JP, Martel-Pelletier J, Labonté F, Beaudoin G, Bloch DA, et al. Meniscal tear and extrusion are strongly associated with progression of symptomatic knee osteoarthritis as assessed by quantitative magnetic resonance imaging. Ann Rheum Dis. 2005;64(4):556–63.

37. Ding C, Martel-Pelletier J, Pelletier JP, Abram F, Raynauld JP, Cicuttini F, et al. Knee meniscal extrusion in a largely non-osteoarthritic cohort: association with greater loss of cartilage volume. Arthritis Res Ther. 2007;9(2):R21.

38. Byrne C, Alkhayat A, Bowden D, Murray A, Kavanagh EC, Eustace SJ. Degenerative tears of the posterior horn of the medial meniscus: correlation between MRI findings and outcome following intra-articular steroid/bupivacaine injection of the knee. Clin Radiol. 2019;74(6):488.e1–8.

39. Coll C, Coudreuse JM, Guenoun D, Bensoussan L, Viton JM, Champsaur P, et al. Ultrasound-guided Perimeniscal injections: anatomical description and feasibility study. J Ultrasound Med. 2022;41(1):217–24.

# Viscosupplementation Agents

Camila Grandberg, Svenja Höger,
and M. Enes Kayaalp

## 9.1 Background

Viscosupplementation is a procedure in which gel-like hyaluronic acid (HA) is injected into affected joints. The aim is to restore the viscoelastic properties of the synovial fluid (SF) and alleviate joint inflammation and pain. The principal ideas behind the use of HA are that it helps to enhance the physiologic viscoelasticity of the SF and it exhibits downregulatory effects on pro-inflammatory factors and enzymes contributing to joint matrix degradation. Presently, many different HA products exist, varying widely in origin, preparation, molecular weight, rheological characteristics, and concentration [1].

The American College of Rheumatology (ACR) recognized viscosupplementation as a therapeutic option for pain management in knee osteoarthritis (OA) in 2000 [2], following several randomized controlled trials (RCTs), which demonstrated that HA yielded superior functional and pain-related outcomes compared to placebo. Since then, the data on the use of HA has largely evolved. After a decade of research, there is now solid evidence both supporting and contradicting its use. Nevertheless, current practice incorporates the use of HA very frequently despite many reputable guidelines advising against it. The increase of studies on HA can be attributed to the increasing prevalence of OA in modern society, the high numbers of unresponsive or unfit patients for other treatments, and the potential biases in treatment selection by care providers [3].

This chapter aims to provide insights into the current literature and shed light on the reasons behind the divergent approaches to the use of HA in the clinical setting.

C. Grandberg
Department of Orthopaedic Surgery, UPMC Freddie
Fu Sports Medicine Center, University of Pittsburgh,
Pittsburgh, PA, USA
e-mail: grandbergc@upmc.edu

S. Höger
Department of Orthopaedic Surgery, UPMC Freddie
Fu Sports Medicine Center, University of Pittsburgh,
Pittsburgh, PA, USA

Department of Sports Orthopaedics, Technical
University of Munich, Munich, Germany
e-mail: hogers@upmc.edu

M. E. Kayaalp (✉)
Department of Orthopaedic Surgery, UPMC Freddie
Fu Sports Medicine Center, University of Pittsburgh,
Pittsburgh, PA, USA

Department of Orthopaedics, Istanbul Kartal Training
and Research Hospital, Istanbul, Turkey
e-mail: mek@mek.md


B. Kocaoglu et al. (eds.), *Musculoskeletal Injections Manual*,
https://doi.org/10.1007/978-3-031-52603-9_9

## 9.2 Mechanism of Action

In the presence of shear stress, the physiological viscosity and elasticity of the SF present in human joints can alter significantly [2]. The concentration of HA in the SF is thought to have a substantial effect on these native properties, with lower concentrations of HA promoting an increase in SF viscosity [2]. In OA, the joint's SF has a characteristically lower concentration of HA, as well as a higher turnover and lower molecular weight. These combined factors generate various adverse biomechanical effects, possibly leading to pain and loss of function [2]. Therefore, viscosupplementation treatment aims to reinstate the physiologic viscoelasticity of human SF and to provide improvement of OA symptoms [1, 4].

Viscosupplementation essentially exerts its action through two main mechanisms: a biochemical effect and a mechanical effect. Mechanically, the HA lubricates the joint, lessening the friction generated by joint movement. It also promotes force distribution and shock absorption, diminishing the effects of weight bearing and movement impact exerted upon the joint [3, 5]. Biochemically, the HA promotes direct analgesia through joint nociceptor inhibition [2, 5, 6] and provides a moderate anti-inflammatory effect by reducing the expression of pro-inflammatory cytokines, via interaction with joint synoviocytes, and suppressing macrophage production of prostaglandin E2, via down-regulation of NF-kappaB [1, 2, 5, 6]. Lastly, HA exerts a chondroprotective effect through proteoglycan and glycosaminoglycan synthesis, stimulation of chondrocyte proliferation, promotion of synoviocyte-mediated endogenous HA production, and inhibition of metalloproteinases production and cartilage degradation [1–3, 5, 6].

## 9.3 Viscosupplementation Agents

HA is a natural compound present in SF, the physiologic solution present in synovial joints with the purpose of lubricating the joint and minimizing friction during joint movement [1]. The medical use of HA, the main component of viscosupplementation, was initiated in the late 1960s, employing purified human umbilical cord and rooster comb as a product source [7]. The physiologic volume of SF present in the human knee is about 2 ml, with a HA concentration of 2.5–4.0 mg/ml and a molecular weight of 4 to $5 \times 10^6$ Da [7]. Exogenous HA may be produced from two origins: avian or non-avian. HAs from avian origin are purified natural products made from rooster combs, while non-avian products are made using fermentation with bacteria (*Streptococcus zooepidemicus*) [5]. These two types of HAs can be further classified as non-cross-linked (hyaluronans) or cross-linked (hylans). The first has long-chain molecules with a molecular weight of 0.5 to $1.8 \times 10^6$ Da. The latter is a chemically modified molecule formed by cross-link connections between hyaluronans or other molecules, such as chondroitin sulfate, sorbitol, and mannitol, in different concentrations. These cross-linked HAs exhibit a liquid phase of molecular weight around $6 \times 10^6$ Da and a solid phase of infinite molecular weight [4, 5, 8].

Currently, there are over 100 HA products available globally. These products have appreciable differences, not only in terms of origin and molecular properties but also in concentrations, dosage, and intervals between injection administration (Table 9.1) [8].

**Table 9.1** Characteristics of the main HA products currently available [1, 4, 5, 9, 10]

| Product name | Origin | Structure | Molecular weight ($\times 10^6$ Da) | Injection interval | Dosage |
|---|---|---|---|---|---|
| Synvisc | Avian | Cross-linked | 6.0 | 1 per week over 3 weeks | 2 ml (16 mg) |
| Euflexxa | Bacterial fermentation | Non-cross-linked | 2.4–3.6 | 1 per week over 3 weeks | 2 ml (20 mg) |
| Supartz Fx | Avian | Non-cross-linked | 0.6–1.2 | 1 per week over 3–5 weeks | 2.5 ml (25 mg) |
| Gelsyn-3 (Sinovial®) | Bacterial fermentation | Non-cross-linked | 1.1 | 1 per week over 3 weeks | 2 ml (16.8 mg) |
| Durolane | Bacterial fermentation | Cross-linked | 1.0-1.8 | Single injection | 3 ml (60 mg) |
| Hyalgan | Avian | Non-cross-linked | 0.5-0.7 | 1 per week over 3–5 weeks | 2 ml (20 mg) |
| Gel-One | Avian | Cross-linked | Not reported | Single injection | 3 ml (30 mg) |
| Synvisc-One | Avian | Cross-linked | 6.0 | Single injection | 6 ml (48 mg) |
| Monovisc | Bacterial fermentation | Non-cross-linked | 1.0–2.9 | Single injection | 4 ml (88 mg) |
| Orthovisc | Bacterial fermentation | Non-cross-linked | 1.0–2.9 | 1 per week over 3–4 weeks | 2 ml (30 mg) |
| Hymovis | Bacterial fermentation | Non-cross-linked | 0.5–0.7 | 1 per week over 2 weeks | 3 ml (24 mg) |

## 9.4 Contraindications and Adverse Events

Viscosupplementation is a generally well-tolerated procedure, with less systemic side effects than other intra-articular injections or more aggressive interventions used for OA. However, adverse events are still a possibility, and local reactions may happen more often than with other treatment options [4]. The literature on possible adverse events associated with viscosupplementation is still scarce and presents a wide variety of incidence rates. Nevertheless, frequently reported adverse events are injection site pain, local skin reactions and erythema, local joint pain and effusion, septic arthritis, and pseudo septic reactions [1, 4].

Therefore, HA injections should be indicated with caution, and contraindications should be fully respected. General contraindications of intra-articular injections are likewise applied to viscosupplementation injections. Absolute contraindications include fracture site, overlying skin infection, severely compromised immune status, and suspected or confirmed bacteremia or infectious arthritis. Relative contraindications comprise diagnosed coagulopathies, joint prosthesis, and poorly controlled diabetes mellitus. Additionally, contraindications to viscosupplementation include hypersensitivity to previous HA products or other avian products, previous lack of efficacy, and pediatric patients [1, 4].

## 9.5 Evidence

There are numerous guidelines available worldwide that provide up-to-date recommendations for health care professionals [11–15]. However, differences in methodology in their preparation, as well as publication timelines, can lead to variations in their recommendations. For instance, the Osteoarthritis Research Society International (OARSI) has provided a conditional recommendation for patients with OA [12]. The American Academy of Orthopaedic Surgeons (AAOS) does not endorse the use of HA injection, with ACR/Arthritis Foundation (AF) indicating conditional recommendation against their use [13]. The European Society for Clinical and Economic Aspects of Osteoporosis, Osteoarthritis and Musculoskeletal Diseases (ESCEO) has issued a weak recommendation for HA use, suggesting it only be used when patients have contraindications to the use of nonsteroidal anti-inflammatory drugs (NSAIDs) or experience insufficient pain relief on NSAID therapy [14]. Similarly, the previous AAOS guidelines advised strongly against HA use for knee OA [15]. However, the latest AAOS guidelines indicated significant advances with high molecular weight cross-linked HA therapy, modifying their recommendation from strongly to moderately against the use of HA [13, 15].

First and foremost, it is crucial to acknowledge that the evidence regarding the use of HA is characterized by inconsistent and conflicting findings. While certain RCTs demonstrated significant pain improvement, others do not yield similar positive outcomes. Moreover, comprehensive systematic reviews and meta-analyses conducted on a larger scale have concluded that the effectiveness of HA may not extend significantly beyond the placebo effect [1]. Additionally, an expert opinion on this matter suggested that comparing HA with placebo may not reflect actual clinical outcomes, as placebo should not be considered a treatment for knee OA [16]. Taking into consideration the significant placebo effect of saline injections, there is some rationale to this approach [17, 18]. In fact, this raises further doubts about the cost-effectiveness and clinical benefits compared to intra-articular saline. Therefore, high-quality RCTs are necessary to assess the effect of HA in osteoarthritic knees to enrich the existing literature.

## 9.6	Joint-Based Applications

The available evidence on the use of HA primarily focuses on knee OA. This section aims to offer a concise overview of the existing evidence, categorized by anatomic location.

### 9.6.1	Shoulder

The guidelines provided by AAOS regarding glenohumeral OA strongly discourage the use of HA, providing robust evidence for this recommendation [19]. OARSI and ACR do not have special recommendations regarding the use of HA in shoulder OA. Additionally, although a meta-analysis conducted in 2019 concluded that HA use in the shoulder was safe and offered pain relief, ultimately it was determined that that the pain improvement observed in both the HA and control groups was due to placebo effect [20]. Therefore, HA seems unsuitable for use in shoulder OA, considering scientific justifications.

### 9.6.2	Ankle

The National Institute for Health and Care Excellence (NICE) guidelines for OA care and management in adults is opposed to the use of HA in ankle OA [11]. Further guidelines did not make any statements on the use of HA in the ankle joint [2]. Furthermore, a systematic review evaluating the use of HA in ankle OA indicated improved pain scores [21]. In addition, a Cochrane systematic review reported that HA use was associated with lower pain score on the Ankle Osteoarthritis Scale at 6 months, although the clinical significance of this observation remains unclear [22].

### 9.6.3	Knee

A systematic review conducted in 2020 analyzed 27 sets of guidelines on intra-articular therapies for knee OA, including guidelines from AAOS and NICE [23]. Among these guidelines, 74% ($n = 20$) provided strong or conditional recommendations in favor of using intra-articular HA; approximately 15% ($n = 4$) offered uncertain or weak recommendations; and 11% ($n = 3$) strongly recommended against its use [23].

In the past, previous studies primarily focused on the pain-reducing effects of HA, known as "symptomatic slow-acting drugs in osteoarthritis."

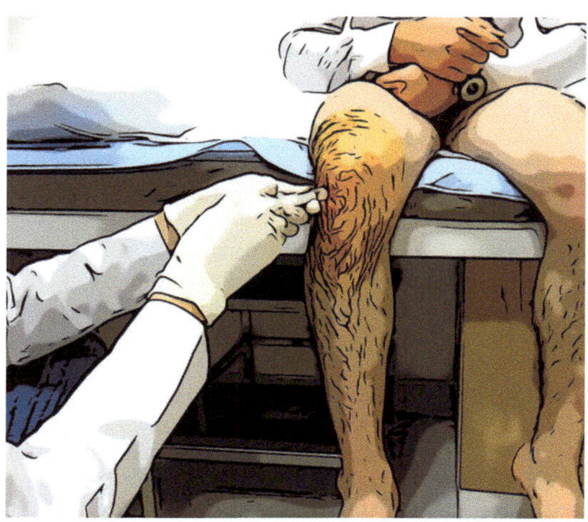

**Fig. 9.1** Intra-articular knee injection using a lateral approach

Currently, emerging evidence suggests that high molecular weight HA may also impact cartilage metabolism itself, suggesting a potential disease-modifying effect (Fig. 9.1). Intra-articular administration of high molecular weight HA has shown promise in reducing C-telopeptide of type II collagen, indicating improved cartilage metabolism [24, 25]. This suggests that HA could be classified as a "disease-modifying osteoarthritis drug" instead of solely as a symptomatic agent [8].

Based on RCTs that compared viscosupplementation with placebo and with no intervention for knee OA treatment, the evidence strongly indicates that viscosupplementation offers only a minor reduction in knee OA pain compared to placebo. However, this difference falls below the minimal clinically important threshold. Furthermore, the RCTs also demonstrate a higher risk of serious adverse events associated with viscosupplementation compared to placebo. Consequently, these findings do not support the widespread use of HA for the treatment of knee OA [26].

### 9.6.4  Hip

Regarding hip OA, there is currently no major international guideline that recommends the use of HA. Recent NICE, AAOS, and ACR guide-lines specifically advise against its use [2]. Several meta-analyses have been conducted on the use of HA in hip OA in the literature. Lieberman et al. found that the observed change in their analysis had uncertain clinical relevance, as the decrease in VAS was only −0.27 in the six RCTs and most studies had a follow-up duration of less than 6 months [27]. Another meta-analysis which included eight RCTs concluded that viscosupplementation for hip OA is not recommended [28]. The analysis revealed limited evidence compared to placebo, scarce evidence showing an efficacy up to 3 months, and suggested no difference at the 5-month mark [28]. Moreover, a third meta-analysis of RCTs showed that the use of HA for hip OA provided only comparable improvements to those achieved with intra-articular saline injections [29].

### 9.6.5  Hand

Recent ACR and NICE guidelines both advise against the use of HA in the hand. Conversely, AAOS and OARSI guidelines do not provide specific recommendations regarding the management of OA in the hand [2]. According to a 2019 meta-analysis analyzing 9 RCTs, with a total of 504 patients, that compared HA, corticosteroid, saline placebo, and dextrose injections, for base of thumb OA, none of the evaluated studies found a therapy superior to placebo [30]. Furthermore, a different meta-analysis found no difference between HA and corticosteroid until 12 weeks posttreatment when comparing 428 patients across six RCTs involving HA, corticosteroid, and placebo injections [31]. However, HA appeared to improve function by pulp pinch force, although HA had less effective outcomes in terms of pain control than corticosteroid injections [2, 31].

## 9.7  Conclusion

In conclusion, viscosupplementation is widely acknowledged as a safe treatment option for OA, not only providing symptomatic management but possibly altering the disease course due to HA's chondroprotective properties. However, clinical benefits remain uncertain, and recommended use varies according to the joint being treated. Furthermore, the variations in product formulations and the inconsistency in study designs present challenges in obtaining high-quality comparative results from the existing data.

# References


1. Peck J, Slovek A, Miro P, et al. A comprehensive review of viscosupplementation in osteoarthritis of the knee. Orthop Rev (Pavia). 2021;13(2):25549. https://doi.org/10.52965/001c.25549.
2. Bowden DJ, Eustace SJ, Kavanagh EC. The value of injectable viscoelastic supplements for joints. Skelet Radiol. 2023;52(5):933–40. https://doi.org/10.1007/s00256-022-04178-3.
3. Pavone V, Vescio A, Turchetta M, Giardina SMC, Culmone A, Testa G. Injection-based management of osteoarthritis of the knee: a systematic review of guidelines. Front Pharmacol. 2021;12:661805. https://doi.org/10.3389/fphar.2021.661805.
4. Provenzano DA, Chandwani K. Joint injections. In: Benzon HT, Rathmell JP, Wu CL, Turk DC, Argoff CE, Hurley RW, editors. Practical Management of Pain. 5th ed. Mosby; 2014. p. 966–80. chap 71.
5. de Rezende MU, de Campos GC. Viscosupplementation. Rev Bras Ortop. 2012;47(2):160–4. https://doi.org/10.1016/S2255-4971(15)30080-X.
6. Legre-Boyer V. Viscosupplementation: techniques, indications, results. Orthop Traumatol Surg Res. 2015;101(1 Suppl):S101–8. https://doi.org/10.1016/j.otsr.2014.07.027.
7. Watterson JR, Esdaile JM. Viscosupplementation: therapeutic mechanisms and clinical potential in osteoarthritis of the knee. J Am Acad Orthop Surg. 2000;8(5):277–84. https://doi.org/10.5435/00124635-200009000-00001.
8. Jerosch J. Conservative treatment options for arthritis of the ankle: what is possible, what is effective? Unfallchirurg. 2022;125(3):175–82. Konservative Optionen zur Therapie der Sprunggelenkarthrose : Was gibt es und was bringt es? https://doi.org/10.1007/s00113-021-01122-3.
9. Nicholls M, Manjoo A, Shaw P, Niazi F, Rosen J. Rheological properties of commercially available hyaluronic acid products in the United States for the treatment of osteoarthritis knee pain. Clin Med Insights Arthritis Musculoskelet Disord. 2018;11:1179544117751622. https://doi.org/10.1177/1179544117751622.
10. Householder NA, Raghuram A, Agyare K, Thipaphay S, Zumwalt M. A review of recent innovations in cartilage regeneration strategies for the treatment of primary osteoarthritis of the knee: intra-articular injections. Orthop J Sports Med. 2023;11(4):23259671231155950. https://doi.org/10.1177/23259671231155950.
11. Osteoarthritis: Care and Management in Adults. 2014. National Institute for Health and Clinical Excellence: Guidance.
12. Bannuru RR, Osani MC, Vaysbrot EE, et al. OARSI guidelines for the non-surgical management of knee, hip, and polyarticular osteoarthritis. Osteoarthr Cartil. 2019;27(11):1578–89. https://doi.org/10.1016/j.joca.2019.06.011.
13. Brophy RH, Fillingham YA. AAOS clinical practice guideline summary: management of osteoarthritis of the knee (nonarthroplasty), third edition. J Am Acad Orthop Surg. 2022;30(9):e721–9. https://doi.org/10.5435/JAAOS-D-21-01233.
14. Bruyere O, Honvo G, Veronese N, et al. An updated algorithm recommendation for the management of knee osteoarthritis from the European Society for Clinical and Economic Aspects of osteoporosis, osteoarthritis and musculoskeletal diseases (ESCEO). Semin Arthritis Rheum. 2019;49(3):337–50. https://doi.org/10.1016/j.semarthrit.2019.04.008.
15. Jevsevar DS. Treatment of osteoarthritis of the knee: evidence-based guideline, 2nd edition. J Am Acad Orthop Surg. 2013;21(9):571–6. https://doi.org/10.5435/JAAOS-21-09-571.
16. Miller LE. Towards reaching consensus on hyaluronic acid efficacy in knee osteoarthritis. Clin Rheumatol. 2019;38(10):2881–3. https://doi.org/10.1007/s10067-019-04597-z.
17. Altman RD, Devji T, Bhandari M, Fierlinger A, Niazi F, Christensen R. Clinical benefit of intra-articular saline as a comparator in clinical trials of knee osteoarthritis treatments: a systematic review and meta-analysis of randomized trials. Semin Arthritis Rheum. 2016;46(2):151–9. https://doi.org/10.1016/j.semarthrit.2016.04.003.
18. Saltzman BM, Leroux T, Meyer MA, et al. The therapeutic effect of intra-articular normal saline injections for knee osteoarthritis: a meta-analysis of evidence level 1 studies. Am J Sports Med. 2017;45(11):2647–53. https://doi.org/10.1177/0363546516680607.
19. Khazzam M, Gee AO, Pearl M. Management of glenohumeral joint osteoarthritis. J Am Acad Orthop Surg. 2020;28(19):781–9. https://doi.org/10.5435/JAAOS-D-20-00404.
20. Zhang B, Thayaparan A, Horner N, Bedi A, Alolabi B, Khan M. Outcomes of hyaluronic acid injections for glenohumeral osteoarthritis: a systematic review and meta-analysis. J Shoulder Elb Surg. 2019;28(3):596–606. https://doi.org/10.1016/j.jse.2018.09.011.
21. Chang KV, Hsiao MY, Chen WS, Wang TG, Chien KL. Effectiveness of intra-articular hyaluronic acid for ankle osteoarthritis treatment: a systematic review and meta-analysis. Arch Phys Med Rehabil. 2013;94(5):951–60. https://doi.org/10.1016/j.apmr.2012.10.030.

22. Witteveen AG, Hofstad CJ, Kerkhoffs GM. Hyaluronic acid and other conservative treatment options for osteoarthritis of the ankle. Cochrane Database Syst Rev. 2015;2015(10):CD010643. https://doi.org/10.1002/14651858.CD010643.pub2.

23. Phillips M, Bhandari M, Grant J, et al. A systematic review of current clinical practice guidelines on intra-articular hyaluronic acid, corticosteroid, and platelet-rich plasma injection for knee osteoarthritis: an international perspective. Orthop J Sports Med. 2021;9(8):23259671211030272. https://doi.org/10.1177/23259671211030272.

24. Conrozier T, Balblanc JC, Richette P, et al. Early effect of hyaluronic acid intra-articular injections on serum and urine biomarkers in patients with knee osteoarthritis: an open-label observational prospective study. J Orthop Res. 2012;30(5):679–85. https://doi.org/10.1002/jor.21580.

25. Henrotin Y, Chevalier X, Deberg M, et al. Early decrease of serum biomarkers of type II collagen degradation (Coll2-1) and joint inflammation (Coll2-1 NO(2) ) by hyaluronic acid intra-articular injections in patients with knee osteoarthritis: a research study part of the Biovisco study. J Orthop Res. 2013;31(6):901–7. https://doi.org/10.1002/jor.22297.

26. Pereira TV, Juni P, Saadat P, et al. Viscosupplementation for knee osteoarthritis: systematic review and meta-analysis. BMJ. 2022;378:e069722. https://doi.org/10.1136/bmj-2022-069722.

27. Lieberman JR, Engstrom SM, Solovyova O, Au C, Grady JJ. Is intra-articular hyaluronic acid effective in treating osteoarthritis of the hip joint? J Arthroplast. 2015;30(3):507–11. https://doi.org/10.1016/j.arth.2013.10.019.

28. Leite VF, Daud Amadera JE, Buehler AM. Viscosupplementation for hip osteoarthritis: a systematic review and meta-analysis of the efficacy on pain and disability, and the occurrence of adverse events. Arch Phys Med Rehabil. 2018;99(3):574–83. e1. https://doi.org/10.1016/j.apmr.2017.07.010.

29. Gazendam A, Ekhtiari S, Bozzo A, Phillips M, Bhandari M. Intra-articular saline injection is as effective as corticosteroids, platelet-rich plasma and hyaluronic acid for hip osteoarthritis pain: a systematic review and network meta-analysis of randomised controlled trials. Br J Sports Med. 2021;55(5):256–61. https://doi.org/10.1136/bjsports-2020-102179.

30. Riley N, Vella-Baldacchino M, Thurley N, Hopewell S, Carr AJ, Dean BJF. Injection therapy for base of thumb osteoarthritis: a systematic review and meta-analysis. BMJ Open. 2019;9(9):e027507. https://doi.org/10.1136/bmjopen-2018-027507.

31. Trellu S, Dadoun S, Berenbaum F, Fautrel B, Gossec L. Intra-articular injections in thumb osteoarthritis: a systematic review and meta-analysis of randomized controlled trials. Joint Bone Spine. 2015;82(5):315–9. https://doi.org/10.1016/j.jbspin.2015.02.002.

# Radiosynovectomy



Goksel Dikmen, Vahit Emre Ozden,
and Kayahan Karaytug

## 10.1  Introduction

Local intra-articular injection of colloidal beta-emitting radionuclides solution, radio synovectomy (RSV), was first described by Fellinger et al. in 1952 for treatment of persistent synovitis of a rheumatoid arthritis (RA) patient with gold-based colloid [1]. RSV is a salvage procedure for chronic synovitis of individual joints after failure of long-term systemic pharmacotherapy and repeated other non-biological injections.

The $^{90}$yttrium citrate ($^{90}$Y), $^{186}$rhenium sulfide ($^{186}$R), and $^{169}$erbium citrate ($^{169}$E) were common approved radiopharmaceuticals for RSV in Europe [2] (Table 10.1). Mechanism of action stars with their high-energy beta participles after injection and homogenous distribution in joint. Beta particles show limited tissue penetration of up to 10 mm to minimize nontarget tissue radiation [3]. The radiation-absorbed dose nearly 100 Gy leads to synovectomy via pronounces cell death, occlusion of capillary system, and further fibrosis and sclerosis of the synovial membrane [4]. Improvement of pain-related restriction, daily living activities, and range of motion with less effusion takes over 3 months after RSV [5].

The best clinical recovery is observed in patients who suffer from highest inflammatory activity and within early phase of their underlying disease. The evaluation of clinical history is the key point before RSV including duration symptoms and underlying disease, duration and doses of systemic pharmacotherapy, previous surgical application, and previous other injections [6]. Swelling, hyperthermia, disability, and physical functions of the effected joints should be evaluated in clinical examination. Clinical history of the patient and disability of the joint are not enough to decide RSV, and a dedicated interdisciplinary team (e.g., hematology, rheumatologist, orthopedic surgeon, and nuclear medicine physician) consisting of referring physician should be necessary to consider RSV application.

Legal issues including indications, approval, and radiation protection protocols of different radionuclides may vary from country to country, and each physician must follow their national health laws. In Europe, $^{169}$E for smaller size joints as the carpometacarpal joints (CMC) and metatarsophalangeal (MTP) joints; $^{186}$R for medium size joints—ankle, shoulder, elbow, wrist, and hip; and $^{90}$Y for larger size joint, e.g., knee and re-RSV cases (Table 10.1).

G. Dikmen (✉)
Acibadem Mehmet Ali Aydinlar University,
Faculty of Medicine, Department of Orthopedics and
Traumatology, Istanbul, Türkiye

International Joint Centre (IJC), Acıbadem Maslak
Hospital, Sarıyer, Istanbul, Türkiye
e-mail: goksel.dikmen@acibadem.edu.tr

V. E. Ozden · K. Karaytug
Acibadem Mehmet Ali Aydinlar University,
Faculty of Medicine, Department of Orthopedics and
Traumatology, Istanbul, Türkiye

**Table 10.1** Common radiopharmaceuticals, according to the European Association of Nuclear Medicine (EANM) guideline for RSV [2]

| | $^{90}$**Yttrium** | $^{186}$**Rhenium** | $^{169}$**Erbium** |
| --- | --- | --- | --- |
| **Half-life (days)** | 2.7 | 3.7 | 9.4 |
| **Average penetration (mm)** | 3.6 | 1.1 | 0.3 |
| **Particles** | Citrate/silicate | Colloid/sulfide | Citrate |
| **Joints** | **Recommended activities (MBq)** | | |
| Knee | 185–222 | | |
| Hip | 74–148 | | |
| Shoulder | 74–148 | | |
| Elbow | 74–111 | | |
| Wrist | 37–74 | | |
| Ankle | 74 | | |
| Subtalar | 74 | | |
| CMC I/SIJ | 20–80 | | |
| MCP others | 20–40 | | |
| PIP/SCJ | 10–20 | | |
| DIP | 10–15 | | |
| MTP | 30–40 | | |
| TMT | 20–40 | | |
| ACJ/TMJ | 20–40 | | |

*CMC* carpometacarpal joints, *SIJ* sacroiliac join, *MCP* metacarpophalangeal joints, *PIP* proximal interphalangeal joint, *SCJ* sternoclavicular joint, *DIP* distal interphalangeal joint, *MTP* metatarsophalangeal joints, *TMT* tarsometatarsal joint, *ACJ* acromioclavicular joint, *TMJ* temporomandibular joint

## 10.2 Indications and Contraindications of Radiosynovectomy

The main indications for RSV in patients with joint pain were persistent synovitis of RA, inflammatory joint diseases (Lyme's borreliosis) and seronegative spondyloarthropathy (psoriatic arthritis, ankylosing spondylitis, reactive arthritis), persistent synovial effusion (e.g., adjuvant therapy after endoprosthesis placement or arthroscopic synovectomy, open synovectomy). (Fig. 10.1), osteoarthritis characterized by secondary synovitis resistant to other treatments, giant cell tumor/pigmented villonodular synovitis (for the prevention of relapse after arthroscopic or open surgery), hemarthrosis, and synovitis associated with hemophilia (for prevention of intra-articular hemorrhage and further arthroplasty) and undifferentiated arthritis characterized by synovitis [5, 7].

- $^{90}$Y is mainly used in adults for persistent synovial hypertrophy of the knee, mono- or oligoarthritis of chronic inflammatory rheumatologic diseases, and hemophilic arthroplasty.
- $^{186}$R is indicated for the mono- or oligoarthritis of the medium-sized joints, RA, hemophilic arthroplasty, and chondrocalcinosis. In Switzerland, $^{186}$R is additionally approved for the knee joint in patients younger than 20 years [2].
- $^{169}$E is used for the treatment of mono- or oligoarthritis of smaller-sized joints, especially in hand and feet small joints after failure of intra-articular corticosteroid infections.

If the first RSV application failed, RSV could be repeated three times within 3 months intervals. Repeated RSV treatments have more chance to decrease synovial hypertrophy than single treatments with higher activity. Pregnancy and postpartum breastfeeding, local and systemic infection, massive hemarthroses of the target joint (except for hemophiliac patients), recent joint surgery or arthroplasty, within ruptured 6 weeks Baker cyst's are the main absolute contra-

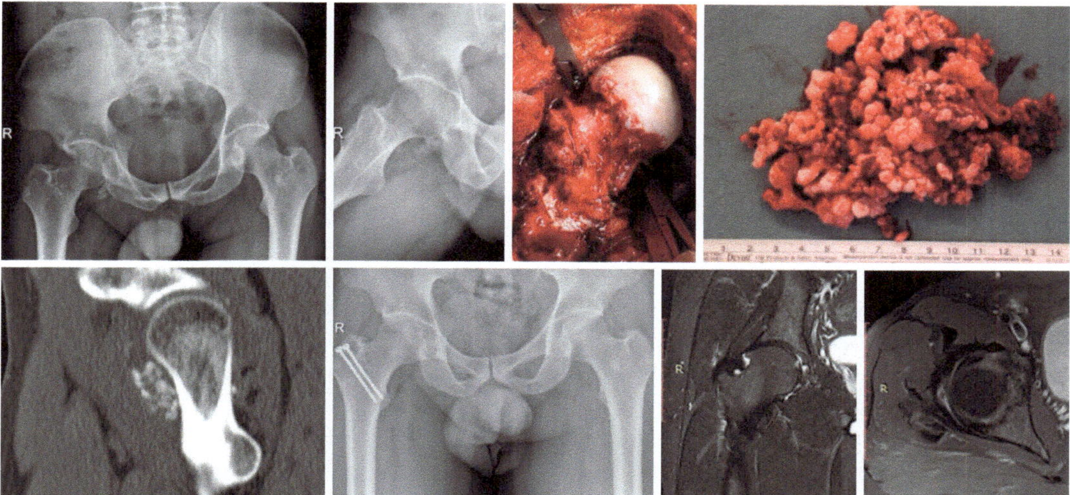

**Fig. 10.1** 30-year male refers our clinic with right hip pain and limited range of motion. Safe dislocation with trans-trochanteric approach was used for open synovectomy due to synovial chondromatosis. [186]Rhenium infusion with subsequent betamethasone and saline infusion 3 cc were used for the RSV after 3 months from index surgery. Patient was not complaining any discomfort, and last radiological MRI showed no relapse at an average 8.7-year follow-up

indications for RSV [5]. Extensive joint instability, high-grade bone destruction, children and younger ager patients (indication should be restricted for the patient's age < 20 years old) are the relative contraindications according to EANM guideline [2].

Patients who are candidates for RSV treatment should have radiograph or MRI of target the joint at no more than 6 months previously. MRI also should be taken for hemophilia patients with younger age group and children due to lack radiation exposure [2]. In addition, joint ultrasonography may be useful to evaluate synovial hypertrophy of the joint and to exclude a ruptured Baker's cyst or massive hemarthrosis [8, 9]. Two- or three-phase bone scan with 99mTc-phosphonate is still the method of choice to evaluate the severity of active soft-tissue inflammation of the mono- and/or oligoarticular involvements in patients with systemic inflammatory disease and activity in bone of the target joints for RSV [10, 11].

## 10.3 Informed Concept and Procedure

RSV decision should be made by a multidisciplinary team consisting of referring physician according to national regulations and national approved indication for the RSV treatment. Finally, specialist nuclear medicine physician is the main responsible for ultimate RSV indication and possible RSV-related complications, who should apply the final intra-articular radiocolloid injections. Patients must be informed about nature of the radioactive treatment, its mechanism, indications, contraindications, and complications including the risks involving puncture of a joint, such as infection, local hemorrhage, extravasation, radiation burden, and the risk of radionecrosis. If it is not restricted intra-articularly, theoretical malignancy risk, postinjection pyrexia or allergy, and risk of thromboembolic events after immobilization of the lower limb for 48 h may occur after RSV. The patient should be informed that the RSV application has recovery of 60–80% chance and may face recurrent synovial attack after at least 6 months [7].

According to EANM, RSV procedure should be administered in a dedicated room equipped for sterile injection procedures and approved for the use of beta emitters [10]. Hygienic requirements of patients, regular infection prophylaxis of the room, and facility requirements should be considered. Sterile disposable cannulas and syringes must be used. Measurement of activity or calculation of activity by volume, the selection of suitable syringe, and shielding device for radionuclides should be prepared by nuclear medicine physician. Intra-articular injections should be done based on the published guidelines and most favorable and simplest access routes to decrease extravasation risk and neurovascular complication.

Ultrasonography or arthrography and fluoroscopy with image documentation are necessary during injection. The contrast agents should be used to evaluate correct needle placement into the joint, and physiological saline may improve the homogeneous distribution of the injected radionuclide [12]. Concomitant use of subsequent glucocorticosteroids, e.g., triamcinolone hexacetonide, triamcinolone acetonide, or betamethasone, increases the effect of RSV [7]. Some authors advice not to use glucocorticosteroids due to possibility of avascular necrosis of the femoral head after RSV treatment.

### 10.3.1 Joints Specific Approaches and Needle Sizes for RSV

- Knee joint preferred approach is generally ultrasonography-guided superolateral approach with positioning of needle to the suprapatellar recess. Removal of synovial fluid even if it is present in small amount is important before radionuclide injection. 21 gauge needle is advanced 40 or 50 mm. A three-way valve should be used not to change the needle's position for subsequent glucocorticosteroids and saline injections.
- RSV for wrist and finger/toe joints should be performed under fluoroscopy control with 25 gauge 25-mm length needle from dorsal into the cruciate fossa perpendicular to the skin.
- Hip joint RSV injection should be performed under fluoroscopy or ultrasonography. Supine position and slightly internal rotation of the hip are helpful. A21 gauge 50 mm or 20 gauge 9–10 mm needle can be used.
- Shoulder joint anterior, superior, and posterior approaches can be used for RSV. A 50 mm or 40 mm 21-gauge needle mostly used diameters to inject in patients'. The most common injection positions used for shoulder joint are supine position, arm in external rotation, and injection through the anterior route. The needle must be positioned in the lower third of the joint space and should be checked with fluoroscopy.
- Elbow joint RSV can be done in patient sitting in an upright position with the elbow flex to 90° and forearm in maximum pronation position. Anatomical triangle of olecranon, the lateral humeral epicondyle, and the radial head is the target for puncture. An 18-gauge needle should be used in the direction of the radial head under fluoroscopy.

### 10.3.2 RSV in Rheumatoid Arthritis

The most common type of inflammatory arthritis is rheumatoid arthritis (RA). RA is a chronic autoimmune disease and is primarily considered an inflammatory joint disease. In the USA and northern Europe, the annual incidence is about 40 per 100,000 [13, 14]. Pharmacological treatments are the first steps for RA-related synovial hypertrophy which include nonsteroidal anti-inflammatory drugs (NSAIDs), systemic or intra-articular glucocorticosteroids (GC), and non-biological and biological disease-modifying antirheumatic drugs (DMARDs) [15]. Persistent synovitis of one or more joints can be treated with intra-articular injections of GC, and RSO is indicated after at least one failed intra-articular injection of GC in RA patients treated with DMARDs [2, 6]. Liepe et al. reported success rates (excellent or good response) in 57% of the treated knees, 63% of shoulders, 60% of wrists, 64% of ankles, 54% of thumb bases, 55% of

MCPs, 54% of PIPs, 53% of DIPs, and 54% of MTPs. The best results for RSO are reported for joints with minimal or moderate joint damage, and RSO should be considered by early in RA in case of persistent synovitis [16].

### 10.3.3 RSV in Hemophilia

Hemophilia is a bleeding disorder due to an inherited X-linked deficiency or absence of clotting factors. As a result, affected patients often experience bleeding that causes synovitis in the musculoskeletal system, particularly in the joints [17]. The pathophysiology is mainly driven by an inflammatory response to the iron load which results in neo-angiogenesis and increased fragility of the synovial membrane and further increased bleeding tendency [18, 19]. RSV is indicated if synovitis can be demonstrated despite coagulation factor substitution or if three joint hemorrhages occur within 6 months [2]. 70% to 90% of patients benefit from RSO concerning bleeding frequency, the intensity of pain, joint function, and thickness of the synovium [17, 19, 20]. The primary goal of RSV, namely, to stop bleeding, is usually achieved. The best results are obtained before the onset of hemarthropathy and are best obtained in the ankle joints, followed by the elbow and knee joints [21]. Therefore, same as the RA disease, RSV should be used as early phase as possible before hemarthropathy develops, and RSV should be used as first-line therapy in chronic synovitis. Finally, if the synovitis persists or recurs after the first RSV, it is recommended that the procedure could be repeated up to three times. If synovitis persists after the third RSO, it should be treated surgically [20, 22].

### 10.3.4 RSV in Osteoarthritis

RSV could be used in selected patients with osteoarthritis after failure of long-term systemic pharmacotherapy and repeated other non-biological injections. The improvement rate after RSV reported from 40% to 89% for knee joint [23]. Patients with early-grade osteoarthritis (Kellgren-Lawrence grade I/II) demonstrated higher functional scores than grade III patients. Success rate could decrease with continuing severe degenerative changes of the joint [6]. Recent two double-blind controlled prospective studies for upper extremity and thumb showed significant reduction of inflammation and symptoms within 1-year follow-up [24]. Szerb et al. reported that RSV could prevent radiologic deterioration among 70.6% of the hip patients and in 79.1% of the treated ankle joint patients at an average 9.2 years of follow-up time [25].

## 10.4 Conclusion

RSV should be considered the initial fast-acting and patient-friendly therapeutic option for the treatment of patients with hemophilic hemarthrosis. RSV therapy should be performing early stages of diseases, before the development of advance stage of cartilage loss. RSV is still classified as "helpful" in patients with chronic synovitis secondary to RA or active osteoarthritis. The interdisciplinary teamworks with nuclear medicine physician are essential for the therapy protochol.

## References


1. Fellinger K, Schmid J. Local therapy of rheumatic diseases. Wiener Zeitschrift fur innere Medizin und ihre Grenzgebiete. 1952;33(9):351.
2. Kampen W, Boddenberg-Pätzold B, Fischer M, Gabriel M, Klett R, Konijnenberg M, Kresnik E, Lellouche H, Paycha F, Terslev L. The EANM guideline for radiosynoviorthesis. Eur J Nucl Med Mol Imaging. 2021;1
3. Ingrand J. Characteristics of radio-isotopes for intra-articular therapy. Ann Rheum Dis. 1973;32(Suppl):3.
4. Bowring C, Keeling D. Absorbed radiation dose in radiation synovectomy. Br J Radiol. 1978;51(610):836.
5. Knut L. Radiosynovectomy in the therapeutic management of arthritis. World J Nucl Med. 2015;14(01):10.
6. Kresnik E, Mikosch P, Gallowitsch H, Jesenko R, Just H, Kogler D, Gasser J, Heinisch M, Unterweger O, Kumnig G. Clinical outcome of radiosynoviorthesis: a meta-analysis including 2190 treated joints. Nucl Med Commun. 2002;23(7):583.
7. Schneider P, Farahati J, Reiners C. Radiosynovectomy in rheumatology, orthopedics, and hemophilia. J Nucl Med. 2005;46(1 suppl):48S.

8. Pirich C, Schwameis E, Bernecker P, Radauer M, Friedl M, Lang S, Kritz H, Wanivenhaus A, Trattnig S, Sinzinger H. Influence of radiation synovectomy on articular cartilage, synovial thickness and enhancement as evidenced by MRI in patients with chronic synovitis. J Nucl Med. 1999;40(8):1277.

9. Takase-Minegishi K, Horita N, Kobayashi K, Yoshimi R, Kirino Y, Ohno S, Kaneko T, Nakajima H, Wakefield RJ, Emery P. Diagnostic test accuracy of ultrasound for synovitis in rheumatoid arthritis: systematic review and meta-analysis. Rheumatology. 2018;57(1):49.

10. Sandrock D, Backhaus M, Burmester G, Munz D, Rheumatologie fdKBVdDGf. Imaging techniques in rheumatology: scintigraphy in rheumatoid arthritis. Z Rheumatol. 2003;62:476.

11. Stollfuss JC, Freudenberg LS, Wieder H. 99mTc-DPD SPECT/CT for localisation of inflammatory and chronic osteoarthritis of the foot and ankle. Nuklearmedizin-NuclearMedicine. 2016;55(04):145.

12. Franssen M, Koenders E, Boerbooms AT, Buijs W, Lemmens J, Van De Putte L. Does application of radiographic contrast medium in radiation synovectomy influence the stability of Yttrium-90 colloid? British Journal of Rheumatology. 1997;36(04);506.

13. Myasoedova E, Crowson CS, Kremers HM, Therneau TM, Gabriel SE. Is the incidence of rheumatoid arthritis rising?: results from Olmsted County, Minnesota, 1955–2007. Arthritis Rheum. 2010;62(6):1576.

14. Hunter TM, Boytsov NN, Zhang X, Schroeder K, Michaud K, Araujo AB. Prevalence of rheumatoid arthritis in the United States adult population in healthcare claims databases, 2004–2014. Rheumatol Int. 2017;37:1551.

15. Fraenkel L, Bathon JM, England BR, St. Clair EW, Arayssi T, Carandang K, Deane KD, Genovese M, Huston KK, Kerr G. 2021 American College of Rheumatology guideline for the treatment of rheumatoid arthritis. Arthritis Rheumatol. 2021;73(7):1108.

16. Liepe K. Efficacy of radiosynovectomy in rheumatoid arthritis. Rheumatol Int. 2012;32(10):3219.

17. Srivastava A, Santagostino E, Dougall A, Kitchen S, Sutherland M, Pipe SW, Carcao M, Mahlangu J, Ragni MV, Windyga J. WFH guidelines for the management of hemophilia. Haemophilia. 2020;26:1.

18. Wyseure T, Mosnier LO, von Drygalski A. Advances and challenges in hemophilic arthropathy. In: Seminars in hematology. Elsevier; 2016. p. 10.

19. van Vulpen LF, Thomas S, Keny SA, Mohanty SS. Synovitis and synovectomy in haemophilia. Haemophilia. 2021;27:96.

20. Haberman B. S2k-Leitlinie Synovitis bei Hämophilie 2018. Gesellschaft für Thrombose-und Hämostaseforschung e V(GTH).

21. Walsh DA, McWilliams DF, Turley MJ, Dixon MR, Fransès RE, Mapp PI, Wilson D. Angiogenesis and nerve growth factor at the osteochondral junction in rheumatoid arthritis and osteoarthritis. Rheumatology. 1852;49(10):2010.

22. Rodriguez-Merchan EC, Heim M, Wallny T. The hemophilic joints. J Blood Disord Transfus. 2011;2(3):111.

23. Szentesi M, Nagy Z, Géher P, Papp I, Kampen WU. A prospective observational study on the long-term results of 90 yttrium citrate radiosynoviorthesis of synovitis in osteoarthritis of the knee joint. Eur J Nucl Med Mol Imaging. 2019;46:1633.

24. Van der Zant F, Jahangier Z, Moolenburgh J, Swen W, Boer R, Jacobs J. Clinical effect of radiation synovectomy of the upper extremity joints: a randomised, double-blind, placebo-controlled study. Eur J Nucl Med Mol Imaging. 2007;34:212.

25. Szerb I, Gál T, Kiss D, Nagy V, Hangody L. Efficacy assessment of radiosynoviorthesis on the progression of radiological osteoarthritic features of hip and ankle joint in patients with osteoarthritis and rheumatoid arthritis. Nuklearmedizin-NuclearMedicine. 2020;59(03):269.

# Deproteinized Hemoderivative of Calf Blood-Natural Botanical and Mineral Extracts

**11**

Berhan Bayram and Baris Kocaoglu

## 11.1 Introduction

Muscle injuries are one of the most common injuries among sports-related injuries, with an incidence of 30–55% [1]. More than 90% of muscle injuries are caused either by excessive contusion or strain of muscle. In professional sports, some of these injuries can cause significant pain and injury, resulting in loss of training and competition time. Muscle strain may be a consequence of eccentric exercise, when the muscle develops tension during this type of lengthening contraction. These types of injuries are more common in sports that require sprinting or jumping [2]. For injuries like this, in professional sports, medical staff face pressure to return the player to training and competition as soon as possible. Physical examination is very important to diagnose muscle injuries. Ultrasonography and magnetic resonance imaging (MRI) can be useful in confirming the diagnosis and helping the clinician make treatment decisions. The muscle healing process consists of several consecutive stages: myofibril degeneration, inflammation, regeneration, and fibrosis [3]. The treatment options for structural muscle injuries are still very limited. There are clinical studies describing the treatment methods of muscle injuries in the literature; however, the underlying mechanisms leading to the accelerated recovery time reported to date have not been fully identified and proven with randomized controlled trials [4]. Rest, immobilization, physical therapy, and sometimes nonsteroidal anti-inflammatory drugs (NSAIDs) are the main components of treatment for grade 1 and 2 muscle injuries [5]. Immobilization can lead to better granulation of injured muscle fibers and shorten the healing process but will cause significant atrophy and joint stiffness in healthy myofibers [1]. The aim of the treatments used in the acute phase is to minimize the formation of a large hematoma that has the potential to affect the size of the scar tissue at the end of the repair process. While some studies have shown that the administration of NSAIDs promotes muscle recovery by reducing degeneration and inflammation [6], another research has shown that NSAIDs are detrimental to the entire healing process [7]. For this reason, studies on treatment methods that lead to faster recovery in the treatment of muscle injuries have increased.

Injection therapy in sports muscle injuries has a history of safe use of approximately 60 years. As a result of studies conducted in recent years, several new injection methods have been reported that shorten the recovery time after muscle strain injuries. For example, growth factor injection therapy has shown good therapeutic results.

B. Bayram
Acibadem Altunizade Hospital, Department of orthopedics and Traumatology, Istanbul, Turkey

B. Kocaoglu (✉)
Acibadem Mehmet Ali Aydinlar University, Faculty of Medicine, Department of Orthopedics and Traumatology, Istanbul, Turkey


B. Kocaoglu et al. (eds.), *Musculoskeletal Injections Manual*,
https://doi.org/10.1007/978-3-031-52603-9_11

However, due to their performance enhancing and anabolic properties, growth factors have been banned by the World Anti-Doping Agency (WADA). Local anesthetics, calf blood compound (CPC) (Actovegin), and homeopathic drug (Tr14) (Traumeel) or injection of platelet-rich plasma (PRP) are the other most commonly reported drugs.

## 11.2 Most Used Deproteinized Hemoderivative of Calf Blood-Natural Botanical and Mineral Extracts

Calf blood compound (CPC) (Actovegin) is a biological drug manufactured from a natural source. It is a calf blood hemodialysate and contains typical salts, trace elements, and high concentrations of various amino acids [8]. It can be administered orally as tablets, topical formulations, intramuscular or intravenous injections, or infusions. Because calf blood compound (CPC) is a complex of blood dialysate with many different active molecules, it is difficult to identify the single pharmacologically active component.

Traumeel (Tr14), on the other hand, is a homeopathic solution with a fixed combination formula of 12 botanical and 2 mineral substances with proven anti-inflammatory, antiedema, as well as anti-exudative properties [9]. It has been on the market in Germany since 1937 and is currently available in approximately 60 countries around the world. The constituents of Traumeel are used traditionally and in homoeopathy for the broad spectrum of symptoms associated with various traumas such as contusions, sprains, wounds, pain, inflammation, neuralgia, etc. Various cellular and biochemical pathways appear to be modulated by the components in Tr14. Homeopathic drug has shown efficacy comparable to NSAIDs in terms of anti-inflammatory property, but it has been suggested that the Tr14 injection solution does not inhibit the cyclooxygenase (COX) or lipoxygenase enzyme pathways as does nonsteroidal anti-inflammatory drugs (NSAIDs) [10]. In vitro studies showed that Tr14 inhibits the production of

proinflammatory cytokines such as TNF-alfa and IL-1beta [11]. The authors suggested that Tr14 injection solution acts by accelerating the healing process rather than blocking the development of edema from the beginning [12]. The oral application of Tr14 has been demonstrated to have beneficial effects on epicondylitis and different musculoskeletal disorders [13]. Although athletes have been reported to be successfully treated in practice, there is limited scientific evidence in the literature to support the general use of these agents in the treatment of muscle injuries. The mechanism of action of calf blood compound (CPC) and Tr14 in the muscle recovery process is still not fully understood.

Another homeopathic solution, Zeel comp. N (Zeel, Heel, GmbH, Baden-Baden, Germany), is a homeopathic medication that has been widely used for many years for the treatment of arthritic disorders in many countries around the world. Zeel's formulation includes a combination of highly diluted extracts from *Arnica montana* (arnica root), *Sanguinaria canadensis* (bloodroot), *Rhus toxicodendron* (poison oak), *Solanum dulcamara* (climbing nightshade), and sulfur. Zeel is available as tablets or injection solution with slightly different compositions. Clinical observational studies have shown that Zeel reduces symptoms of osteoarthritis, including stiffness and pain, and is generally well tolerated with a good safety profile [14]. Initial preclinical data suggested that Zeel has an inhibitory effect on cartilage degradation; however, the mechanism of action of Zeel is still not fully understood. The mechanism of action of Zeel and its efficacy in the treatment of osteoarthritis will be better understood with future studies with a large number of patients.

The use of deproteinized hemoderivative of calf blood-natural botanical and mineral extracts in training regimes by high profile athletes has led to the anecdotal opinion that the blood product is ergogenic, enhancing athlete performance. Calf blood compound (CPC) has been used clinically for decades to improve blood circulation in the brain after ischemic injury caused by impaired peripheral blood circulation. In vitro studies have suggested that these types of drugs have membrane-stabilizing effects to interrupt oxidative stress and cell death processes while increasing the efficiency of energy balance in cells during postischemic metabolic events. Actovegin contains physiological components, electrolytes, and essential trace elements. 30% of organic components are amino acids, nucleosides, intermediates of carbohydrates, and fat metabolites. Ultrafiltered up to 6000 daltons, therefore it contains no growth factors or hormone-like substances. It can be administered as tablets, topical formulations, injections, or infusions by intramuscular or intravenous routes [15]. For example, a recent in vitro model of cell injury has shown that Actovegin (a biological drug produced by Nycomed GmbH, Linz, Austria, which in 2015 was taken over by Takeda Pharmaceutical Ltd., Japan) improves intrinsic mitochondrial respiration capacity in injured human skeletal muscle fibers. It was concluded that the findings of this study support and explain the reported ergogenic properties [16]. The mechanism of action with Actovegin can potentially protect ischemic cells, modulate the inflammatory process, and help the initial phase of recovery from acute muscle injuries become more efficient.

Skeletal muscle is a very vascular structure and is highly energy dependent. Once injured, blood flow often is disturbed, which leads to cell ischemia and energy imbalance. Therefore, Actovegin injection therapy to the injury site might aid recovery and limit further ischemic effects and control cellular damage from initial injuries. Another publication reported that Actovegin may inhibit the release of inflammatory mediators and therefore potentially speed up the muscle recovery process [17]. In a study with football players, the Actovegin injection therapy regimen described for grade 1 hamstring injuries appears to significantly reduce the number of days to return to play, with an average reduction of 8 days compared to rehabilitation therapy alone [15]. The career life span for the professional elite athlete is often short lived, and shortened recovery time could mean continuing with training, increased game play, and benefit to the team and club. And some agents administered after muscle injuries have been found to be myotoxic when administered intramuscularly, but there is no information in the studies that calf blood compound (CPC) and Tr14 are myotoxic. This feature is one of the most important advantages that distinguish CPC and Tr14 from other agents. In addition, no other side effects were reported in the studies on Actovegin and Traumeel. However, previously published data have suggested beneficial effects of CPC on numerous diseases, including acute and chronic wounds and circulatory disorders [18]. CPC has neuroprotective and antioxidative properties and is intended to have ergogenic qualities and play an important role in muscle tissue metabolism [19]. CPC improves muscle cell proliferation and has been successfully applied clinically in combination with Tr14 for the treatment of muscle injuries. A result of another in vivo clinical study showed that Actovegin has no effect on peak aerobic capacity in humans [20].

## 11.3  Conclusion

In conclusion, injection therapies with deproteinized hemoderivative of calf blood-natural botanical and mineral extracts are frequently used as a possible treatment option for acute muscle injuries and osteoarthritis. The use of these drugs in the acute phase of muscle injuries suggests that it may be a promising intervention for athletes who experience muscle injuries by shortening the recovery period. Based on previous studies, these agents appear to be safe and well tolerated. There is no new evidence to question the long-standing, good safety profile of these drugs. Compared with conventional conservative RICE and NSAID therapy, these agents propose an exciting and legal alternative for high-performance athletes. Although athletes have been reported to be successfully treated in practice, there is only limited scientific evidence to support the general use of the abovementioned agents in the treatment of skeletal muscle injuries in athletes. The limited number of high-quality studies makes it difficult to compare the effectiveness of injection therapy with these agents in muscle injuries and osteoarthritis.

Further research should be encouraged to investigate the effects of these drugs. Also, there are no established doses and its frequency to use in the treatment of acute muscle injury and osteoarthritis. It is also unclear when and which solution was considered. Furthermore, Actovegin can be shown to improve muscle cell proliferation [8]. This may help explain the positive effects of Actovegin on muscle injuries. Another important point is that the intramuscular use of such drugs is not prohibited by the regulatory authority, the World Anti-Doping Agency (WADA). Based on the most recent literature, it suggests that deproteinized hemoderivative of calf blood-natural botanical and mineral extracts is a safe injectable therapy that has demonstrated some efficacy in the treatment of muscle injuries and is unlikely to be ergogenic. It should be noted that an injection therapy that will prove to shorten the healing process could potentially revolutionize the treatment of muscle injuries. In addition, with studies on the use of these drugs in the treatment of osteoarthritis, they may be an alternative in the clinical follow-up of these patients.

## References


1. Järvinen TAH, Kääriäinen M, Järvinen M, Kalimo H. Muscle strain injuries. Curr Opin Rheumatol. 2000;12(2):155–61. https://doi.org/10.1097/00002281-200003000-00010.
2. Garrett J. Muscle strain injuries. Am J Sports Med. 1996;24(SUPPL) https://doi.org/10.1177/036354659602406S02/ASSET/036354659602406S02.FP.PNG_V03.
3. Harmon KG. Muscle injuries and PRP: what does the science say? Br J Sports Med. 2010;44(9):616–7. https://doi.org/10.1136/BJSM.2010.074138.
4. Wright-Carpenter T, Klein P, Schäferhoff P, Appell HJ, Mir LM, Wehling P. Treatment of muscle injuries by local administration of autologous conditioned serum: a pilot study on sportsmen with muscle strains. Int J Sports Med. 2004;25(8):588–93. https://doi.org/10.1055/S-2004-821304.
5. Kasemkijwattana C, et al. Use of growth factors to improve muscle healing after strain injury. Clin Orthop Relat Res. 2000;370(370):272–85. https://doi.org/10.1097/00003086-200001000-00028.
6. Abramson SB, Weissmann G. The mechanisms of action of nonsteroidal antiinflammatory drugs. Arthritis Rheum. 1989;32(1):1–9. https://doi.org/10.1002/ANR.1780320102.
7. Shen W, Li Y, Tang Y, Cummins J, Huard J. NS-398, a cyclooxygenase-2-specific inhibitor, delays skeletal muscle healing by decreasing regeneration and promoting fibrosis. Am J Pathol. 2005;167(4):1105–17. https://doi.org/10.1016/S0002-9440(10)61199-6.
8. Reichl FX, et al. Comprehensive analytics of Actovegin® and its effect on muscle cells. Int J Sports Med. 2017;38(11):809–18. https://doi.org/10.1055/S-0043-115738/ID/R6256-0028.
9. Vanden Bossche L, Vanderstraeten G. A multi-center, double-blind, randomized, placebo-controlled trial protocol to assess Traumeel injection vs dexamethasone injection in rotator cuff syndrome: the TRAumeel in ROtator cuff syndrome (TRARO) study protocol. BMC Musculoskelet Disord. 2015;16:1. https://doi.org/10.1186/S12891-015-0471-Z.
10. Schneider C. Traumeel - an emerging option to nonsteroidal anti-inflammatory drugs in the management of acute musculoskeletal injuriesInt J Gen Med. 2011;25(4):225–34. https://doi.org/10.2147/IJGM.S16709.
11. Porozov S, Cahalon L, Weiser M, Branski D, Lider O, Oberbaum M. Inhibition of IL-1beta and TNF-alpha secretion from resting and activated human immunocytes by the homeopathic medication Traumeel S. Clin Dev Immunol. 2004;11(2):143–9. https://doi.org/10.1080/10446670410001722203.
12. Lussignoli S, Bertani S, Metelmann H, Bellavite P, Conforti A. Effect of Traumeel S, a homeopathic formulation, on blood-induced inflammation in rats. Complement Ther Med. 1999;7(4):225–30. https://doi.org/10.1016/S0965-2299(99)80006-5.
13. Schneider C, Schneider B, Hanisch J, van Haselen R. The role of a homoeopathic preparation compared with conventional therapy in the treatment of injuries: an observational cohort study. Complement Ther Med. 2008;16(1):22–7. https://doi.org/10.1016/J.CTIM.2007.04.004.
14. Sanchez C, et al. Reduction of matrix metallopeptidase 13 and promotion of chondrogenesis by Zeel T in primary human osteoarthritic chondrocytes. Front Pharmacol. 2021;12 https://doi.org/10.3389/FPHAR.2021.635034/FULL.
15. Lee P, Rattenberry A, Connelly S, Nokes L. Our experience on Actovegin, is it cutting edge? Int J Sports Med. 2011;32(4):237–41. https://doi.org/10.1055/S-0030-1269862.
16. Søndergård SD, Dela F, Helge JW, Larsen S. Actovegin, a non-prohibited drug increases oxidative capacity in human skeletal muscle. Eur J Sport Sci. 2016;16(7):801–7. https://doi.org/10.1080/17461391.2015.1130750.

17. Lee P, Kwan A, Nokes L. Actovegin—cutting-edge sports medicine or 'voodoo' remedy? Curr Sports Med Rep. 2011;10(4):186–90. https://doi.org/10.1249/JSR.0B013E318223CD8A.

18. Buchmayer F, Pleiner J, Elmlinger MW, Lauer G, Nell G, Sitte HH. Actovegin®: a biological drug for more than 5 decades. Wien Med Wochenschr. 2011;161(3–4):80–8. https://doi.org/10.1007/S10354-011-0865-Y.

19. Elmlinger MW, Kriebel M, Ziegler D. Neuroprotective and anti-oxidative effects of the hemodialysate actovegin on primary rat neurons in vitro. doi: https://doi.org/10.1007/s12017-011-8157-7.

20. Lee P, Nokes L, Smith PM. No effect of intravenous Actovegin® on peak aerobic capacity. Int J Sports Med. 2012;33(4):305–9. https://doi.org/10.1055/S-0031-1291322.

# Ortho-biologic Agents for Injections

Paola De Luca, Michela Maria Taiana,
and Laura de Girolamo

## 12.1 Introduction

Orthobiologic therapy for regenerative medicine applications has gained increasing interest in recent years in the orthopedic field from the scientific and clinical community due to the promising results obtained in preclinical and clinical studies. Defined as biological materials and substrates derived from the body that promote bone, ligament, muscle, and tendon healing in musculoskeletal injuries and degeneration [1], orthobiologics represent a category of innovative products that exploit the biological properties of their constituent elements to enhance the reparative and regenerative capacity of damaged tissue. These products open the frontiers to innovative therapeutic strategies, and they can be both used as conservative injectable treatments and/or in combination with surgical procedures providing a viable alternative to more conventional treatments.

Orthobiologics are heterogeneous products including bioactive molecules such as cytokines, growth factors, and/or cells and their released products with anti-inflammatory, reparative

immunomodulatory, and regenerative properties that can promote the restoration of tissue homeostasis counteracting the inflammatory microenvironment or increase tissue repair native biologic potential through multifactorial mechanisms [2–4]. Their mechanisms of action are based on the therapeutic potential of soluble factors as well as exosomes and micro-vesicles carrying DNA, proteins/peptides, lipids, organelles, mRNA, and miRNA [5] that exert antiapoptotic, anti-catalytic, trophic, and immunomodulatory activities. Indeed, inflammation is a frequent hallmark of musculoskeletal diseases such as osteoarthritis, tendinopathy, and muscle injuries and is often responsible of the degenerative cascade onset that results in long-term tissue destruction. Based on these premises, the promotion of cell proliferation and inhibition of pro-inflammatory mechanisms in the early stage of tissue homeostatic alteration result in improvement of the pain status of patients with musculoskeletal diseases due to tissue damage. Therefore, the use of orthobiologics represents a promising strategy also for the development of new pain therapies.

P. De Luca · M. M. Taiana · L. de Girolamo (✉)
Orthopaedic Biotechnology Laboratory, Ospedale
Galeazzi Sant'Ambrogio, Milan, Italy
e-mail: deluca.paola@grupposandonato.it;
michelamaria.tiana@grupposandonato.it;
laura.degirolamo@grupposandonato.it

## 12.2 Types and Effects of Orthobiologics

Orthobiologics come in many different forms derived from blood or adult tissue cells obtained from adipose tissue, bone marrow, or fetal annexes (placenta, cord blood, amniotic membrane) through procedures with variable complexity. Because of regulatory matter, the most used orthobiologics are those prepared by minimal manipulation at the point to care. This allows to avoid fall into the more complex regulations of the advanced therapy medicinal products (ATMPs). Among the most used minimally manipulated orthobiologics, blood-derived products (BDPs) are widely employed for the injective treatment of musculoskeletal conditions. This category includes platelet derivatives products such as platelet-rich plasma (PRP), growth factors-rich plasma (PRGF), autologous protein solution (APS), autologous conditioned plasma (ACP), and other blood products such as autologous conditioned serum (ACS) and α2-macroglobulin (A2M).

Platelet concentrate was first proposed as a strategy to promote tissue healing in regenerative medicine in 1998 by Marx and colleagues for bone healing in maxillofacial surgery [6]. After this publication, numerous clinical trials and case series have provided promising results in terms of safety and efficacy of platelet-based products in patients with osteoarthritis demonstrated by reduction of symptoms and tissue degeneration. Some studies have shown an opposite trend, but it should be considered that procedures for PRP production and administration in different trials were not univocal and also some studies suffer from intrinsic flaws. However, despite some controversial results, the current literature seems to report more and more positive findings. BDP injection has been proposed also as therapy for tendinopathy. Indeed, neovascularization promoted by BDPs in tendon tissue could be particularly relevant, although also in this case conflicting results are reported, likely due to the aforementioned reasons. Recent studies showed positive effect of BDP injection in treatment of muscles injuries too, without negative side effects reported.

However, particularly in case of extended damage, muscle recovery is not clearly evident, and further investigations are needed to validate the efficacy of BDP in muscle healing [7].

In BDPs, the regenerative potential derives mainly from the high content of growth factors, cytokines, and bioactive molecules stored in platelet α-granules that are released after degranulation and that promote healing process and modulate inflammation, ECM synthesis, and new blood vessel formation by paracrine mechanisms. The procedure for obtaining these products is usually rapid and technically simple. After being harvested, the blood usually undergoes a centrifugation cycle—differing in speed and duration—and directly injected into the site of treatment. The centrifugation procedure leads to a platelet concentration of three to six times than whole blood, and based on the preparation method, the final product composition can vary significantly. BDPs can be classified according to different platelet and/or leukocyte content. More in detail, PRGF is a type of plasma enriched of proteins, circulating growth factors, and coagulative factors that are pre-activated exogenously before injection [8]. It can improve the healing process by forming blood clots that release their content at the site of the injury. According to different centrifugation methods, it is possible to obtain platelet-rich plasma products with different leucocyte content, labeled as leucocyte-poor PRP (LP-PRP) or leucocyte-rich PRP (LR-PRP) [9]. Leukocytes can secrete several molecules involved in inflammation and in wound healing; however, the optimal formulation of platelet and leukocyte ratio in PRP has not been defined and could vary for different pathologies. To date, clinical recommendation suggests LR-PRP injection for lateral epicondylitis and LP-PRP for osteoarthritis of the knee [10].

APS is a novel LR-PRP containing high concentration of leukocytes, cytokines involved in inflammation, and anabolic cytokines, and it has been proposed as an autologous treatment for patients with joint conditions specifically osteoarthritis [11].

Autologous conditioned serum (Orthokine) is another autologous blood product whose peculiarity is the enrichment in the interleukin-1 receptor antagonist (IL-1Ra); therefore, its application is suitable for pathologies resulting from altered levels of IL-1β such as osteoarthritis, degenerative joint diseases, but it is also indicated for muscle regenerative treatment [12]. It is obtained after blood monocytes selection, platelets degranulation, and release of anti-inflammatory proteins such as IL-1Ra [13].

A2M is the most abundant blood proteinase inhibitor able to bind and to inhibit metalloproteinase and endoproteases included many cartilage catabolic factors; for these reasons, it would seem to attenuate cartilage degeneration and osteoarthritis disease [14]. It is obtained after whole blood centrifugation to separate the PRP from the platelet poor plasma (PPP) from which larger molecules including A2M (720 kDa) are isolated and concentrated [15].

In addition to BDPs, orthobiologics include also product whose efficacy relies on cells, mainly mesenchymal stem/stromal cells (MSCs), obtained through one-step procedures which have pro-regenerative ability crucial for the treatment of damaged musculoskeletal tissues [16]. MSCs can produce and release active molecules able to stimulate the resident cell population to cellular repair, as well as to act as immunomodulators on the local immune system, reducing fibrous scarring processes and cell apoptosis, and to stimulate angiogenesis [17]. MSCs have been more recently described as a subtype of pericytes, quiescent cells wrapped around vasculature network present in all the vascularized tissue, including adipose tissue and in bone marrow. In fact, those two tissues are the most common sources of autologous MSCs for clinical use in regenerative medicine.

Bone marrow has been the first and widely used source of isolation of MSCs (BMSCs, bone marrow stem/stromal cells) due to the high yield. For practical reason related to a simpler procedure and smother regulations, to date autologous concentrate of bone marrow aspirate (BMAC) is the most common form of bone marrow-dried products [18]. BMAC is obtained by harvesting and centrifuging the bone marrow aspirate to concentrate the mononuclear cell phase containing BMSCs as well as leukocytes and platelet [19]. As for other cell therapies, BMAC composition, and consequently its effectiveness, depends on the donor characteristics and on the harvest site. Indeed, BMAC harvested from the iliac crest has the highest concentration of mononuclear cells compared to other sites. Given the preclinical and clinical promising results, BMAC therapy is often chosen for the treatment of several musculoskeletal and spinal conditions. In particular, BMAC has resulted to be efficacious in treatment of knee osteoarthritis, for which a significant pain relief and function improvement have been reported, as well as in osteochondral repair mostly when associated with a scaffold. Moreover, in rotator cuff repair setting, BMAC resulted to be more efficacious than PRP [20, 21]. Standardized reproducible procedures for clinical application have not yet been defined, and long-term studies are needed to clearly establish the therapeutic efficacy of BMAC injection.

More recently, it has become evident that also adipose tissue, which is easier to obtain than bone marrow, represents a relevant resource of MSCs, namely, adipose-derived MSCs (ASCs).

Moreover, from adipose tissue, it is possible to obtain the stromal vascular fraction (SVF), a heterogeneous cell population comprising endothelial cells, pericytes, lymphocytes, and pre-adipocytes, excluding mature adipocytes.

Several studies demonstrate the SVF efficacy in inhibiting inflammation, promoting repair of cartilage damage, and reducing pain in osteoarthritis patients through paracrine mechanisms. However, different patients showed variable outcome probably due to individual differences in donors and different product preparations that can affect the therapeutic effectiveness [22].

Alternatively, adipose tissue can be minced to obtain microfragmented adipose tissue (microfat or MFAT) through the mechanical fragmentation of lipoaspirate using devoted medical devices [23]. The resulting product maintains the adipose tissue microarchitecture and preserves the so-called stem cell niche therefore respecting the physiological cell environment [24].

Also, in this case, the different devices for obtaining micro-fat, including harvesting, processing washing, and centrifugation steps, in addition to donor variability led to different products [25, 26]. Extensive in vitro evidence encourages the use of bone marrow and adipose tissue-derived products for the treatment of musculoskeletal conditions [16], but several issues have still limited translation of cell-based products into clinical practice. These treatments can be used in one-step procedure through minimally manipulated procedures as for BMAC, SVF, or micro-fat or can be expanded in vitro, generating more characterized products that are however more expensive and complex to manage from a regulatory perspective. Indeed, when MSCs undergo consistent manipulation, the resulting products fall within the ATMPs category of therapeutics. ATMPs are subject to very stringent regulation that classifies and categorizes them as sterile drugs, and therefore, they must be produced in accordance with the "Good Manufacturing Practices" (GMPs) outlined in the Regulation (EC) No 1394/2007, Directive 2001/83/EC on Sterile Medicines. To facilitate the MSC translation into orthopedic clinical practice, several solutions have been proposed with a low degree of tissue and cell manipulation, therefore not recognized as ATMPs, which have given rise to the category of the products prepared at the point of care (POC).

## 12.3 Conclusion

Because of the strong rationale behind their use, the clinical satisfactory evidence, the safety profile, and the minimal invasiveness and easiness of the procedures, orthobiologics are rapidly advancing and are considered to be able to significantly influence modern orthopedic practice, being a valid alternative to conventional treatments for the management of musculoskeletal problems. This would allow to reduce or postpone the need for more invasive approaches with benefit of both the patients and the socioeconomical tissue. However, although the basic science findings support their application in musculoskeletal conditions, a better understanding of the mechanisms of action and efficacy is needed to

define standardized protocols and provide concrete guidelines for clinicians according to scientific findings and recommendation.

It is not possible to identify yet the best orthobiologic injective treatment for each given pathology, as different orthobiologics can exert different efficacies in specific patients. Literature, although rich of reports about these therapeutic procedures, often also of good level of evidence (level I studies), suffers from some inconsistencies due to the different preparation methods resulting in different efficacy and results. Moreover, orthobiologics prepared at the POC do not undergo specific quality control analysis or evaluation of their potential clinical effectiveness prior to administration, making more difficult to find a consistent correlation between the product quality and the clinical outcomes. The literature reports a stable percentage of patients who does not respond to treatment as expected, despite the indication for the treatment is in line with the current knowledge. In this view, it is essential to understand how variable factor as age, nutrition state, lifestyle, medical comorbidities, and site or time of harvest can influence the composition and then the outcome of orthobiological treatments [27]. Therefore, study of biomarker profile correlated with the clinical outcome is ongoing to define potential predictive elements of potency and biologic activity that should be considered in order to choose the optimal personalized therapy.

## References

1. Calcei JG, Rodeo SA. Orthobiologics for bone healing. Clin Sports Med. 2019;38:79–95.
2. Jeyakumar V, Niculescu-Morzsa E, Bauer C, et al. Platelet-rich plasma supports proliferation and redifferentiation of chondrocytes during in vitro expansion. Front Bioeng Biotechnol. 2017;6(5):75.
3. Zhou Y, Wang JHC. PRP treatment efficacy for tendinopathy: a review of basic science studies. Biomed Res Int. 2016;2016:9103792.
4. Zhu Y, Yuan M, Meng HY, et al. Basic science and clinical application of platelet-rich plasma for cartilage defects and osteoarthritis: a review. Osteoarthr Cartil. 2013;21(11):1627–37.3
5. Qiu G, Zheng G, Ge M, et al. Mesenchymal stem cell-derived extracellular vesicles affect disease outcomes

via transfer of microRNAs. Stem Cell Res Ther. 2018;9(1):320.

6. Marx RE, Carlson ER, Eichstaedt RM, et al. Platelet-rich plasma: growth factor enhancement for bone grafts. Oral Surg. Oral med. Oral Pathol Oral Radiol Endodontol. 1998;85:638–46.

7. Mariani E, Pulsatelli L. Platelet concentrates in musculoskeletal medicine. Int J Mol Sci. 2020;21(4):1328.

8. Giannini S, Cielo A, Bonanome L. Comparison between PRP, PRGF and PRF: lights and shadows in three similar but different protocols. Eur Rev Med Pharmacol Sci. 2015;19(6):927–30.

9. Lin KY, Chen P, Chen AC, et al. Leukocyte-rich platelet-rich plasma has better stimulating effects on tenocyte proliferation compared with leukocyte-poor platelet-rich plasma. Orthop J Sports Med. 15. 2022;10(3):23259671221084706.

10. Le ADK, Enweze L, DeBaun MR, et al. Current clinical recommendations for use of platelet-rich plasma. Curr Rev Musculoskelet Med. 2018;11(4):624–34.

11. Wakayama T, Saita Y, Kobayashi Y, et al. Quality comparison between two different types of platelet-rich plasma for knee osteoarthritis. Regen Med Res. 2020;8:3.

12. Evans CH, Chevalier X, Wehling P. Autologous conditioned serum. Phys Med Rehabil Clin N Am. 2016;27(4):893–908.

13. Meijer H, Reinecke J, Becker C, et al. The production of anti-inflammatory cytokines in whole blood by physico-chemical induction. Inflamm Res. 2003;52(10):404–7.

14. Li S, Xiang C, Wei X, et al. Early supplemental α2-macroglobulin attenuates cartilage and bone damage by inhibiting inflammation in collagen II-induced arthritis model. Int J Rheum Dis. 2019;22(4):654–65.

15. Montesano P, Cuellar J, Scuderi G. Injection of an autologous Alpha-2-macroglobulin (A2M) concentrate alleviates Back pain in FAC-positive patients. Orthop Rheum Open Access J. 2017;4:2.

16. Muthu S, Jeyaraman M, Jain R, et al. Accentuating the sources of mesenchymal stem cells as cellular therapy for osteoarthritis knees-a panoramic review. Stem Cell Investig. 2021;9(8):13.

17. Caplan AI, Correa D. The MSC: an injury drugstore. Cell Stem Cell. 2011;9(1):11–5. 1726829

18. Gupta PK, Chullikana A, Rengasamy M, et al. Efficacy and safety of adult human bone marrow-derived, cultured, pooled, allogeneic mesenchymal stromal cells (Stempeucel®): preclinical and clinical trial in osteoarthritis of the knee joint. Arthritis Res Ther. 2016;18(1):301.

19. Chahla J, Alland JA, Verma NN. Bone marrow aspirate concentrate for orthopaedic use. Orthop Nurs. 2018;37(6):379–81.

20. Lamplot JD, Rodeo SA, Brophy RH. A practical guide for the current use of biologic therapies in sports medicine. Am J Sports Med. 2020;48(2):488–503.

21. Manchikanti L, Centeno CJ, Atluri S, et al. Bone marrow concentrate (BMC) therapy in musculoskeletal disorders: evidence-based policy position statement of American Society of Interventional Pain Physicians (ASIPP). Pain Physician. 2020;23(2):E85–E131.

22. Tang Q, Zhao XS, Guo A, et al. Therapeutic applications of adipose-derived stromal vascular fractions in osteoarthritis. World J Stem Cells. 2022;14(10):744–55.

23. Ulivi M, Meroni V, Viganò M, et al. Micro-fragmented adipose tissue (mFAT) associated with arthroscopic debridement provides functional improvement in knee osteoarthritis: a randomized controlled trial. Knee Surg Sports Traumatol Arthrosc. 2022;30:1–12.

24. Vezzani B, Shaw I, Lesme H, et al. Higher pericyte content and secretory activity of microfragmented human adipose tissue compared to enzymatically derived stromal vascular fraction. Stem Cells Transl Med. 2018;7(12):876–86.

25. Iyyanki T, Hubenak J, Liu J, et al. Harvesting technique affects adipose-derived stem cell yield. Aesthet Surg J. 2015;35(4):467–76.

26. Trivisonno A, Alexander RW, Baldari S, et al. Intraoperative strategies for minimal manipulation of autologous adipose tissue for cell- and tissue-based therapies: concise review. Stem Cells Transl Med. 2019;8(12):1265–71.

27. Lattermann C, Leite CBG, Frisbie D, et al. Orthobiologics in orthopedic applications: a report from the TMI Havemeyer meeting on orthobiologics. J Cartil Joint Preserv. 2022;2(3):100055. ISSN 2667-2545

# Platelet-Rich Plasma for Osteoarthritis

Trifon Totlis and Angelo V. Vasiliadis

## 13.1 Introduction

Osteoarthritis (OA) is considered the most common type of arthritis and the most prevalent joint disease in adults [1]. It is characterized by articular cartilage degeneration that affects patient's mobility and quality of life [2]. The World Health Organization (WHO) estimates that approximately 10% of all men and 13% of all women aged over 60 have OA [3]. Articular cartilage degradation results from a disruption in homeostasis, due to an imbalance between chondrocyte catabolic and anabolic pathways, driven by local production of metalloproteinases and multiple inflammatory mediators [4]. The most commonly affected joint is the knee followed by the hip, hands, and spine [3, 5].

A stepwise approach is recommended for OA management, including non-pharmacological, pharmacological, and surgical treatment for patients with a more advanced stage of disease [6, 7]. Recently, there has been an increasing interest in non-operative treatment options for knee OA, primarily focusing on intra-articular injectable therapies [8]. Intra-articular injections traditionally included corticosteroids and hyaluronic acid (HA) with integral roles in improving joint lubrication and anti-inflammatory effects [8–10]. More recently, orthobiologics have emerged as potential adjunctive therapies to treat mild to moderate OA. Agents such as platelet-rich plasma (PRP), bone marrow aspirate concentrate (BMAC), adipose tissue, and allogenic amniotic fluid products are more commonly being used as injectable therapies for OA [7, 9].

PRP is an autologous formulation derived from centrifugation of the patient's whole blood. It is defined as a volume of autologous plasma with a platelet concentration greater than the average in peripheral blood (150,000–350,000 platelets/µl) [11, 12]. PRP preparation protocols vary widely. There are essentially three different methods for PRP production: single centrifugation, double centrifugation, or blood filtration and plateletpheresis. This is done on manual or automatic systems operated in open or closed circuits. The high number of variables involved has led to the development of several custom protocols and to a large number and variability of commercially available PRP systems. These devices vary in PRP collection volumes and preparation protocols that result in distinctive PRP bioformulations and properties.


T. Totlis (✉)
The-MIS Orthopaedic Center, St. Luke's Hospital, Thessaloniki, Greece

School of Medicine, Faculty of Health Sciences, Aristotle University of Thessaloniki, Thessaloniki, Greece
e-mail: totlis@auth.gr

A. V. Vasiliadis
Orthopaedic Department, St. Luke's Hospital, Thessaloniki, Greece

Research Fellow Hôpital de la Croix-Rousse, Lyon, France

The critical variables to characterize PRPs usually include the following [13]:

1. The proportion of platelets in PRP to platelets in whole blood called platelet enrichment factor (PEF).
2. The presence or absence of WBCs. Leukocyte rich-PRP (LR-PRP) is defined as having a greater number of leukocytes when compared to whole blood. In leukocyte poor PRP (LP-PRP), the concentration of leukocytes is equivalent or reduced when compared to baseline.
3. The method of activation. Platelet activation can be triggered by exogenous agents, such as calcium chloride or thrombin. Alternatively, direct injection of nonactivated PRP allows local tissue factors to endogenously activate platelets.

Based on those three variables and other PRP characteristics, several classification systems have been put forward over the years for a more complete and standardized description of different PRP categories (Table 13.1) [12, 14].

The rationale for PRP use in patients with OA is based on evidence that intra-articular PRP injection may reduce pain and inflammation as well as influence joint homeostasis promoting both the healing process and immunoregulation (Fig. 13.1a, b) [12, 15, 16]. PRP contains biologically active proteins, cytokines, and growth factors derived from platelets a-granules, which have inflammation reduction and cell regenerative properties (Table 13.2) [14, 17]. Once injected and activated, platelets degranulation releases a great number of GFs that act through autocrine, paracrine, or endocrine mechanisms by binding to receptors on the cell membrane [17]. This results in a cascade of events, such as chondrocyte proliferation, chemotaxis, cell migration and differentiation, matrix production stimulation, angiogenesis, and inflammation modulation [12]. This mechanism favors the restoration of a homeostatic balance in degenerative joints, slowing down the inflammatory, catabolic, and degenerative processes, thus offering benefits in terms of symptom relief, functional improvement, and potentially on disease progression. Moreover, PRP is considered to be cost-effective and convenient for patients [18–20]. Very rarely leads to complications, it is easy to prepare and administer, and it is less invasive compared to other orthobiologic therapeutic options [21, 22].

**Table 13.1** Platelet-rich plasma classification systems

| Classification system | Criteria |
| --- | --- |
| Ehrenfest classification | Presence of cell content |
| | Fibrin architecture |
| PAW classification | Platelets |
| | Activation |
| | White blood cells |
| PLRA classification | Platelet count |
| | Leucocyte content |
| | Red blood cell content |
| | Activation |
| DEPA classification | Dose of injected platelets |
| | Efficiency of production |
| | Purity of PRP |
| | Activation process |
| MARSPILL classification | Method |
| | Activation |
| | Red blood cells |
| | Spin |
| | Platelet concentration |
| | Image guided |
| | Leucocyte concentration |
| | Light activation |

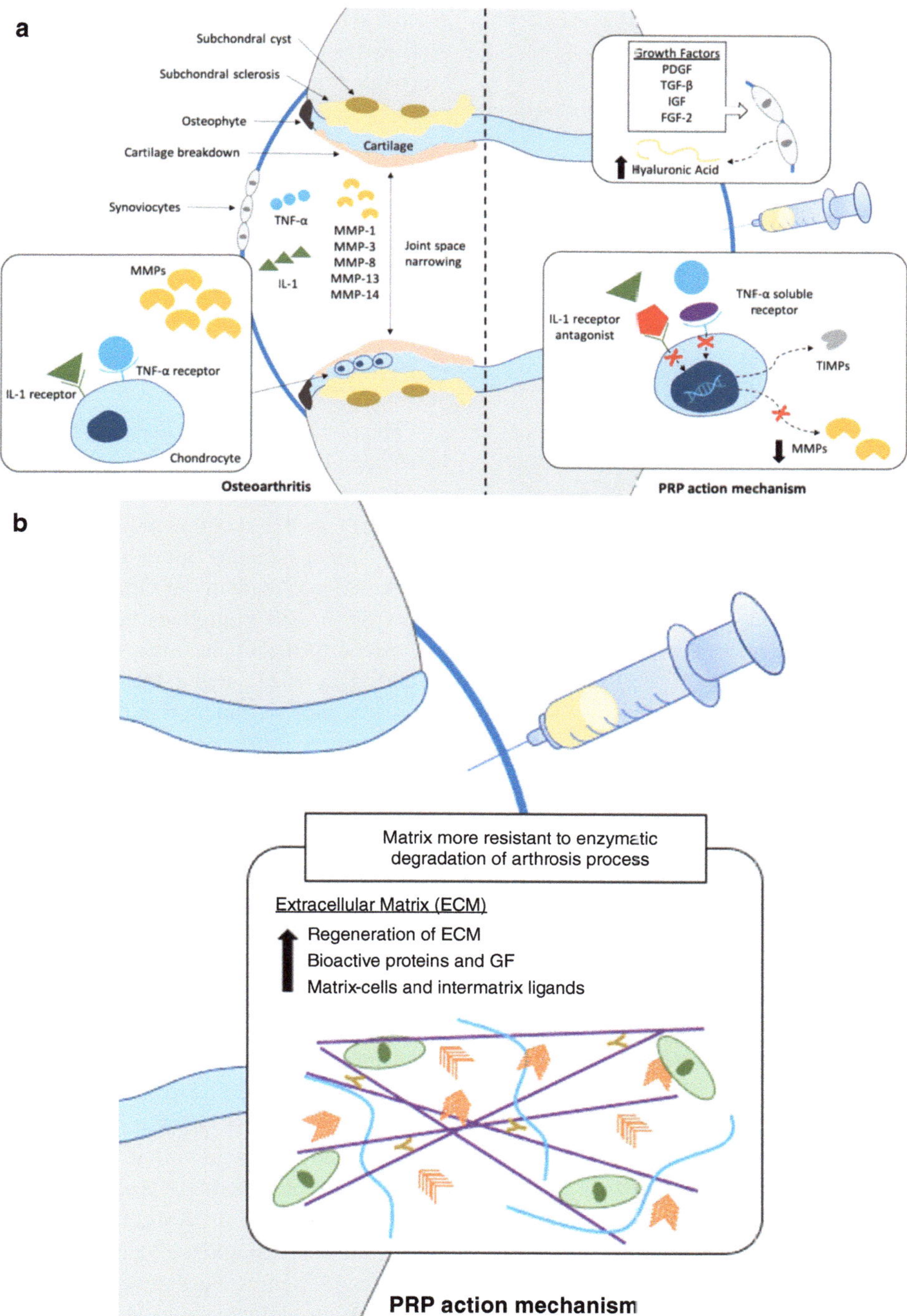


**Fig. 13.1** (**a**) Pro-inflammatory factors interleukin-1 (IL-1) and tumor necrosis factor alpha (TNF-α) bind to receptors on chondrocytes and promote synthesis of matrix metalloproteinases (MMPs) and cartilage breakdown, contributing to osteoarthritis pathogenesis (Fig. **a**, left). Intra-articular platelet-rich plasma (PRP) injection has immunoregulatory effects and creates a regenerative microenvironment. A number of anti-inflammatory factors including IL-1 receptor antagonist and TNF-α soluble receptors block the action of IL-1 and TNF-α by preferentially binding to the receptor on chondrocytes and inhibit production of MMPs. The growth factors (GF) produced from activated platelets stimulate synoviocytes to synthesize hyaluronic acid. (**b**) PRP promotes cell migration and proliferation but also enhances the extracellular matrix remodeling (proteoglycans, fibronectin, type II collagen) via the stimulation of angiogenesis

**Table 13.2** Platelet-rich plasma-based growth factors

| Growth factor | Function |
| --- | --- |
| PDGF | Stimulates chemotaxis and mitogenesis in fibroblast |
| | Regulates collagenase secretion and collagen synthesis |
| TGF-β | Stimulates MSC proliferation |
| | Regulates mitogenesis |
| | Regulates collagen synthesis and collagenase secretion |
| | Stimulates chemotaxis and angiogenesis |
| VEGF | Increases angiogenesis and vessel permeability |
| | Stimulates mitogenesis |
| FGF | Promotes growth and differentiation of chondrocytes and osteoblasts |
| | Regulates mitogenesis for MSC, chondrocytes, and osteoblasts |
| IGF-1 | Stimulates chemotaxis |
| | Regulates proliferation and maturation of chondrocytes |

*PDGF* platelet-derived growth factor, *TGF* transforming growth factor, *VEGF* vascular endothelial growth factor, *FGF* fibroblast growth factor, *IGF* insulin-like growth factor, *MSC* mesenchymal stem cells

## 13.2    PRP Outcomes for OA

The role of PRP in alleviating pain and improving functional outcomes in patients with OA has been well established both in basic science and clinical studies over the last two decades [23, 24]. There are many randomized controlled trials (RCTs) and meta-analyses about the efficacy of intra-articular PRP injections, documenting positive results in terms of pain relief and functional improvement up to 12 months following administration [23, 25–27]. Compared with placebo (normal saline injections) in patients with mild to moderate knee osteoarthritis, PRP showed significant superiority in providing symptomatic relief [27, 28]. Furthermore, intra-articular PRP injections provide superior outcomes compared to hyaluronic acid and corticosteroids for symptomatic management of knee OA, including pain relief, improving joint function and participation in physical activities at 12 months follow-up [26, 28–30]. Particularly, a recent meta-analysis performed on 34 RCTs reported that these benefits increase over time, being not significant at earlier follow-ups, but becoming clinically significant through 6 and 12 months [27]. However, these improvements remain partial, and the level of evidence for some of the outcomes is still low [27]. Whether PRP can affect cartilage regeneration and OA progression is yet to be confirmed in clinical studies, although 68% of relevant animal studies showed disease-modifying effect [31]. There are also RCTs and meta-analyses which do not support the use of autologous PRP as an effective treatment for the management of knee OA [32, 33].

The plethora of RCTs favoring PRP injections for knee OA [27] provides adequate evidence for PRP use in daily practice. Nevertheless, this treatment is not recommended by most of the international societies' guidelines. Only recently, the American Academy of Orthopaedic Surgeons (AAOS) in their guidelines for the management of osteoarthritis of the knee (non-arthroplasty), released in 2021, stated that PRP may reduce pain and improve function in patients with symptomatic osteoarthritis of the knee. To address this issue, the European Society of Sports Traumatology, Knee Surgery and Arthroscopy (ESSKA) launched a consensus study in 2022 and concluded that there is currently enough preclinical and clinical evidence to recommend the use of PRP in knee OA, mainly in mild and moderate degrees. The principal drawback of the relevant literature is the high heterogeneity and lack of standardization in terms of PRP preparation and content. To overcome these issues, further double-blinded RCTs are needed, in which the preparation of PRP should be analyzed with platelets and cells counting and standardized. The sample size needs to be adequate based on a priori sample size analysis. Patient compliance and follow-up should be maintained at a high level for 12 months. Furthermore, it is essential to establish strict inclusion and exclusion criteria with highly specific indications and adhere to specific reporting requirements.

Several studies have consistently found that PRP injections showed better efficacy in younger patients with mild to moderate structural changes on plain radiographs (Kellgren-Lawrence grade

1, 2, and 3) than in older patients with severe knee OA (Kellgren-Lawrence grade 4) [13, 34–36]. In patients with moderate to severe knee OA [37], PRP treatment did not present a statistically significant improvement in pain and function, although some authors suggested a possible beneficial effect despite the lack of statistical difference [38, 39]. In recent years, the application of biological treatments, such as PRP injections, in moderate knee OA has been reported to decrease the rate of cartilage loss and may contribute to delay or avoid the need of total knee replacement [23, 35, 40]. One of the main advantages of PRP therapy is that the procedure uses an autologous blood product with no known systemic side effects but only local adverse effects, such as bleeding, bruising, swelling, stiffness, and soreness [24, 34]. These symptoms, when present, are transient and usually resolved within a few hours to days without any specific treatment [34]. Interestingly, the incidence of adverse effects after intra-articular PRP injections has been associated with the concentration of leukocytes, with the LR-PRP increasing the risk [24].

As for the clinical superiority of LP-PRP versus LR-PRP, it remains controversial. The role of leukocytes has been a subject of debate because of their positive as well as negative properties. In a meta-analysis by Riboh et al., LP-PRP showed significantly better WOMAC scores than did HA, but LR-PRP did not [41]. However, the latter study [41], a comparative clinical study [42] and two recent meta-analyses [24, 43] found no significant difference in clinical outcomes when LP-PRP was directly compared to LR-PRP. The LP-PRP has been suggested to be preferable to LR-PRP for the treatment of knee OA [43]; however, further comparative studies are necessary to confirm this.

Delivering PRP with hyaluronic acid has also gained interest in recent years under the assumption that the combined application could provide a synergistic effect resulting in inflammatory inhibition and inducing cartilage regeneration through targeting both agents' biological pathways [44]. The relevant literature has shown greater improvements in pain and function for the combined injection compared with both HA alone [7] and PRP alone [45]. The potentially synergistic effects of PRP and HA may be considered in the management of knee OA to maximize outcomes; however, whether combination therapy is cost-effective remains unclear [46].

## 13.3 Application Protocol

Several factors in the PRP application process and injection protocol may influence the outcome. The optimal number and interval between PRP injections have not been clearly defined yet. Several randomized controlled trials [47–49] comparing single to multiple PRP injections have been reported. Moreover, a recent meta-analysis concluded that within a 6-month interval, a single injection was as effective as multiple ($n = 2$ or 3) PRP injections in pain improvement; however, multiple injections ($n = 3$) were more effective than a single injection in terms of knee functionality improvement [50]. Interval between injections vary among clinical trials between 1 week and 1 month [42, 47, 51–56]. However, the majority of studies have shown that protocols with more than one injection (up to three injections) are better than a single injection with the most common regimen being the application of three weekly PRP injections (1-week intervals) [51–53, 57, 58].

Injection techniques are also very diverse. Various approaches have been used including lateral suprapatellar and lateral mid-patellar with the patient supine or anterolateral and anteromedial with the patient seated. The skin is typically sterilely prepped with an alcohol or iodine-based solution, and optionally a sterile drape is placed over the prepped skin. The use of local anesthetics during PRP injections has been associated with a possible detrimental effect on platelet aggregation [59]. However, it is unclear whether a local subcutaneous injection of a local anesthetic without penetrating the capsule, as opposed to an intra-articular injection, would have similar adverse effects on platelet function. A 21 to 23G needle is inserted into the joint under ultrasound guidance or blindly with surface landmark technique. Aspiration of synovial fluid confirms proper needle placement and could relieve effusion when present. If an effusion is present, it is recommended to aspirate/evacuate it before injecting the PRP in order to avoid dilution of the PRP.

Postinjection recommendations also vary. A frequent approach includes: The patient is asked to flex and extend the knee several times to allow the PRP to spread throughout the joint before PRP transformed to a more solid material. After 10–15 min of rest, the patient is discharged home with instructions to restrict the use of the leg, regularly apply cold therapy, avoid nonsteroidal or steroidal medication or medications influencing platelet count or function for at least 7 days, and use only acetaminophen, dipyrone, or similar for pain management if necessary. Subsequently, the patient is recommended to continue the mild-to-moderate level of activities and gradually increase the level as tolerated. A postinjection rehabilitation program aiming to improve mobility, muscle strength, and stability may be considered. Gradual resumption of normal sport or recreational activities is allowed as tolerated [27, 56].

## 13.4 Conclusion

The management of knee OA is complex, and the use of orthobiologics is ever growing for this indication. PRP is defined as a volume of autologous plasma, derived from centrifugation of the patient's whole blood, with a platelet concentration greater than the average in peripheral blood. PRP contains biologically active proteins, cytokines, and growth factors derived from platelets a-granules. Intra-articular PRP injection may reduce pain and inflammation as well as influence joint homeostasis promoting both the healing process and immunoregulation. It is a safe and easy to perform procedure for patients. PRP has been shown to provide superior clinical outcomes compared to placebo, steroids, and HA in most studies. The majority of studies have shown protocols with multiple injections (up to three) are preferable to single-injection protocols. PRP could be considered a valid first-line injectable treatment option for nonoperative management of knee OA, mainly for KL grades 1–3.

## References

1. Suri P, Morgenroth DC, Hunter DJ. Epidemiology of osteoarthritis and associated comorbidities. PM R. 2012;4:S10–9.
2. Dong Y, Zhang B, Yang Q, Zhu J, Sun X. The effects of platelet-rich plasma injection in knee and hip osteoarthritis: a meta-analysis of randomized controlled trials. Clin Rheumatol. 2021;40(1):263–77.
3. Zhang Y, Jordan JM. Epidemiology of osteoarthritis. Clin Geriatr Med. 2010;26(3):355–69.
4. Garner M, Alshameeri Z, Khanduja V. Osteoarthritis: genes, nature-nurture interaction and the role of leptin. Int Orthop. 2013;37(12):2499–505.
5. Park JH, Hong JY, Han K, Suh SW, Park SY, Yang JH, Han SW. Prevalence of symptomatic hip, knee, and spine osteoarthritis nationwide health survey analysis of an elderly Korean population. Medicine (Baltimore). 2017;96(12):e6372.
6. Ebell MH. Osteoarthritis: rapid evidence review. Am Fam Physician. 2018;97(8):523–6.
7. Karasavvidis T, Totlis T, Gilat R, Cole BJ. Platelet-rich plasma combined with hyaluronic acid improves pain and function compared with hyaluronic acid alone in knee osteoarthritis: a systematic review and meta-analysis. Arthroscopy. 2021;37(4):12277–1287.
8. Yaftali NA, Weber K. Corticosteroids and hyaluronic acid injections. Clin Sports Med. 2019;38(1):1–15.
9. Campbell KA, Erickson BJ, Saltzman BM, et al. Is local viscosupplementation injection clinically superior to other therapies in the treatment of osteoarthritis of the knee: a systematic review of overlapping meta-analyses. Arthroscopy. 2015;31(10):2036–45.e14
10. Waddell DD. Viscosupplementation with hyaluronans for osteoarthritis of the knee: clinical efficacy and economic implications. Drugs Aging. 2007;24(8):629–42.
11. Cole BJ, Seroyer ST, Filardo G, Bajaj S, Fortier LA. Platelet-rich plasma: where are we now and where are we going? Sports Health. 2010;2(3):203–10.
12. Kon E, Filardo G, DiMartino A, Marcacci M. Platelet-rich plasma (PRP) to treat sports injuries: evidence to support its use. Knee Surg Sport Traumatol Arthrosc. 2011;19(4):516–27.
13. Gato-Calvo L, Magalhaes J, Ruiz-Romero C, Blanco FJ, Burguera EF. Platelet-rich plasma in osteoarthritis treatment: review of current evidence. Ther Adv Chronic Dis. 2019;10:1–18.
14. Collins T, Alexander D, Barkatali B. Platelet-rich plasma: a narrative review. EFFORT Open Rev. 2021;6(4):225–35.

15. Cole BJ, Karas V, Hussey K, Pilz K, Fortier LA. Hyaluronic acid versus platelet-rich plasma: a prospective, double-blind randomized controlled trial comparing clinical outcomes and effects on intra-articular biology for the treatment of knee osteoarthritis. Am J Sports Med. 2017;45(2):339–46.

16. Halpern B, Chaudhury S, Rodeo SA, et al. Clinical and MRI outcomes after platelet-rich plasma treatment for knee osteoarthritis. Clin J Sport Med. 2013;23(3):238–9. https://doi.org/10.1097/JSM.0b013e31827c3846.

17. Everts P, Onishi K, Jayaram P, Lana JF, Mautner K. Platelet-rich plasma: new performance understanding and therapeutic considerations in 2020. Int J Mol Sci. 2020;21(20):7794.

18. Kennedy MI, Whitney K, Evans T, LaPrade RF. Platelet-rich plasma and cartilage repair. Curr Rev Musculoskelet Med. 2018;11(4):573–82.

19. Marx RE. Platelet-rich plasma (PRP): what is PRP and what is not PRP? Implant Dent. 2001;10(4):225–8.

20. Foster TE, Puskas BL, Mandelbaum BR, Gerhardt MB, Rodeo SA. Platelet-rich plasma: from basic science to clinical applications. Am J Sports Med. 2009;37(11):2259–72.

21. Forogh B, Mianehsaz E, Shoaee S, et al. Effect of single injection of platelet-rich plasma in comparison with corticosteroid on knee osteoarthritis: a double-blind randomized clinical trial. J Sports Med Phys Fitness. 2016;56:901–8.

22. Ayhan E, Kesmezacar H, Akgun I. Intraarticular injections (corticosteroid, hyaluronic acid, platelet rich plasma) for the knee osteoarthritis. World J Orthop. 2014;5:351–61.

23. Paterson KL, Hunter DJ, Metcalf BR, et al. Efficacy of intra-articular injections of platelet-rich plasma as a symptom- and disease-modifying treatment for knee osteoarthritis o the RESTORE trial protocol. BMC Musculoskelet Disord. 2018;19(1):272.

24. Kim J-H, Park Y-B, Ha C-W, et al. Adverse reactions and clinical outcomes for leukocyte-poor versus leukocyte-rich platelet-rich plasma in knee osteoarthritis. A systematic review and meta-analysis. Orthop J Sports Med. 2021;9(6):23259671211011948.

25. Dai WL, Zhou AG, Zhang H, et al. Efficacy of platelet-rich plasma in the treatment of knee osteoarthritis: a meta-analysis of randomized controlled trials. Arthroscopy. 2017;33(3):659–70.

26. Aw ALL, Leeu JJ, Tao X, et al. Comparing the efficacy of dual platelet-rich plasma (PRP) and hyaluronic acid (HA) therapy with PRP-alone therapy in the treatment of knee osteoarthritis: a systematic review and meta-analysis. J Exp Orthop. 2021;8(1):101.

27. Filardo G, Previtali D, Napoli F, et al. PRP injections for the treatment of knee osteoarthritis: a meta-analysis of randomized controlled trials. Cartilage. 2021;13(1_suppl):364S–75S.

28. Lin KY, Yang CC, Hsu CJ, et al. Intra-articular injection of platelet-rich plasma is superior to hyaluronic acid or saline solution in the treatment of mild to moderate knee osteoarthritis: a randomized, double-blind, triple-parallel, placebo-controlled clinical trial. Arthroscopy. 2016;35(1):106–17.

29. Huang Y, Liu X, Xu X, et al. Intra-articular injections of platelet-rich plasma, hyaluronic acid or corticosteroids for knee osteoarthritis: a prospective randomized controlled study. Orthopade. 2019;48(3):239–47.

30. McLarnon M, Heron N. Intra-articular platelet-rich plasma injections versus intra-articular corticosteroid injections for symptomatic management of knee osteoarthritis: systematic review and meta-analysis. BMC Musculoskelet Disord. 2021;22(1):550.

31. Boffa A, Salerno M, Merli G, et al. Platelet-rich plasma injections induce disease-modifying effects in the treatment of osteoarthritis in animal models. Knee Surg Sports Traumatol Arthrosc. 2021;29:4100–21.

32. Bennell KL, Paterson KL, Metcalf BR, et al. Effect of intra-articular platelet-rich plasma vs placebo injection on pain and medial tibial cartilage volume in patients with knee osteoarthritis: the RESTORE randomized clinical trial. JAMA. 2021;326(20):2021–30.

33. Han SB, Seo IW, Shin YS. Intra-articular injections of hyaluronic acid or steroids associated with better outcomes than platelet-rich plasma, adipose mesenchymal stromal cells, or placebo in knee osteoarthritis: a network meta-analysis. Arthroscopy. 2021;37(1):292–306.

34. Eymard F, Ornetti P, Maillet J, et al. Intra-articular injections of platelet-rich plasma in symptomatic knee osteoarthritis: a consensus statement from French-speaking experts. Knee Surge Sports Traumatol Arthrosc. 2021;29(10):3195–210.

35. Sánchez M, Jorquera C, Sánchez P, et al. Platelet-rich plasma injections delay the need for knee arthroplasty: a retrospective study and survival analysis. Int Orthop. 2021;45(2):401–10.

36. Kuffler DP. Variables affecting the potential efficacy of PRP in providing chronic pain relief. J Pain Res. 2019;12:109–16.

37. Dorio M, Rodrigues Pereira RM, Branco Luz AG, et al. Efficacy of platelet-rich plasma and plasma for symptomatic treatment of knee osteoarthritis: a double-blinded placebo-controlled randomized clinical trial. BMC Musculoskelet Disord. 2021;22:822.

38. Jubert NJ, Rodriguez L, Reverte-Vinaixa MM, Navarro A. Platelet-rich plasma injections for advanced knee osteoarthritis: a prospective, randomized, double-blinded clinical trial. Orthop J Sports Med. 2017;5(2):2325967116698386.

39. Vilchez-Cavazos F, Blazquez-Saldana J, Gamboa-Alonso AA, et al. The use of platelet-rich plasma in studies with early knee osteoarthritis versus advanced stages of the disease: a systematic review and meta-analysis of 31 randomized clinical trials. Arch Orthop Trauma Surg. 2022; https://doi.org/10.1007/s00402-021-04304-1.

40. Bansal H, Leon J, Pont JL, et al. Platelet-rich plasma (PRP) in osteoarthritis (OA) knee: correct dose critical for long term clinical efficacy. Sci Rep. 2021;11(1):3971.

41. Riboh JC, Saltzman BM, Yanke AB, et al. Effect of leukocyte concentration on the efficacy of platelet-rich plasma in the treatment of knee osteoarthritis. Am J Sports Med. 2016;44(3):792–800.

42. Filardo G, Kon E, Pereira Ruiz MT, et al. Platelet-rich plasma intra-articular injections for cartilage degeneration and osteoarthritis: single- versus double-spinnng approach. Knee Surg Sports Traumatol Arthrosc. 2012;20(10):2082–91.

43. Abbas A, Tong Du J, Dhotar HS. The effect of leukocyte concentration on platelet-rich plasma injections for the knee osteoarthritits: A network meta-analysis. 2021;https://doi.org/10.2106/JBJS.20.02258.

44. Laver L, De Girolamo L. Injectable orthobiologics for osteoarthritis: current and future advances. ESSKA. 2021;

45. Zhao J, Huang H, Liang G, et al. Effects and safety of the combination of platelet-rich plasma (PRP) and hyaluronic acid (HA) in the treatment of knee osteoarthritis: a systematic review and meta-analysis. BMC Musculoskelet Disord. 2020;21(1):224.

46. Dwyer T, Chahal J. Editorial commentary: injections for knee osteoarthritis: doc, you gotta help me! Arthroscopy. 2021;37(4):1288–9.

47. Görmeli G, Görmeli CA, Ataoglu B, et al. Multiple PRP injections are more effective than single injections and hyaluronic acid in knees with early osteoarthritis: a randomized, double-blind, placebo-controlled trial. Knee Surg Sports Traumatol Arthrosc. 2017;25:958–65.

48. Patel S, Dhillon MS, Aggarwal S, et al. Treatment with platelet-rich plasma is more effective than placebo for knee osteoarthritis: a prospective, double-blind, randomized trial. Am J Sports Med. 2013;41:356–64.

49. Uslu Güvendi E, Askin A, Güvendi G, et al. Comparison of efficiency between corticosteroid and platelet rich plasma injection therapies in patients with knee osteoarthritis. Arch Rheumatol. 2018;33:273–81.

50. Vilches-Cavazos F, Millan-Alanis JM, Blazquez-Saldana J, et al. Comparison of the clinical effectiveness of single versus multiple injections of platelet-rich plasma in the treatment of knee osteoarthritis. A sys-

51. Vaquerizo V, Angel Plasencia M, et al. Comparison of intra-articular injections of plasma rich in growth factors (PRGF-Endoret) versus durolane hyaluronic acid in the treatment of patients with symptomatic osteoarthritis: a randomized controlled trial. Arthroscopy. 2013;29:1635–43.

52. Sanchez M, Fiz N, Azofra J, et al. A randomized clinical trial evaluating plasma rich in growth factors (PRGF-Endoret) versus hyaluronic acid in the short-term treatment of symptomatic knee osteoarthritis. Arthroscopy. 2012;28:1070–8.

53. Spakova T, Rosocha J, Lacko M, et al. Treatment of knee joint osteoarthritis with autologous platelet-rich plasma in comparison with hyaluronic acid. Am J Phys Med Rehabil. 2012;91:411–7.

54. Li M, Zhang C, Ai Z, et al. Therapeutic effectiveness of intra-knee-articular injection of platelet-rich plasma on knee articular cartilage degeneration. Zhongguo Xiu Fu Chong Jian Wai Ke Za Zhi. 2011;25:1192–6.

55. Kon E, Mandelbaum B, Buda R, et al. Platelet- rich plasma intra-articular injection versus hyaluronic acid viscosupplementation as treatments for cartilage pathology: from early degeneration to osteoarthritis. Arthroscopy. 2011;27:1490–501.

56. Raeissadat SA, Rayegani SM, Hassanabadi H, et al. Knee osteoarthritis injection choices: platelet- rich plasma (PRP) versus hyaluronic acid (a one-year randomized clinical trial). Clin Med Insights Arthritis Musculoskelet Disord. 2015;8:1–8.

57. Sanchez M, Anitua E, Azofra J, et al. Intra- articular injection of an autologous preparation rich in growth factors for the treatment of knee OA: a retrospective cohort study. Clin Exp Rheumatol. 2008;26:910–3.

58. Paterson KL, Nicholls M, Bennell KL, et al. Intra-articular injection of photo-activated platelet-rich plasma in patients with knee osteoarthritis: a double-blind, randomized controlled pilot study. BMC Musculoskelet Disord. 2016;17:67.

59. Bausset O, Magalon J, Giraudo L, et al. Impact of local anaesthetics and needle calibres used for painless PRP injections on platelet functionality. Muscles Ligaments Tendons J. 2014;4(1):18–23.

# Platelet-Rich Plasma Treatment for Meniscal Tears

## 14

Yosef Sourugeon, Yaniv Yonai, Yaron Berkovich, and Lior Laver

## 14.1 The Anatomy of the Menisci

The menisci are semilunar, fibrocartilaginous structures that primarily seek to transform axial loads from the femoral condyles more evenly along the tibial plateaus. They are also responsible for lubrication, stress reduction, stabilization, and congruency in the knee joint [1, 2]. Traditionally, the meniscus is divided into three zones based on its vascularization (from outside to inside): (1) the red zone, (2) the red-white or pink zone, and (3) the white zone. Only the lateral and medial peripheral areas of the menisci are supplied directly by blood vessels (branches of the genicular artery), while the remaining portions receive nutrition via diffusion. As demonstrated by Arnoczky and Warren, only 10 to 25 percent of the periphery of the meniscus is vascularized in adults, which reduces with age [3]. This lack of vascularity is the principal challenge in meniscal restoration.

## 14.2 Meniscal Injuries and Rationale for Use of PRP

Several inherent factors contribute to the unfavorable healing environment of a meniscus tear. These include the avascular nature of the meniscus, the presence of synovial fluid and pro-inflammatory cytokines, and the repetitive load subjected to the meniscus. Arnoczky and Warren previously showed that only the peripheral 10–30% of the meniscus is vascularized [3]. Moreover, the synovial fluid and presence of pro-inflammatory cytokines has been shown to have a catabolic effect on meniscal healing [4].

The improved understanding of the role of the meniscus in knee preservation in the last two decades has brought much focus toward preserving and saving the meniscus, exploring various methods to improve healing when meniscal injuries occur. As a result, being one of the main directions explored for improved meniscal healing, there has been growing interest in the role of orthobiologics in the treatment of meniscal


Y. Sourugeon
Department of Orthopedic Surgery, Chaim Sheba Medical Center, Rama Gan, Israel

Y. Yonai · L. Laver (✉)
Department of Orthopedic Surgery and Sports Medicine Unit, Hillel Yaffe Medical Center (HYMC), Hadera, Israel

Rappaport Faculty of Medicine, Technion (Israel Institute of Technology), Haifa, Israel

Y. Berkovich
Rappaport Faculty of Medicine, Technion (Israel Institute of Technology), Haifa, Israel

Department of Orthopedic Surgery, Hillel Yaffe Medical Center (HYMC), Hadera, Israel
e-mail: YARONB@hymc.gov.il

pathology. Techniques to improve meniscal healing based on enhancing one's own biologic properties have been explored in the past, such as meniscal trephination (to promote bleeding in the red zone of the meniscus) and microfractures to the notch area during arthroscopy. Several other modalities for biologic augmentation of meniscal repair include the use of a fibrin clot, cytokines and growth factors, PRP, and cell-based therapies. Several animal studies have explored the potential of PRP use for nonoperative management of meniscal tears. Xiao et al. reported that the application of PRP alone or in combination with BMSCs promoted the healing rate of meniscal white-white zone injury in a dog model [5]. Shin et al. found no significant differences in meniscal healing between the LR-PRP group and controls when applied to horizontal medial meniscus tears in a rabbit model [6]. In an in vitro and an in vivo study in a rabbit model, Ishida et al. demonstrated increased healing with meniscal defect filling using a gelatin hydrogel delivery system for PRP [7].

## 14.3 PRP Use for Meniscal Tears: Clinical Data

Treatment strategy for meniscal tears is dictated by a plethora of variables including tear type and pattern, zone involvement, age, tear location and extent, time from injury (acute or chronic), previous meniscus injuries, additional injuries (i.e., ligamentous, cartilage), and symptoms (i.e., existence of mechanical symptoms). While for many years surgical management has been the centerpiece of treatment of meniscal tears, primarily arthroscopic partial meniscectomy (APM) [8], the management goal has shifted in recent years toward preservation rather than resection of the meniscus, and there is a growing interest in orthobiologic treatments, including PRP.

Only a handful of studies examined PRP injections as a sole treatment for meniscal tears, while the majority of studies focused on the use of PRP as augmentation for meniscus repair. Blanke et al. used percutaneous PRP injections under fluoroscopic guidance for intra-substance meniscal tears (grade 2) in ten recreational ath-

letes. Each patient received three sequential injections with 7-day intervals. They reported six patients (60%) showed clinical improvement and improved sports activity [9]. Delen et al. examined the clinical effect of PRP injections on symptomatic meniscal tears in 41 patients (12 males, 29 females; mean age 38.2 + 8.37 years; range 21 to 50 years) with grade 2 or 3 meniscal tears. Three PRP injections were administered (2.7 mL, 1.2–1.5 million platelets per mL) at 1-week intervals, in a lateral patellofemoral approach. Authors reported at posttreatment weeks 1 and 4, both VAS and Lequesne Index scores significantly decreased, suggesting that PRP injections may improve short-term pain and disability in patients with meniscal tears [10].

In another small retrospective study with 6 months follow-up, Guenoun et al. examined the effect of ultrasound-guided intra-meniscal PRP injection (4 mL, 1999 ± 616 million platelets, 2 ± 2 million leukocytes) in ten patients with degenerative meniscal tears of the knee (grades 1–3), without knee osteoarthritis. They reported KOOS score was significantly improved and that all patients that were regularly practicing sports were able to return to competition or training activities, and VAS wasn't significantly improved; however, a decrease from the baseline was described [11].

Another option for utilizing of PRP for meniscal tears management is augmentation of meniscus repair. A randomized controlled trial comparing PRP augmentation of repaired vertical tears vs. isolated suture repair showed favorable results in the PRP-augmented group, with statistically significant functional outcome improvement, lower failure rates, and better healing on second look arthroscopy at 42 months post-surgery [12]. Another study by Everhart et al. reported that PRP augmentation of isolated meniscus repairs resulted in significantly decreased failure rates at 3 years post-surgery [13]. However, a number of smaller studies reported more variable results, with some showing modest benefits in functional outcomes, while others finding no benefits when compared to placebo [14–17]. While showing promising potential, it is still difficult to draw clear conclusion with regard to the full potential of PRP use for meniscal tears due to the relatively small number of existing studies, especially high level ones, the significant heterogeneity in PRP preparation techniques used in the different studies, the injection timings, as well as the type of tear being treated, and therefore future well-designed studies are required to better understand the role of PRP use in meniscal tears management.

> **Tip Box 1 PRP Use for Meniscal Tears**
> - Can be considered for stable, non-displaced meniscal tears.
> - Should not be used for displaced tears (i.e., bucket-handle tears, flap tears, complete root tears), tears with mechanical symptoms.
> - In acute tears extending to the periphery or acute horizontal tears, intra-articular injections could be combined with intra-meniscal injections performed under ultrasound guidance from the periphery of the meniscus with multiple penetrations in various trajectories (an outside-in trephination) to stimulate the red-zone and menisco-capsular area.
> - In acute horizontal tears, PRP could be injected intra-meniscally in a gel form; however, high volumes should be avoided intra-meniscally to prevent extending the tear.

> **Tip Box 2 PRP Use Following Meniscus Repair**
> - A PRP clot or PRP gel could be injected into and around the repair site intraoperatively.
> - Intra-articular injection at the end of surgery is not ideal as there is often remnant fluid and bleeding which could dilute the PRP and affect its activity and effectiveness.
> - If considering PRP injections postoperatively following a repair, injections should start at around 2 weeks post-surgery when swelling has subsided.

## 14.4 Conclusion

PRP treatment for meniscal tears shows promising results in promoting tissue repair and regeneration, potentially reducing the need for invasive surgical procedures and fostering quicker recoveries. The use of PRP, a concentrated solution of autologous platelets and growth factors derived from the patient's blood, has emerged as a potential therapeutic option to enhance the healing process and improve clinical outcomes. The potential of PRP use for the management of meniscal tears has been shown in several animal model studies as well as in several clinical studies exploring its use in nonoperative management of meniscal tears as well as for augmentation of meniscus repairs. By promoting natural healing mechanisms, PRP treatment may offer a nonsurgical and minimally invasive alternative, and this can be particularly advantageous for patients who wish to avoid the risks and prolonged recovery associated with surgery.

## References


1. Fox AJS, Bedi A, Rodeo SA. The basic science of human knee menisci: structure, composition, and function. Sports Health. 2012;4(4):340–51.
2. Patel H, Skalski MR, Patel DB, White EA, Tomasian A, Gross JS, Vangsness CT, Matcuk GR. Illustrative review of knee meniscal tear patterns, repair and replacement options, and imaging evaluation. Clin Imaging. 2021;69:4–16.
3. Arnoczky SP, Warren RF. Microvasculature of the human meniscus. Am J Sports Med. 1982;10(2):90–5.
4. Taylor SA, Rodeo SA. Augmentation techniques for isolated meniscal tears. Curr Rev Musculoskelet Med. 2013;6(2):95–101.
5. Xiao W, Yang Y, Xie W, He M, Liu D, Cai Z, Yu D, Li Y, Wei L. Effects of platelet-rich plasma and bone marrow mesenchymal stem cells on meniscal repair in the White-White zone of the meniscus. Orthop Surg. 2021;13(8):2423–32.
6. Shin KH, Lee H, Kang S, Ko YJ, Lee SY, Park JH, Bae JH. Effect of leukocyte-rich and platelet-rich plasma on healing of a horizontal medial meniscus tear in a rabbit model. Biomed Res Int. 2015; https://doi.org/10.1155/2015/179756.
7. Ishida K, Kuroda R, Miwa M, Tabata Y, Hokugo A, Kawamoto T, Sasaki K, Doita M, Kurosaka M. The regenerative effects of platelet-rich plasma on meniscal cells in vitro and its in vivo application

with biodegradable gelatin hydrogel. Tissue Eng. 2007;13(5):1103–12.

8. Bhan K. Meniscal tears: current understanding, diagnosis, and management. Cureus. 2020; https://doi.org/10.7759/CUREUS.8590.

9. Blanke F, Vavken P, Haenle M, von Wehren L, Pagenstert G, Majewski M. Percutaneous injections of platelet rich plasma for treatment of intrasubstance meniscal lesions. Muscles Ligaments Tendons J. 2015;5(3):162–6.

10. Delen V, Ediz L, Alpaycı M. The clinical effect of platelet-rich plasma injections on symptomatic meniscal tears of the knee. East J Med. 2021;26(3):367–70.

11. Guenoun D, Magalon J, de Torquemada I, Vandeville C, Sabatier F, Champsaur P, Jacquet C, Ollivier M. Treatment of degenerative meniscal tear with intrameniscal injection of platelets rich plasma. Diagn Interv Imaging. 2020;101(3):169–76.

12. Kaminski R, Kulinski K, Kozar-Kaminska K, Wielgus M, Langner M, Wasko MK, Kowalczewski J, Pomianowski S. A prospective, randomized, double-blind, parallel-group, placebo-controlled study evaluating meniscal healing, clinical outcomes, and safety in patients undergoing meniscal repair of unstable, complete vertical meniscal tears (bucket handle) aug-mented with platelet-rich plasma. Biomed Res Int. 2018;2018:9315815.

13. Everhart JS, Cavendish PA, Eikenberry A, Magnussen RA, Kaeding CC, Flanigan DC. Platelet-rich plasma reduces failure risk for isolated meniscal repairs but provides no benefit for meniscal repairs with anterior cruciate ligament reconstruction. Am J Sports Med. 2019;47(8):1789–96.

14. Dai W-L, Zhang H, Lin Z-M, Shi Z-J, Wang J. Efficacy of platelet-rich plasma in arthroscopic repair for discoid lateral meniscus tears. BMC Musculoskelet Disord. 2019;20:113.

15. Griffin JW, Hadeed MM, Werner BC, Diduch DR, Carson EW, Miller MD. Platelet-rich plasma in meniscal repair: does augmentation improve surgical outcomes? Clin Orthop Relat Res. 2015;473(5):1665–72.

16. Kemmochi M, Sasaki S, Takahashi M, Nishimura T, Aizawa C, Kikuchi J. The use of platelet-rich fibrin with platelet-rich plasma support meniscal repair surgery. J Orthop. 2018;15(2):711–20.

17. Pujol N, Salle De Chou E, Boisrenoult P, Beaufils P. Platelet-rich plasma for open meniscal repair in young patients: any benefit? Knee Surg Sports Traumatol Arthrosc. 2015;23(1):51–8.

Ferran Abat, Ignacio De Rus Aznar,
Federico Ibañez, and Charlotte Rafé


## 15.1 Introduction

Currently, the treatment of tendinopathies encompasses various therapeutic options. However, their standardized use as well as the way in which they should be combined remains uncertain. It is important to know the basic pathophysiology of tendinopathy in order to understand the overall process of tendon degeneration and regeneration. A comprehensive understanding of the tendon would allow to apply the most appropriate therapy and to know how to combine biological and mechanical stimulation as both are of fundamental importance in the quest for tendon regeneration.

This chapter will examine platelet-rich plasma (PRP) treatment in tendinopathies. It will also explain how it is suggested to be combined with other biological and mechanical therapies.

## 15.2 Fundamentals of PRP in Tendinopathy

The term PRP refers to a preparation obtained by centrifuging peripheral blood in a small volume of plasma where platelet concentration is increased over baseline [1]. The erythrocytes present in these preparations must be discarded. However, the need to include leukocytes and their quantity is currently being questioned [2].

The healing response triggered at the tendon level when it is injured consists of three well-studied phases. The first is referred as an inflammatory phase, followed by the proliferative and, finally, a maturation phase [3]. During that first phase, platelets play a key role, since their appearance and activation trigger a regulatory response of homeostasis, inflammation, neovascularization, and tissue remodeling [4]. Specifically, to control homeostasis, various growth factors capable of modulating the process of tissue repair and remodeling are released. It has been proven, via in vitro studies, that these factors increase their concentration in injured


F. Abat (✉) · F. Ibañez
Sports Orthopaedic Department, ReSport Clinic
Barcelona, Universitat Pompeu Fabra, Escola
Superior de Ciències de la Salut TecnoCampus, Grup
de recerca GRACIS (GRC 01604), Barcelona, Spain

I. D. R. Aznar
Hospital Universitario de Torrejón: Torrejon de
Ardoz, Madrid, Spain

Hospital Olympia Quironsalud, Madrid, Spain

C. Raflé
Investigation Department, ReSport Clinic Barcelona i
Mataró, Universitat Pompeu Fabra, Escola Superior
de Ciències de la Salut TecnoCampus, Grup de
recerca GRACIS (GRC 01604), Barcelona, Spain

tendons when PRP is applied [5]. One of them is the transforming factor β (TGF-β1). It is present in all stages of tendon healing and has mitogenic properties for fibroblasts and favors the production of extracellular matrix. There is also the group of growth factors derived from platelets (platelet-derived growth factors, PDGF) that are key in the activation and recruitment of macrophages and fibroblasts as well as in the synthesis of collagen [6]. It has been proven that the concentration of deposited platelets is relevant to inducing the mechanisms they mediate [7].

Conversely, PRP concentrates are capable of inducing the differentiation of tenocyte progenitor cells into active tenocytes [4]. As seen in animal studies, PRP seems to shorten the healing time of tendon, favors the organization of collagen fibers, and reduces proinflammatory macrophages, which confirms the gene modulation observed in in vitro studies [8].

## 15.3 Role of Leukocytes in the Treatment of Tendinopathies with PRP

The leukocytes present in PRP may have a proinflammatory, immunomodulatory, and nociceptive effect, eventually contributing to tissue repair and remodeling [1]. The optimal concentration of leukocytes in PRP preparation is currently a controversial issue [2]. The leukocyte population contains lymphocytes, neutrophils, and monocytes. In leukocyte-rich PRP (LR-PRP), the predominant population is lymphocytes. They produce two types of cytokines, interferon gamma (IFN-Đ) and interleukin 4 (IL-4), both involved in the noninflammatory cellular response through the modulation of macrophages.

Neutrophils are also well represented in a mixed leukocytes population. The need for their presence in PRP concentrates is controversial. Although they have been shown to play a role in angiogenesis and tissue regeneration, which could play a key role in the management of tendinopathy, it has been reported that they trigger an inflammatory reaction that have led to the application of LR-PRP not being recommended in pathologies such as osteoarthritis [9].

Finally, the monocytes, precursors of macrophages, can be classified into classically activated M1 (phenotype 1) and alternatively activated M2 (phenotype 2) in the microenvironment of a lesion. M1 have a pathogen-killing function through the production of IFN-Đ and nitric oxide, whereas M2 are capable of repairing damaged tissue and have anti-inflammatory properties. They produce components of the extracellular matrix, interleukin 10 (IL-10), and angiogenic factors. The presence of the two phenotypes depends on the physiopathology of the environment, and the switching process from M1 to M2 or vice versa (macrophage polarization) is affected by the type of PRP used [10]. It seems justifiable to try to get PRP concentrates able to polarize M1 toward M2.

## 15.4 Leukocyte-Rich PRP and Leukocyte-Poor PRP in Tendinopathy

There are numerous works in the literature that show variable efficacy when using PRP in the treatment of tendinopathy [11–13]. However, it is difficult to find studies that specifically compare the two treatment modalities: leukocyte-rich PRP (LR-PRP) and leukocyte-poor PRP (LP-PRP).

A recent meta-analysis [14] presents the available in vitro studies related to the application of *leukocyte-rich PRP* (LR-PRP) in tendinopathy. There are two works [15, 16] that analyze the results of samples of the supraspinatus tendon of the shoulder. Rubio-Azpeitia et al. [16] describe an increase in cell migration and proliferation, a regulation of the genes associated with the remodeling of the extracellular matrix, as well as an increase in inflammatory proteins when compared to the control group at 96 h. Cross et al. [15] report an increase in the expression of catabolic genes and remodeling of the extracellular matrix (COL1:COL3 ratio), the latter data differing from those presented by Rubio-Azpeitia [16].

When specifically comparing LR-PRP to LP-PRP, it can be seen that the release of growth factors, the proliferation of the tenocytes, and/or the gene expression of genes related to the tenocytes are quantitatively greater in the former [17]. In clinical studies, the injection of LR-PRP at the level of the lesion in gluteus medius tendinopathy significantly improves pain and function compared to corticosteroid injection at the 2-year follow-up [18]. However, when this data is looked at closely, the confidence intervals of the scores obtained in both groups show that the clinical differences might not be so relevant even though significant. Conversely, in a randomized multicenter trial focusing on patellar tendinopathy [19], no differences were found in the scores on the functional or pain scales with different treatment groups (LR-PRP, LP-PRP, or saline solution associated with an exercise program) at 1-year follow-up.

Finally, a systematic review compares the effectiveness of both types of preparations in lateral elbow tendinopathy [20]. Although patients see improvement in terms of pain and functional results with PRP treatment, no differences were found between the two preparations, even though an increase in the rate of complications in LR-PRP was notable [21].

## 15.5 Importance of the Activation of the Inflammatory Process in the Use of PRP

The interaction between PRP and the inflammatory milieu is thought to be essential to the therapeutic efficacy of PRP-based treatment options. It has been determined that PRP administration in the absence of an active inflammatory reaction may additionally yield suboptimal results. Conversely, the presence of an active inflammatory process at the time of PRP application has been associated with better tissue healing and regenerative responses [1, 2].

The body's natural response to an acute injury is to start an inflammatory process. The removal of injured tissue, the prevention of infection, and the beginning of the healing process are all made easier via acute inflammation. By directly providing a significant concentration of platelets and growth factors to the wounded region, PRP can improve the healing process in this situation [4, 5, 8].

However, the inflammatory response may not occur or may not be sufficiently potent in chronic tendon disorders [22, 23]. The essential signaling and cellular responses that PRP depends on to promote tissue regeneration may be slowed down in the absence of inflammation. The recruitment of platelets, growth factors, and immune cells to the affected region may be reduced as a result of the decreased inflammatory environment in chronic tendon disorders.

To solve this problem, scientists have investigated several methods to trigger or intensify the inflammatory response in chronic tendinopathy [24]. The ultrasound-guided galvanic electrolysis technique (USGET), which applies galvanic current to the afflicted tendon, is one that promotes this inflammation activation. This infiltrative act induces a regulated localized inflammatory response that encourages the recruitment of platelets and growth factors and speeds up tissue repair [25, 26].

During USGET, a small needle electrode is inserted into the affected tendon under ultrasound guidance. A localized chemical reaction occurs within the tissue once a galvanic current is introduced after the electrode is suitably positioned. As a result of this reaction, certain elements are produced, including hydrogen and hydroxide ions, which support localized inflammation in the treated region [25].

It is thought that the regulated inflammatory response brought on by electrolysis procedures has several beneficial effects on tendon recovery [26]. It can encourage angiogenesis (formation of new blood vessels) and the production of growth factors, which are essential for tissue regeneration. The inflammatory reaction can also encourage the recruitment of cells necessary for the healing process and aid in the breakdown of scar tissue [26]. The extracellular matrix (ECM) of connective tissue at the tendon level supports most of the physical loads of the body. Those ECM, like collagen, fibronectin, and proteoglycans, are produced by tenocytes to maintain tendon homeostasis and repair injured tendons [27].

The mechanical stimulus of exercise on the tendons can be translated into chemical signals, triggering multiple cellular responses [28]. McBeath et al. showed that mechanical stimulation, in combination with specific growth factors, can act as a switch that controls the differentiation of mesenchymal stem cells [29]. Zhang et al. [30] discovered that mice subjected to intensive use of a wheel in their cage had a greater number of myofibroblasts in the patellar tendons than the control group without this stimulus. Myofibroblasts are activated fibroblasts and are involved in the repair and remodeling of injured tissues [31]. Therefore, applying moderate tension to the tendons is beneficial to their healing.

## 15.6 Conclusion

Tendinopathy is a highly prevalent tendon disorder affecting a wide range of individuals, regardless of age and activity level, with a consistent socioeconomic impact. Although the mechanisms of tendinopathy are not completely understood, in the last years, it has emerged a clear role of inflammation. PRP is a popular autologous therapy used worldwide to treat a variety of conditions, including tendinopathy (Fig. 15.1). Basic science studies have repeatedly demonstrated a positive effect of PRP on tendon cell proliferation, increase of expression of anabolic genes and proteins, and reduction of tendon inflammation.

Nevertheless, the literature shows controversial results, with some RCTs showing very good outcomes whereas other poor or no results. This conflicting evidence is due to differences in terms of PRP type, protocol of applications and indications, as well as flaws in study methodology. A relevant factor affecting study result is PRP composition. LR- and LP-PRP are often considered as the same products, although the basic science studies show consistent differences among them. Often the same PRP type is administered to patients regardless of age, gender, disease history, and others, and this may result in conflicting or unclear results. In some cases, tendinopathy would require a more personalized approach. In example, the combination of PRP and electrolysis may offer a synergistic strategy that makes use of the positive effect of inflammation to promote tendon regeneration and recovery.

Currently available evidence is still insufficient to indicate PRP as a totally effective treat-

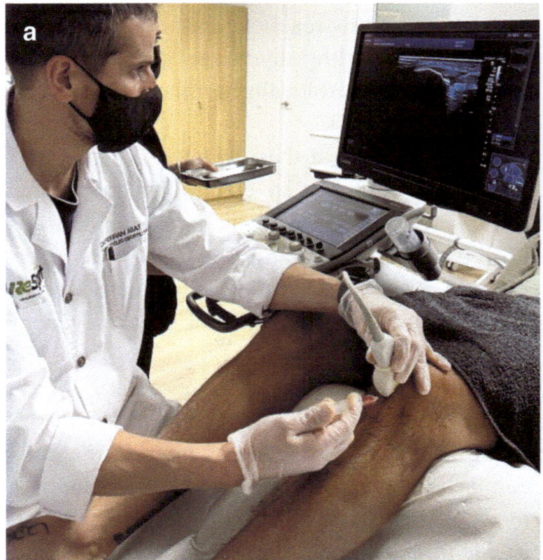
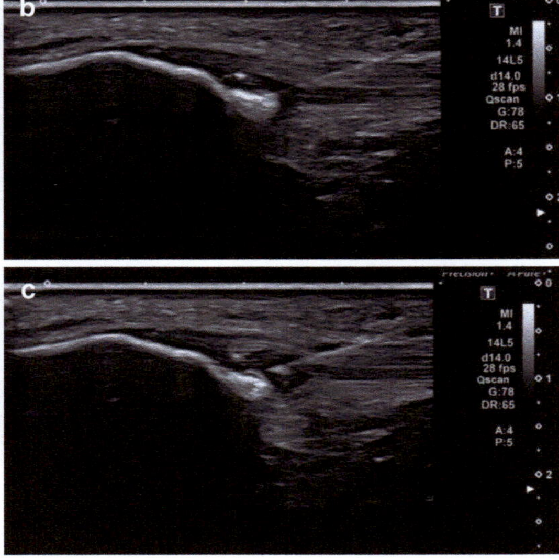

**Fig. 15.1** Ultrasound-guided injection of L-PRP in the patellar tendon. (**a**) Infiltration was made in the long axis with a linear probe in the zone of tendon injury. Observe in the ultrasound images in longitudinal view (**b**, **c**) how the PRP penetrate the tissue in the injured area

ment for tendinopathy. Further studies with a greater impact are necessary to establish PRP as the gold standard. In addition, detailed protocols relative to PRP preparation and its application as well as the posttreatment (exercise/mechanical stimulation) are called for.

# References


1. Everts PA, Mazzola T, Mautner K, Randelli PS, Podesta L. Modifying orthobiological PRP therapies are imperative for the advancement of treatment outcomes in musculoskeletal pathologies. Biomedicine. 2022;10(11):2933. https://doi.org/10.3390/biomedicines10112933.
2. Everts P, Onishi K, Jayaram P, Lana JF, Mautner K. Platelet-rich plasma: new performance understandings and therapeutic considerations in 2020. Int J Mol Sci. 2020;21(20):7794. https://doi.org/10.3390/ijms21207794.
3. Dragoo JL, Wasterlain AS, Braun HJ, Nead KT. Platelet-rich plasma as a treatment for patellar tendinopathy: a double-blind, randomized controlled trial. Am J Sports Med. 2014;42(3):610–8. https://doi.org/10.1177/0363546513518416.
4. Chalidis B, Givissis P, Papadopoulos P, Pitsilos C. Molecular and biologic effects of platelet-rich plasma (PRP) in ligament and tendon healing and regeneration: a systematic review. Int J Mol Sci. 2023;24(3):2744. https://doi.org/10.3390/ijms24032744.
5. Geaney LE, Arciero RA, DeBerardino TM, Mazzocca AD. The effects of platelet-rich plasma on tendon and ligament: basic science and clinical application. Oper Tech Sports Med. 2011;19(3):160–4.
6. McCarrel T, Fortier L. Temporal growth factor release from platelet-rich plasma, trehalose lyophilized platelets, and bone marrow aspirate and their effect on tendon and ligament gene expression. J Orthop Res. 2009;27(8):1033–42. https://doi.org/10.1002/jor.20853.
7. Giusti I, Rughetti A, D'Ascenzo S, Millimaggi D, Pavan A, Dell'Orso L, Dolo V. Identification of an optimal concentration of platelet gel for promoting angiogenesis in human endothelial cells. Transfusion. 2009;49(4):771–8. https://doi.org/10.1111/j.1537-2995.2008.02033.x.
8. Zhou Y, Zhang J, Wu H, Hogan MV, Wang JH. The differential effects of leukocyte-containing and pure platelet-rich plasma (PRP) on tendon stem/progenitor cells - implications of PRP application for the clinical treatment of tendon injuries. Stem Cell Res Ther. 2015;6(1):173. https://doi.org/10.1186/s13287-015-0172-4.
9. Eymard F, Ornetti P, Maillet J, Noel É, Adam P, Legré-Boyer V, Boyer T, Allali F, Gremeaux V, Kaux JF, Louati K, Lamontagne M, Michel F, Richette P, Bard H, GRIP (Groupe de Recherche sur les Injections de PRP, PRP Injection Research Group). Knee Surg Sports Traumatol Arthrosc. 2021;29(10):3211–2. https://doi.org/10.1007/s00167-020-06331-8.
10. Ferrante CJ, Leibovich SJ. Regulation of macrophage polarization and wound healing. Adv Wound Care. 2012;1(1):10–6. https://doi.org/10.1089/wound.2011.0307.
11. Peng Y, Li X, Wu W, Ma H, Wang G, Jia S, Zheng C. Effect of mechanical stimulation combined with platelet-rich plasma on healing of the rotator cuff in a murine model. Am J Sports Med. 2022;50(5):1358–68. https://doi.org/10.1177/03635465211073339.
12. Tanpowpong T, Thepsoparn M, Numkarunarunrote N, Itthipanichpong T, Limskul D, Thanphraisan P. Effects of platelet-rich plasma in tear size reduction in partial-thickness tear of the supraspinatus tendon compared to corticosteroids injection. Sports Med Open. 2023;9(1):11. https://doi.org/10.1186/s40798-023-00556-w.
13. Arthur Vithran DT, Xie W, Opoku M, Essien AE, He M, Li Y. The efficacy of platelet-rich plasma injection therapy in the treatment of patients with Achilles tendinopathy: a systematic review and meta-analysis. J Clin Med. 2023;12(3):995. https://doi.org/10.3390/jcm12030995.
14. Liu X, Li Y, Shen L, Yan M. Leukocyte and platelet-rich plasma (L-PRP) in tendon models: a systematic review and meta-analysis of *in vivo/in vitro* studies. Evid Based Complement Alternat Med. 2022;2022:5289145. https://doi.org/10.1155/2022/5289145.
15. Cross JA, Cole BJ, Spatny KP, Sundman E, Romeo AA, Nicholson GP, Wagner B, Fortier LA. Leukocyte-reduced platelet-rich plasma normalizes matrix metabolism in torn human rotator cuff tendons. Am J Sports Med. 2015;43(12):2898–906. https://doi.org/10.1177/0363546515608157.
16. Rubio-Azpeitia E, Bilbao AM, Sánchez P, Delgado D, Andia I. The properties of 3 different plasma formulations and their effects on tendinopathic cells. Am J Sports Med. 2016;44(8):1952–61. https://doi.org/10.1177/0363546516643814.
17. Lin KY, Chen P, Chen AC, Chan YS, Lei KF, Chiu CH. Leukocyte-rich platelet-rich plasma has better stimulating effects on tenocyte proliferation compared with leukocyte-poor platelet-rich plasma. Orthop J Sports Med. 2022;10(3):23259671221084706. https://doi.org/10.1177/23259671221084706.
18. Fitzpatrick J, Bulsara MK, O'Donnell J, Zheng MH. Leucocyte-rich platelet-rich plasma treatment of gluteus medius and minimus tendinopathy: a double-blind randomized controlled trial with 2-year follow-up. Am J Sports Med. 2019;47(5):1130–7. https://doi.org/10.1177/0363546519826969.
19. Scott A, LaPrade RF, Harmon KG, Filardo G, Kon E, Della Villa S, Bahr R, Moksnes H, Torgalsen T, Lee J, Dragoo JL, Engebretsen L. Platelet-rich plasma for patellar tendinopathy: a randomized controlled trial

of leukocyte-rich PRP or leukocyte-poor PRP versus saline. Am J Sports Med. 2019;47(7):1654–61. https://doi.org/10.1177/0363546519837954.

20. Li S, Yang G, Zhang H, Li X, Lu Y. A systematic review on the efficacy of different types of platelet-rich plasma in the management of lateral epicondylitis. J Shoulder Elb Surg. 2022;31(7):1533–44. https://doi.org/10.1016/j.jse.2022.02.017.

21. Moraes VY, Lenza M, Tamaoki MJ, Faloppa F, Belloti JC. Platelet-rich therapies for musculoskeletal soft tissue injuries. Cochrane Database Syst Rev. 2013;12:CD010071. https://doi.org/10.1002/14651858.CD010071.pub2. Update in: Cochrane Database Syst Rev. 2014;(4):CD010071.

22. Alfredson H, Forsgren S, Thorsen K, Lorentzon R. In vivo microdialysis and immunohistochemical analyses of tendon tissue demonstrated high amounts of free glutamate and glutamate NMDAR1 receptors, but no signs of inflammation, in Jumper's knee. J Orthop Res. 2001;19(5):881–6. https://doi.org/10.1016/S0736-0266(01)00016-X.

23. Nirschl RP. Elbow tendinosis/tennis elbow. Clin Sports Med. 1992;11(4):851–70.

24. Millar NL, Murrell GA, McInnes IB. Inflammatory mechanisms in tendinopathy - towards translation. Nat Rev Rheumatol. 2017;13(2):110–22. https://doi.org/10.1038/nrrheum.2016.213.

25. Abat F, Sánchez-Sánchez JL, Martín-Nogueras AM, Calvo-Arenillas JI, Yajeya J, Méndez-Sánchez R, Monllau JC, Gelber PE. Randomized controlled trial comparing the effectiveness of the ultrasound-guided galvanic electrolysis technique (USGET) versus conventional electro-physiotherapeutic treatment on patellar tendinopathy. J Exp Orthop. 2016;3(1):34. https://doi.org/10.1186/s40634-016-0070-4.

26. Abat F, Alfredson H, Cucchiarini M, Madry H, Marmotti A, Mouton C, Oliveira JM, Pereira H, Peretti GM, Spang C, Stephen J, van Bergen CJA, de Girolamo L. Current trends in tendinopathy: consensus of the ESSKA basic science committee. Part II: treatment options. J Exp Orthop. 2018;5(1):38. https://doi.org/10.1186/s40634-018-0145-5.

27. Tan JL, Tien J, Pirone DM, Gray DS, Bhadriraju K, Chen CS. Cells lying on a bed of microneedles: an approach to isolate mechanical force. Proc Natl Acad Sci U S A. 2003;100(4):1484–9. https://doi.org/10.1073/pnas.0235407100.

28. Davies PF. Flow-mediated endothelial mechanotransduction. Physiol Rev. 1995;75(3):519–60. https://doi.org/10.1152/physrev.1995.75.3.519.

29. McBeath R, Pirone DM, Nelson CM, Bhadriraju K, Chen CS. Cell shape, cytoskeletal tension, and RhoA regulate stem cell lineage commitment. Dev Cell. 2004;6(4):483–95. https://doi.org/10.1016/s1534-5807(04)00075-9.

30. Szczodry M, Zhang J, Lim C, Davitt HL, Yeager T, Fu FH, Wang JH. Treadmill running exercise results in the presence of numerous myofibroblasts in mouse patellar tendons. J Orthop Res. 2009;27(10):1373–8. https://doi.org/10.1002/jor.20878.

31. Tomasek JJ, Gabbiani G, Hinz B, Chaponnier C, Brown RA. Myofibroblasts and mechano-regulation of connective tissue remodelling. Nat Rev Mol Cell Biol. 2002;3(5):349–63. https://doi.org/10.1038/nrm809.

# Platelet-Rich Plasma (PRP) for Rotator Cuff Tears

# 16


Ron Gilat, Ilan Y. Mitchnik, Derrick Knapik, Grant Garrigues, Nikhil Verma, and Brian J. Cole


## 16.1 Introduction

The rotator cuff is composed of four tendons and muscles that surround the shoulder joint and aid in glenohumeral joint mobility and stability. Rotator cuff injuries are a common cause of significant shoulder pain and functional limitations [1], leading to muscle atrophy, fatty infiltration, and degenerative changes if not properly diagnosed and managed [2]. Rotator cuff injuries are classified based on acuity (acute versus chronic), severity (partial versus full-thickness), location (bursal-sided versus articular-sided), the extent of tendon retraction or muscle atrophy, as well as tear size. These factors may reflect the extent of damage, which dictate appropriate treatment when evaluating patient symptoms, examination findings, imaging, results, and patient expectations. Nonoperative treatments include physiotherapy, nonsteroidal anti-inflammatory drugs (NSAIDs), and/or injections [3, 4]. When indicated, potential operative interventions include rotator cuff repair, superior capsular reconstruction, reverse total shoulder arthroplasty, tendon transfers, debridement, and subacromial decompression [1]. Recent investigations have suggested that operative management may not be superior to conservative treatment when examining outcomes based on strength, range of motion, functional outcomes, and recurrent tear rates [2, 3, 5]. This controversy has sparked a growing interest in the use of orthobiologics, namely, platelet-rich plasma (PRP), as a biologic augment for the treatment of rotator cuff injuries, especially for patients with smaller tears [4, 6–8].

The purpose of this chapter is to provide a comprehensive overview evaluating the use of PRP injections for the treatment of rotator cuff injuries. By helping promote healing and reducing inflammation, the use of PRP is theorized to stimulate the growth of new tissue, offering a viable alternative to operative intervention in select cases. However, the effectiveness of PRP injections in the treatment of patients with rotator cuff injuries is still a subject of ongoing research and debate in the medical community.


R. Gilat (✉)
Midwest Orthopaedics at Rush University Medical Center, Chicago, IL, USA

Department of Orthopaedic Surgery, Shamir Medical Center and Tel Aviv University, Tel Aviv, Israel

I. Y. Mitchnik
Department of Orthopaedic Surgery, Shamir Medical Center and Tel Aviv University, Tel Aviv, Israel

D. Knapik
Department of Orthopaedic Surgery, Washington University Sports Medicine, St. Louis, MO, USA

G. Garrigues · N. Verma · B. J. Cole
Midwest Orthopaedics at Rush University Medical Center, Chicago, IL, USA
e-mail: Grant.Garrigues@rushortho.com;
Nikhil.Verma@rushortho.com;
Brian.Cole@rushortho.com



© The Author(s), under exclusive license to Springer Nature Switzerland AG 2024
B. Kocaoglu et al. (eds.), *Musculoskeletal Injections Manual*,
https://doi.org/10.1007/978-3-031-52603-9_16

## 16.2	Healing Rotator Cuff Injuries with PRP Therapy

Following injury, the rotator cuff tendons possess a minimal degree of intrinsic healing, undergoing a three-phase healing process that includes *inflammation*, *proliferation*, and *remodeling* [4, 8] (Fig. 16.1). During the *inflammatory phase*, lasting around a week, vascular permeability increases, and immune cells enter the healing site. This triggers the production of growth and cellular factors such as insulin-like growth factor 1 (IGF-1), which promotes the accumulation of macrophages and the proliferation of tenocytes [9, 10]. In addition, inflammatory expression of tissue growth factor beta (TGF-B) promotes fibrosis and scar formation by increasing type I and III collagen [10, 11]. This scar tissue is important for the healing of tendon to bone at the enthesis where the tendons contact with the greater tuberosity [12]. The *proliferative phase* then lasts several weeks, during which an increase in myofibroblasts and regenerative tissues occurs as a result of the presence of platelet-derived growth factor (PDGF) and vascular endothelial growth factor (VEGF) [4]. PDGF supports the proliferation of fibroblasts and matrix deposition while also modulating the proliferation of vascular smooth muscle [10, 11]. This phase also includes upregulation of the receptor for the angiogenic VEGF [10, 13]. Finally, the *remodeling phase*, which may last several months or years, involves the continuous reshaping of the regenerated tissue under the influence of external forces. Unfortunately, when the rotator cuff tear retracts more than a few millimeters, this process is not enough to bridge the tear gap, and consequentially structural healing tends not to occur.

PRP consists of autologous blood, centrifuged to create a supraphysiologic concentration of platelets containing various cytokines and chemokines, including IGF-1, TGF-B, VEGF, and PDGF, the growth factors involved in the three healing phases of rotator cuff injuries [4, 14]. The procedure involves the extraction of peripheral blood from the patient, which is then processed

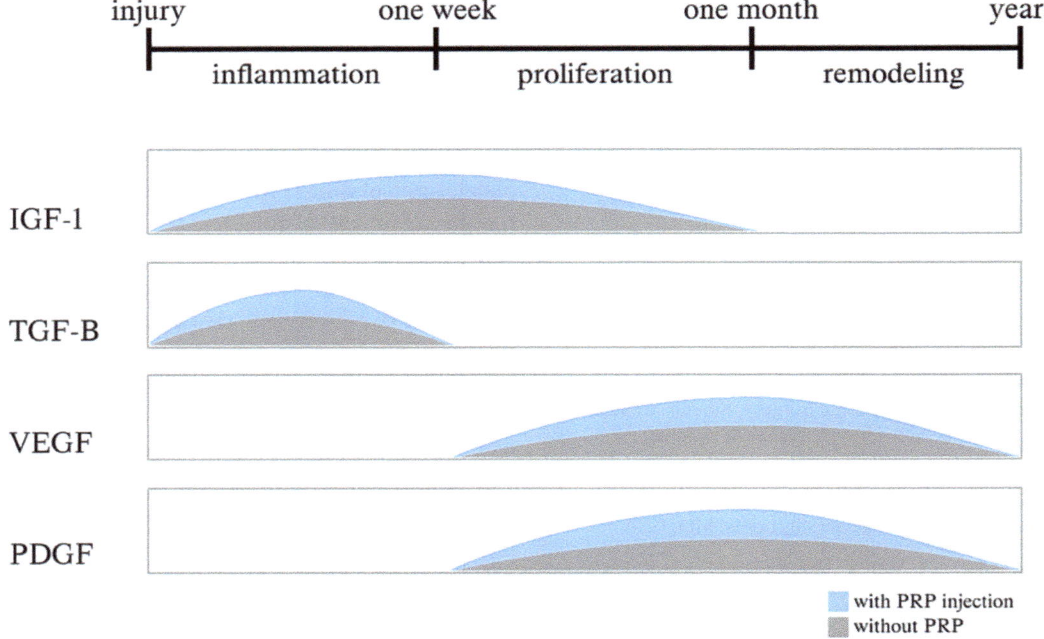


**Fig. 16.1** Three-phase healing of rotator cuff tendons. *IGF-1* insulin-like growth factor 1, *TGF-B* tissue growth factor beta, *VEGF* vascular endothelial growth factor; *PDGF* platelet-derived growth factor, *PRP* platelet-rich plasma

via centrifugation to separate the platelets [8]. PRP can be applied in either a gel or liquid form. The gel state of PRP allows for it to be secured in a specific area of injury along the rotator cuff, believed to result in an extended effect period [4]. When PRP is applied to injured areas, it is believed to have both anabolic and anti-inflammatory effects. Specifically, the high concentration of IGF-1 in PRP stimulates cell proliferation and matrix synthesis, while TGF-B increases collagen production improving tissue strength; additionally, VEGF and PDGF promote angiogenesis, additional cell proliferation, and matrix synthesis.

Using a rabbit model, Chung et al. have shown that the use of PRP for rotator cuff repairs enhanced the tendon to bone healing [15]. At the 4-week mark, vascularity and cellularity were increased in PRP-treated rabbits. After 8 weeks, collagen fibers were more regularly arranged and continuous with PRP treatment. In a murine model, Peng et al. have also shown that PRP improved tendon-to-bone healing [16]. After 4 and 8 weeks, PRP was associated with a larger fibrocartilaginous layer and increased subchondral bone trabeculae number and thickness. Using a rat model, Beck et al. also showed increased vascularity and fibroblast proliferation [17]. Furthermore, collagen fibers were oriented more linearly toward the tendon-to-bone interface. These studies have all shown increased biomechanical strength for rotator cuffs repaired with PRP addition.

## 16.3 The Effectiveness of PRP for Rotator Cuff Tears

Multiple meta-analyses have demonstrated that intraoperative PRP application may significantly reduce pain levels, although measured functional outcome improvements have generally failed to demonstrate the achievement of minimal clinically important differences (MCID) [18–22]. As an intermediate treatment option between nonoperative and operative management, PRP injections have shown promising results in improving

functionality compared to standard corticosteroid injections [23–26].

### 16.3.1 PRP Injections as Isolated Treatment for Rotator Cuff Tears

In cases with rotator cuff tendinitis, subacromial impingement, and partial rotator cuff tears, patients are generally initially treated with a trial of nonoperative management, which may include the use of PRP. However, the benefits of PRP injections for shoulder injuries have been a subject of debate. PRP injections may be performed with intervals reported from 1 week to 1 month, consisting of two to four consecutive injections [19, 20, 24]. PRP may be injected into the subacromial space, or intralesionally into injured tendons [19, 20, 24]. When compared to modalities such as physiotherapy or corticosteroid injections, the use of PRP has been reported to be associated with a potentially more steady and sustained response. Lin et al. and Feltri et al. reported no added benefit in shoulder pain or function up to 3 months following PRP injection [19, 27]. Meanwhile, additional studies have reported worse functional results on 3-month follow-ups [23–25]. However, Jiang et al. observed that after 3 months, PRP injections resulted in improved functional outcomes [24–26]. Adra et al., Jiang et al., and Lin et al. also reported PRP injections to be associated with less shoulder pain after 6 months [19, 24, 26]. While PRP injections may be considered safe based on the low rate of reported adverse events [20], they have also been reported to be associated with fewer repeat shoulder interventions [25].

### 16.3.2 PRP as an Adjunct to Rotator Cuff Repairs

Large and massive rotator cuff tears have a high retear rate of up to 94% [28]. PRP has been utilized in an attempt to augment the healing biology with intraoperative PRP application during rotator cuff repair. PRP use as an adjuvant during surgery has been described as being applied either as a gel over the site of injury or injected as a liquid [18, 29]. Intraoperative application of PRP demonstrates several early clinical benefits. At 3- to 6-month follow-up, Chen et al., Xu et al., and Yang et al. reported the incorporation of PRP to be associated with less shoulder pain and better functional outcomes [18, 21, 22]. At longer follow-ups, PRP augmentation has shown continued improvement in functional outcomes [18, 21]. Importantly, Xu et al. reported that the addition of PRP was not associated with an increase in adverse events [21].

### 16.3.3 Effects on the Rotator Cuff's Structural Integrity

Most structural failures of the rotator cuff occur at the tendon-bone interface, and application of PRP to this site may be more beneficial [30]. Although structural failure may bring temporary pain relief, structural integrity is important to preserve muscle mass and shoulder function [31]. Thus, assessing short-term pain level outcomes is often not enough. It is also important to remember that while most full-thickness tears are likely to progress over time, partial-thickness tears tend to remain unchanged [32, 33]. Unfortunately, not all the meta-analysis described thus far have controlled for these two factors. Despite these limitations, a large body of evidence exists to support that PRP use is associated with less retear rates [18, 21, 22, 27, 29]. However, it is important to note that Vavken et al. have suggested that this may not be a cost-effective indication for the use of intraoperative PRP [34]. Carr et al. were the first to show that PRP injections alter the rotator cuffs tendon cellular tissue in such a way that may increase retear rates due to reduced vascularity and increased apoptosis [35]. However, this may be attributed to the use of a leukocyte-rich PRP preparation. Pandey et al. have shown increased vascularity at PRP injection sites up to 1 year and decreased retear rates, as assessed by ultrasound (US), for large tears [36]. Using magnetic resonance imaging (MRI), both Dukan et al. and Malavolta et al. have shown that PRP gel and liquid, respectively, conserve the structural integrity of the rotator cuff on follow-ups [37, 38].

## 16.4 Discussion

Despite increasing interest in the use of PRP as an isolated and adjunct treatment for patients with rotator cuff tears, additional investigations are warranted to better understand the indications, outcomes, ideal method, and timing of application, as well as the potential risks associated with PRP use (Table 16.1). Namely, there remains a lack of standardization in the preparation of PRP, with multiple systematic reviews discussing various methods with varying centrifugation protocols and the use of activating agents. As such, PRP preparation likely consists of differing platelet concentrations, leukocyte compositions, and fibrin networks, introducing a substantial degree of heterogeneity in the final PRP product utilized between studies [39–43]. Furthermore, the addition of local anesthetic injections at the site of injury may reduce the effects of PRP, with some studies suggesting that the function of platelet function in PRP compounds being compromised secondary to changes in the local pH levels which is decreased by the anesthetic [44, 45]. The clinical benefit of PRP is typically observed when the platelet concentration is two to eight times greater than that found in native blood [20]. PRP can be prepared as leukocyte-poor (LP) or leukocyte-rich (LR), depending on the concentration of leukocytes present [42, 43, 46]. The presence of leukocytes in PRP can prompt fibroblasts to release matrix metalloproteinases (MMP) [47, 48]. Thus, LR-PRP may have tissue-degrading effects on the recovering rotator cuff [4, 49]. Indeed, Cross et al. demonstrated that LR-PRP is associated with an increased expression of tissue-degrading MMP-9 [50]. Furthermore, several meta-analyses have shown that LP-PRP lowers retear rates following tendon repair [51–54]. LP-PRP has also been reported to improve postoperative pain and functional outcomes [51, 52]. However, the reported

**Table 16.1** Effectiveness of PRP injections for rotator cuff injuries

| Treatment type | Pain | Function | Retear | Adverse effects |
|---|---|---|---|---|
| Intraoperative PRP augmentation | Early pain reduction that may not be clinically important | Early general improvement that may not be clinically important | May reduce retear rates | No difference from standard operative treatment |
| PRP as an alternative to other conservative treatment | Promising results, improvement may be more gradual and sustained | Early benefits may be less significant compared to corticosteroid injections | Unknown | Fewer adverse effects than corticosteroid injections |
| LP-PRP compared to LR-PRP | Better short-term reduction in postoperative pain | Better short-term improvement in functional scores | Reduced retear rates | LR-PRP may have tissue-degrading effects on the recovering rotator cuff |

*PRP* platelet-rich plasma, *LP* leukocyte poor, *LR* leukocyte rich

effects of LP-PRP are mostly limited to studies with short-term follow-up [53], where modest improvement has not surpassed the MCID threshold [54]. The current lack of consensus on the optimal preparation and dosing of PRP for treating rotator cuff tears represents a substantial limitation impacting the validity of the currently reported evidence [18, 43, 55, 56]. Larger, well-designed randomized controlled trials are necessary to provide more definitive evidence on the effectiveness of homogenous PRP injections.

PRP injection or augmentation has been shown to be an overall safe procedure [20, 21]. The primary concern, as with any injection, is for infection; therefore, PRP injection should always be performed under sterile conditions. Although allergic reactions are possible, there are few reported complications or adverse events related to the use of PRP. Specific adverse events that have been reported include short-term pain, as well as frozen shoulder and tear extension [20]. Patients should be informed of these risks and carefully evaluated prior to undergoing PRP injection or augmentation.

## 16.5 Conclusion

Current literature suggests that PRP injections may improve shoulder function and potentially reduce shoulder pain, with few reported side effects or adverse events. PRP injections may be used during nonoperative management or to augment operative management of rotator cuff tears

to enhance tendon healing and recovery. However, the effectiveness of PRP injections in treating rotator cuff injuries remains a subject of debate due to the high heterogeneity in PRP preparation methods and variable outcomes in clinical studies, warranting further investigations.

## References


1. Yanik EL, Chamberlain AM, Keener JD. Trends in rotator cuff repair rates and comorbidity burden among commercially insured patients younger than the age of 65 years, United States 2007-2016. JSES Rev Rep Tech. 2021;1(4):309–16.
2. Chalmers PN, Ross H, Granger E, Presson AP, Zhang C, Tashjian RZ. The effect of rotator cuff repair on natural history. JBJS Open Access. 2018;3(1):e0043.
3. Piper CC, Hughes AJ, Ma Y, Wang H, Neviaser AS. Operative versus nonoperative treatment for the management of full-thickness rotator cuff tears: a systematic review and meta-analysis. J Shoulder Elb Surg. 2018;27(3):572–6.
4. Zhang C, Wu J, Li X, Wang Z, Lu WW, Wong T-M. Current biological strategies to enhance surgical treatment for rotator cuff repair. Front Bioeng Biotechnol. 2021;9:657584.
5. Khatri C, Ahmed I, Parsons H, et al. The natural history of full-thickness rotator cuff tears in randomized controlled trials: a systematic review and meta-analysis. Am J Sports Med. 2019;47(7):1734–43.
6. Viswanath A, Monga P. Trends in rotator cuff surgery: research through the decades. J Clin Orthop Trauma. 2021;18:105–13.
7. Hitchen J, Wragg NM, Shariatzadeh M, Wilson SL. Platelet rich plasma as a treatment method for rotator cuff tears. SN Compr Clin Med. 2020;2(11):2293–9.
8. Gowd AK, Cabarcas BC, Frank RM, Cole BJ. Biologic augmentation of rotator cuff repair:

the role of platelet-rich plasma and bone marrow aspirate concentrate. Oper Tech Sports Med. 2018;26(1):48–57.

9. Oliva F, Gatti S, Porcellini G, Forsyth NR, Maffulli N. Growth factors and tendon healing. Med Sport Sci. 2012;57:53–64. Available from: https://www.karger.com/Article/FullText/328878.

10. Molloy T, Wang Y, Murrell GAC. The roles of growth factors in tendon and ligament healing. Sports Med. 2003;33(5):381–94. https://doi.org/10.2165/00007256-200333050-00004.

11. Lee YS, Kim JY, Ki SY, Chung SW. Influence of smoking on the expression of genes and proteins related to fat infiltration, inflammation, and fibrosis in the rotator cuff muscles of patients with chronic rotator cuff tears: a pilot study. Arthroscopy. 2019;35(12):3181–91. Available from: http://www.arthroscopyjournal.org/article/S0749806319306152/fulltext.

12. Kovacevic D, Fox AJ, Bedi A, et al. Calcium-phosphate matrix with or without TGF-β3 improves tendon-bone healing after rotator cuff repair. Am J Sports Med. 2011;39(4):811–9. https://doi.org/10.1177/0363546511399378.

13. Pufe T, Petersen WJ, Mentlein R, Tillmann BN. The role of vasculature and angiogenesis for the pathogenesis of degenerative tendons disease. Scand J Med Sci Sports. 2005;15(4):211–22. https://doi.org/10.1111/j.1600-0838.2005.00465.x.

14. Boswell SG, Cole BJ, Sundman EA, Karas V, Fortier LA. Platelet-rich plasma: a milieu of bioactive factors. Arthroscopy. 2012;28(3):429–39.

15. Chung SW, Song BW, Kim YH, Park KU, Oh JH. Effect of platelet-rich plasma and porcine dermal collagen graft augmentation for rotator cuff healing in a rabbit model. Am J Sports Med. 2013;41(12):2909–18. https://doi.org/10.1177/0363546513503810.

16. Peng Y, Li X, Wu W, et al. Effect of mechanical stimulation combined with platelet-rich plasma on healing of the rotator cuff in a murine model. Am J Sports Med. 2022;50(5):1358–68. https://doi.org/10.1177/03635465211073339.

17. Beck J, Evans D, Tonino PM, Yong S, Callaci JJ. The biomechanical and histologic effects of platelet-rich plasma on rat rotator cuff repairs. Am J Sports Med. 2012;40(9):2037–44. https://doi.org/10.1177/0363546512453300.

18. Chen X, Jones IA, Togashi R, Park C, Vangsness CT. Use of platelet-rich plasma for the improvement of pain and function in rotator cuff tears: a systematic review and meta-analysis with bias assessment. Am J Sports Med. 2020;48(8):2028–41.

19. Lin M-T, Wei K-C, Wu C-H. Effectiveness of platelet-rich plasma injection in rotator cuff tendinopathy: a systematic review and meta-analysis of randomized controlled trials. Diagnostics. 2020;10(4):189.

20. Hamid MS, Sazlina SG. Platelet-rich plasma for rotator cuff tendinopathy: a systematic review and meta-analysis. PLoS One. 2021;16(5):e0251111.

21. Xu W, Xue Q. Application of platelet-rich plasma in arthroscopic rotator cuff repair: a systematic review and meta-analysis. Orthop J Sports Med. 2021;9(7):232596712110168.

22. Yang FA, De Liao C, Wu CW, Shih YC, Wu LC, Chen HC. Effects of applying platelet-rich plasma during arthroscopic rotator cuff repair: a systematic review and meta-analysis of randomised controlled trials. Sci Rep. 2020;10(1):1–10. Available from: https://www.nature.com/articles/s41598-020-74341-0.

23. Peng Y, Li F, Ding Y, et al. Comparison of the effects of platelet-rich plasma and corticosteroid injection in rotator cuff disease treatment: a systematic review and meta-analysis. J Shoulder Elb Surg. 2023;32(6):1303–13.

24. Adra M, El Ghazal N, Nakanishi H, et al. Platelet-rich plasma versus corticosteroid injections in the management of patients with rotator cuff disease: a systematic review and meta-analysis. J Orthop Res. 2023;41(1):7–20.

25. Pang L, Xu Y, Li T, Li Y, Zhu J, Tang X. Platelet-rich plasma injection can be a viable alternative to corticosteroid injection for conservative treatment of rotator cuff disease: a meta-analysis of randomized controlled trials. Arthroscopy. 2023;39(2):402–21.

26. Jiang X, Zhang H, Wu Q, Chen Y, Jiang T. Comparison of three common shoulder injections for rotator cuff tears: a systematic review and network meta-analysis. J Orthop Surg Res. 2023;18(1):272. https://doi.org/10.1186/s13018-023-03747-z.

27. Feltri P, Gonalba GC, Boffa A, et al. Platelet-rich plasma does not improve clinical results in patients with rotator cuff disorders but reduces the retear rate. A systematic review and meta-analysis. Knee Surg Sports Traumatol Arthrosc. 2022;60(6):465–75.

28. Galatz LM, Ball CM, Teefey SA, Middleton WD, Yamaguchi K. The outcome and repair integrity of completely arthroscopically repaired large and massive rotator cuff tears. J Bone Joint Surg Am. 2004;86(2):219–24. Available from: https://pubmed.ncbi.nlm.nih.gov/14960664/.

29. Trantos IA, Vasiliadis ES, Giannoulis FS, Pappa E, Kakridonis F, Pneumaticos SG. The effect of PRP augmentation of arthroscopic repairs of shoulder rotator cuff tears on postoperative clinical scores and retear rates: a systematic review and meta-analysis. J Clin Med. 2023;12(2):581.

30. Warth RJ, Dornan GJ, James EW, Horan MP, Millett PJ. Clinical and structural outcomes after arthroscopic repair of full-thickness rotator cuff tears with and without platelet-rich product supplementation: a meta-analysis and meta-regression. Arthroscopy. 2015;31(2):306–20. Available from: https://pubmed.ncbi.nlm.nih.gov/25450417/.

31. Zumstein MA, Jost B, Hempel J, Hodler J, Gerber C. The clinical and structural long-term results of open repair of massive tears of the rotator cuff. J Bone Joint Surg Am. 2008;90(11):2423–31. Available from: https://pubmed.ncbi.nlm.nih.gov/18978411/.

32. Ranebo MC, Björnsson Hallgren HC, Norlin R, Adolfsson LE. Clinical and structural outcome 22 years after acromioplasty without tendon repair

in patients with subacromial pain and cuff tears. J Shoulder Elb Surg. 2017;26(7):1262–70. Available from: https://pubmed.ncbi.nlm.nih.gov/28131687/.

33. Zingg PO, Jost B, Sukthankar A, Buhler M, Pfirrmann CWA, Gerber C. Clinical and structural outcomes of nonoperative management of massive rotator cuff tears. J Bone Joint Surg. 2007;89(9):1928–34.

34. Vavken P, Sadoghi P, Palmer M, et al. Platelet-rich plasma reduces retear rates after arthroscopic repair of small- and medium-sized rotator cuff tears but is not cost-effective. Am J Sports Med. 2015;43(12):3071–6.

35. Carr AJ, Murphy R, Dakin SG, et al. Platelet-rich plasma injection with arthroscopic acromioplasty for chronic rotator cuff tendinopathy: a randomized controlled trial. Am J Sports Med. 2015;43(12):2891–7. Available from: https://pubmed.ncbi.nlm.nih.gov/26498958/.

36. Pandey V, Bandi A, Madi S, et al. Does application of moderately concentrated platelet-rich plasma improve clinical and structural outcome after arthroscopic repair of medium-sized to large rotator cuff tear? A randomized controlled trial. J Shoulder Elb Surg. 2016;25(8):1312–22. Available from: http://www.jshoulderelbow.org/article/S1058274616300295/fulltext.

37. Dukan R, Bommier A, Rousseau MA, Boyer P. Arthroscopic knotless tape bridging with autologous platelet-rich fibrin gel augmentation: functional and structural results. Phys Sportsmed. 2019;47(4):455–62. Available from: https://pubmed.ncbi.nlm.nih.gov/31136263/.

38. Malavolta EA, Gracitelli MEC, Assunção JH, Ferreira Neto AA, Bordalo-Rodrigues M, de Camargo OP. Clinical and structural evaluations of rotator cuff repair with and without added platelet-rich plasma at 5-year follow-up: a prospective randomized study. Am J Sports Med. 2018;46(13):3134–41. Available from: https://pubmed.ncbi.nlm.nih.gov/30234999/.

39. Baksh N, Hannon CP, Murawski CD, Smyth NA, Kennedy JG. Platelet-rich plasma in tendon models: a systematic review of basic science literature. Arthroscopy. 2013;29(3):596–607.

40. Fice MP, Miller JC, Christian R, et al. The role of platelet-rich plasma in cartilage pathology: an updated systematic review of the basic science evidence. Arthroscopy. 2019;35(3):961–976.e3.

41. Kunze KN, Hannon CP, Fialkoff JD, Frank RM, Cole BJ. Platelet-rich plasma for muscle injuries: a systematic review of the basic science literature. World J Orthop. 2019;10(7):278–91.

42. Chahla J, Cinque ME, Piuzzi NS, et al. A call for standardization in platelet-rich plasma preparation protocols and composition reporting: a systematic review of the clinical orthopaedic literature. J Bone Joint Surg. 2017;99(20):1769–79. Available from: https://journals.lww.com/jbjsjournal/Fulltext/2017/10180/A_Call_for_Standardization_in_Platelet_Rich_Plasma.9.aspx.

43. Hitchen J, Wragg NM, Shariatzadeh M, Wilson SL. Platelet rich plasma as a treatment method for rotator cuff tears. SN Comprehensive Clin Med. 2020;2(11):2293–9. https://doi.org/10.1007/s42399-020-00500-z.

44. Ersen A, Demirhan M, Atalar AC, Kapicioğlu M, Baysal G. Platelet-rich plasma for enhancing surgical rotator cuff repair: evaluation and comparison of two application methods in a rat model. Arch Orthop Trauma Surg. 2014;134(3):405–11. https://doi.org/10.1007/s00402-013-1914-3.

45. Carofino B, Chowaniec DM, McCarthy MB, et al. Corticosteroics and local anesthetics decrease positive effects of platelet-rich plasma: an in vitro study on human tendon cells. Arthroscopy. 2012;28(5):711–9. Available from: http://www.arthroscopyjournal.org/article/S0749806311011935/fulltext.

46. Bhan K, Singh B, Bhan K, Singh B. Efficacy of platelet-rich plasma injection in the management of rotator cuff tendinopathy: a review of the current literature. Cureus. 2022;14:6. Available from: https://www.cureus.com/articles/102283-efficacy-of-platelet-rich-plasma-injection-in-the-management-of-rotator-cuff-tendinopathy-a-review-of-the-current-literature.

47. Pifer MA, Maerz T, Baker KC, Anderson K. Matrix metalloproteinase content and activity in low-platelet, low-leukocyte and high-platelet, high-leukocyte platelet rich plasma (PRP) and the biologic response to PRP by human ligament fibroblasts. Am J Sports Med. 2014;42(5):1211–8. https://doi.org/10.1177/0363546514524710.

48. Zhou Y, Zhang J, Wu H, Hogan MCV, Wang JHC. The differential effects of leukocyte-containing and pure platelet-rich plasma (PRP) on tendon stem/progenitor cells - implications of PRP application for the clinical treatment of tendon injuries. Stem Cell Res Ther. 2015;6:1. Available from: https://pubmed.ncbi.nlm.nih.gov/26373929/.

49. Rossi LA, Piuzzi N, Tanoira I, Brandariz R, Huespe I, Ranalletta M. Subacromial platelet-rich plasma injections produce significantly worse improvement in functional outcomes in patients with partial supraspinatus tears than in patients with isolated tendinopathy. Arthroscopy. 2023;39(9):2000–8.

50. Cross JA, Cole BJ, Spatny KP, et al. Leukocyte-reduced platelet-rich plasma normalizes matrix metabolism in torn human rotator cuff tendons. Am J Sports Med. 2015;43(12):2898–906.

51. Peng Y, Guanglan W, Jia S, Zheng C. Leukocyte-rich and leukocyte-poor platelet-rich plasma in rotator cuff repair: a meta-analysis. Int J Sports Med. 2022;43(11):921–30. Available from: http://www.thieme-connect.com/products/ejournals/html/10.1055/a-1790-7982.

52. Hurley ET, Colasanti CA, Anil U, et al. The effect of platelet-rich plasma leukocyte concentration on arthroscopic rotator cuff repair: a network meta-analysis of randomized controlled trials. Am J Sports Med. 2020;49(9):2528–35. https://doi.org/10.1177/0363546520975435.

53. Ahmad Z, Ang S, Rushton N, et al. Platelet-rich plasma augmentation of arthroscopic rotator cuff

repair lowers retear rates and improves short-term postoperative functional outcome scores: a systematic review of meta-analyses. Arthrosc Sports Med Rehabil. 2022;4(2):823–33.

54. Zhao D, Han Y, Hong, Pan J, et al. The clinical efficacy of leukocyte-poor platelet-rich plasma in arthroscopic rotator cuff repair: a meta-analysis of randomized controlled trials. J Shoulder Elb Surg. 2021;30(4):918–28.

55. Parisien RL, Ehlers C, Cusano A, Tornetta P, Li X, Wang D. The statistical fragility of platelet-rich plasma in rotator cuff surgery: a systematic review and meta-analysis. Am J Sports Med. 2021;49(12):3437–42.

56. DeClercq MG, Fiorentino AM, Lengel HA, et al. Systematic review of platelet-rich plasma for rotator cuff repair: are we adhering to the minimum information for studies evaluating biologics in orthopaedics? Orthop J Sports Med. 2021;9(12):232596712110419.

# Platelet-Rich Plasma Treatment for Muscle Injuries

# 17


Yosef Sourugeon, Yaniv Yonai, Yaron Berkovich, and Lior Laver


## 17.1 Introduction

Muscle injuries (MI) are common, especially in sports. For instance, MIs represent more than 30% of all injuries in professional soccer and more than 20% in basketball [1, 2]. Muscle injuries in the athlete can be classified into intrinsic and extrinsic injuries with intrinsic muscle injuries occurring most commonly at the myotendinous junction during eccentric contraction with various degrees of tearing of the muscle fibers, while extrinsic muscle injuries in the athlete occur most commonly as a result of a direct trauma such as a contusion injury [3]. These injuries are graded using clinical, sonographic, and magnetic resonance parameters or a combination of clinical and imaging parameters [4]. Conservative/nonsurgical management has been the mainstay of treatment for most muscle injuries and usually consists of protection, rest, ice, compression, and elevation (PRICE protocol); physiotherapy; NSAIDs; and time.

Traditionally, NSAIDs treatment was avoided in the acute phase of injury due to fear of prolonged recovery; however, recent works suggest the NSAIDs may be beneficial in reducing strength loss, soreness, and blood creatine kinase levels [5]. Despite the excellent results with nonsurgical treatment, such treatment failure may be devastating for the injured athlete, leading to reinjury and postponing the return to physical activity for weeks and even many months in some instances [6, 7].

New treatment modalities have been explored and introduced into this field with the aim to promote early return to play, decrease recurrence rates, and minimize fibrosis and subsequent muscle weakness. Such new treatment modalities which have been introduced in recent years in the treatment of muscle injuries are orthobiologic therapies such as PRP and cell-based therapies (bone marrow based or adipose tissue based) [3, 8, 9]. This chapter will focus on the use of PRP for muscle injuries.


Y. Sourugeon
Department of Orthopedic Surgery, Chaim Sheba Medical Center, Rama Gan, Israel

Y. Yonai · L. Laver (✉)
Department of Orthopedic Surgery and Sports Medicine Unit, Hillel Yaffe Medical Center (HYMC), Hadera, Israel

Rappaport Faculty of Medicine, Technion (Israel Institute of Technology), Haifa, Israel

Y. Berkovich
Rappaport Faculty of Medicine, Technion (Israel Institute of Technology), Haifa, Israel

Department of Orthopedic Surgery, Hillel Yaffe Medical Center (HYMC), Hadera, Israel
e-mail: YARONB@hymc.gov.il

## 17.2 Muscle Healing Pathophysiology and Rationale for Using PRP

Skeletal muscle is a well-organized tissue, composed of cylindrical syncytial cells that are covered by an endomysium tissue which constitutes the single muscle fiber. Parallel muscle fibers are then bundled into groups (fascicles) that are surrounded by perimysium. These fascicles are further grouped up, and finally they are enclosed in another connective tissue layer, the epimysium. Satellite cells function as adult muscle stem cells that lie in opposition to the muscle fibers. These cells have a limited ability to differentiate into myoblasts; thus, the muscle fibers have a limited ability to regenerate [10].

Muscle healing follows a consistent path, regardless of the injury category. This path consists of three phases, with an overlap between the end of the previous phase and the start of the following phase: (1) destruction of myofibrils, formation of hematoma, and proliferation of inflammatory cells; (2) phagocytosis of necrotic tissue upon arrival of platelets, formation of scar tissue, blood vessels, and neural growth; and (3) remodeling of scar tissue and myofibrils [6, 11].

Muscle tissue regeneration is mainly limited by scar tissue formation rather than by muscle regeneration rate [12, 13]. Therefore, the rationale and potential benefit of PRP use for muscle injuries are not only aimed at early return to sports but also improved tissue healing with improved structural properties, thus potentially reducing the risk of recurrence. However, most clinical studies have only focused on return to sports rates and durations rather than assessing the healed tissue quality as well. Hence, using PRP in the treatment of muscle injuries allows a simple, easily accessible, nonsurgical method to introduce an abundance of naturally occurring growth factors to the site of injury, in order to enhance the natural healing process as well as improve the healing properties of the tissue [14].

In a preclinical study assessing muscle healing of contusion-injured tibialis anterior muscle in mice with combined treatment of an oral antifibrotic agent (Losartan) and PRP, Terada et al. reported increased muscle regeneration and func-

tion, along with decreased fibrosis in the experimental group [15]. In an in vitro study, Li et al. have shown that PRP use can lead to myoblast proliferation, but not to myoblast differentiation, which is important in producing muscle tissue [16].

Dimauro et al. have examined the effect of PRP injection on the early phases of muscle regeneration in rats. They have found that PRP injection enhanced the number of myogenic precursor cells and their proliferation rate. On the other hand, PRP administration raised the number of catabolic cytokines and pro-fibrotic growth factors such as TGF-β1 and myostatin that may induce fibrotic healing instead of muscular regeneration [17]. Enhanced catabolism and pro-fibrotic recovery of muscle tissue may affect recovery and performance and increase the risk of injury recurrence. This double-edged sword nature of PRP treatment in muscle injuries must be taken into consideration when choosing which PRP to use for muscle injuries or even which fraction to use (i.e., a solution with relatively low levels of TGF-β1). In addition, the timing of injection should also be taken into consideration—with the initial local tissue damage still forming even in the first 24–48 h from injury with local bleeding/hematoma formation still taking place; it is recommended that the administration of PRP should therefore be avoided in the first 24 and even 48 h post injury.

Current understanding of PRP biology and the exact role of growth factors (GFs) in the setting of muscle injuries is limited, especially due to the fact that the healing process requires a degree of inflammation to progress. GFs and cytokines influence chemotaxis and proliferation, and artificially adding those factors may greatly impact the healing process in an unexpected manner. The exact composition, concentration, and timing of the PRP application in MIs are subject to further in vivo research, but a new and more targeted strategy is being developed for PRP application in MIs.

A recent work by Tsani et al. from 2021 has examined the effect of a combined treatment of PRP and Suramin (an antifibrotic agent, a TGF-β inhibitor) on MI model of rats and compared it with a PRP monotherapy. They have found that Suramin successfully reduced fibronectin expression, without significant differences between the groups in muscle histological of myofibril evaluation and injured muscle strength. However, both groups had significantly better results than the untreated group, and the combination therapy group had the best overall results [18].

This study joins previous studies from 2016 by Li et al. and from 2013 by Satoshi et al., examining the effect of combined PRP and antifibrotic agents that block TGF-$\beta$, on MI models (rats and mice, respectively). Both of these studies have found that the combined therapy group had significantly less fibrosis when compared to the PRP monotherapy group, without significant histological results regarding myofibril regeneration. Li et al. have also found that there were more satellite cells and higher infiltration of M2 macrophages, contributing to the overall better efficacy of the augmented PRP treatment [15, 19].

## 17.3 Platelet-Rich Plasma for Muscle Injuries: Clinical Experience

Several studies have reported positive outcomes with the use of PRP for the treatment of muscle strains. Sanchez et al. analyzed the use of PRP in different grades of muscle injuries in 21 professional soccer players, reporting the PRP group required half the time to resume normal training activities compared to matched historical controls [8]. Rossi et al. performed a randomized controlled trial comparing a rehabilitation program plus a PRP injection vs. a rehabilitation program alone for muscle injury (hamstrings, quadriceps, and gastrocnemius), reporting significantly earlier full recovery and significantly lower pain scores in the PRP group [20].

Hamstrings injuries are one of the most common injuries in athletes, usually resulting in a prolonged rest period and delayed return to sport even in mild injuries. In a randomized controlled trial of 28 patients comparing PRP with a rehabilitation program for hamstrings injury vs. a rehabilitation program alone, Hamid et al. reported a significantly shorter time to return to play in the PRP group (26.7 days) when compared to the rehabilitation alone group (42.5 days) [21]. In another prospective study, Bezuglov et al. reported similar results in 40 soccer players [22]. However, several other studies have shown contradictory results for PRP use for acute hamstrings tears [23–25]. In a

double-blind, placebo-controlled, randomized study on 80 professional and recreational athletes with acute hamstrings injuries treated with two intramuscular injections of PRP or isotonic saline, Reurink et al. reported no benefit for PRP injections [26].

In a systematic review from 2018, Grassi et al. examined six randomized controlled trial studies (RCTs) that included mostly professional athletes that suffered injuries to different locations (hamstrings, rectus femoris, quadriceps, gastrocnemius, thigh, foot and ankle, and shoulder), reporting statistically significant shorter time to return to sports in the PRP-treated groups. However, they highlighted the fact that not all RCTs were of the highest quality and also the large variability and heterogeneity in the various regarding injury type and PRP preparation method [27].

It is also important that it is quite challenging to perform high-level studies in professional athletes due to their limited availability for continuous repeated assessments. Another important issue is that while it is not easy to cite statistical significance in shorter return to sport timings, even 2–3 days would make a difference in competitive athletes who often compete two to three times a week, which means that even 2–3 days could make the difference of playing or missing the next competition/match.

## 17.4 Tips for PRP Use for Muscle Injuries

When considering the use of PRP for muscle injuries, the extent of muscle injury should be taken into consideration, especially in cases where true and significant muscle fiber disruption is confirmed with imaging studies. If a hematoma or a seroma is present, it is recommended to evacuate it under ultrasound guidance to decompress the area of injury and approximate the injured muscle fibers; once the hematoma is evacuated, a platelet poor fraction (i.e., fraction F1 in PRGF) could be injected into the injury site and adjacent peripheral healthy muscle (Fig. 17.1). The recommendation is to use either a PRP product with low levels of TGF$\beta$-1 or to use the platelet-poor fraction, since it has a reduced concentration of

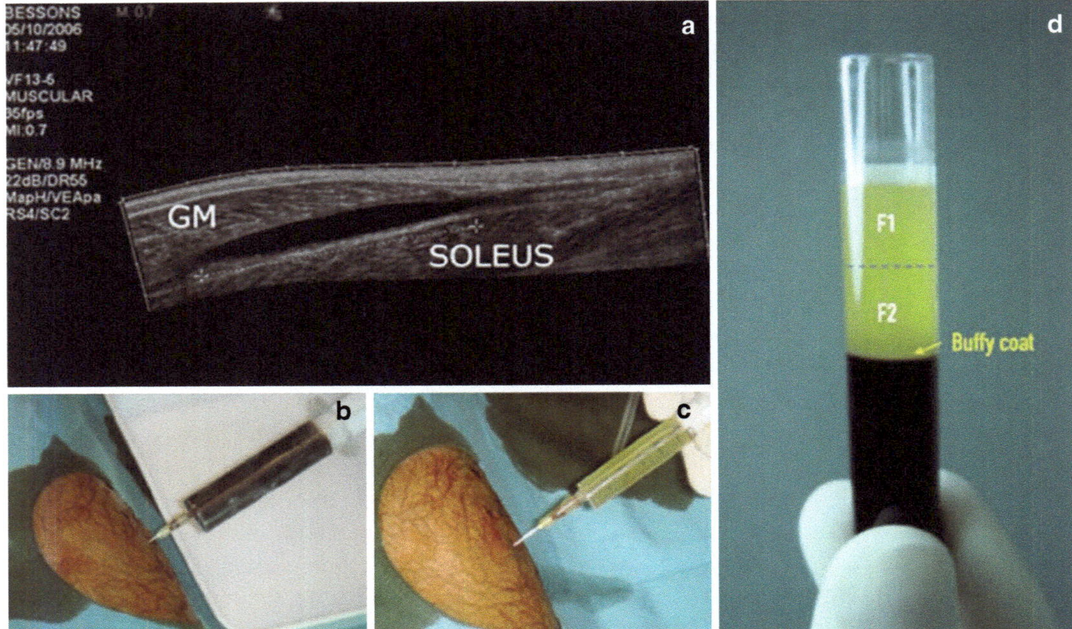

**Fig. 17.1** (**a**) Ultrasound image of an extensive soleus muscle injury and the area of surrounding hematoma (arrow; black area inside the muscle); (**b**) hematoma evacuation using a 10-cc syringe; (**c**) injection of platelet poor plasma (PRGF F1 fraction, Endoret® System, Spain) intramuscular injection using a 10 cc syringe; (**d**) PRGF fractions distribution following centrifugation

the pro-fibrotic factor TGFβ-1, unlike the platelet-rich fraction, which is adjacent to the buffy coat layer or leukocytes sediment; repeated ultrasound (US) or MRI imaging can be used to follow healing progression and assess for fibrosis levels which may predispose to reinjury. Repeated injections may be applied (often considered in higher grades of injury) at a minimum of 1-week intervals, and decision should be based on injury grade, US imaging (to assess muscle tissue damage and healing progression), and symptoms.

## 17.5 Conclusion

Despite promising biological reasoning, positive preclinical findings, and early clinical success, the efficacy of PRP for the management of muscle injuries has not been strongly established as of yet. Further higher quality studies are necessary to properly explore the full potential of PRP for this indication.

**Tip**

- When injecting PRP for acute muscle injury, it is preferable to wait 3–4 days after the injury and not inject immediately after the injury.
- If a hematoma or seroma is present, it is recommended to evacuate/aspirate the hematoma/seroma under US guidance prior to PRP injection.
- It is recommended to use either a PRP product with low levels of TGFβ-1 or to use the platelet-poor fraction, since it has a reduced concentration of the pro-fibrotic factor TGFβ-1.
- Repeated injections are usually considered in higher injury grades and should be performed at a minimum of 1-week intervals, and decision should be based on injury grade, US imaging (to assess muscle tissue damage and healing progression), and symptoms.

## References

1. Ekstrand J, Hägglund M, Waldén M. Epidemiology of muscle injuries in professional football (soccer). Am J Sports Med. 2011;39:1226–32. https://doi.org/10.1177/0363546510395879.

2. Rodas G, Bove T, Caparrós T, et al. Ankle sprain versus muscle strain injury in professional men's basketball: a 9-year prospective follow-up study. Orthop J Sports Med. 2019;7:2325967119849035. https://doi.org/10.1177/2325967119849035.

3. Colio S, McAuliffe M, Uribe Y, Bodor M. Regenerative medicine for muscle and ligament problems: technical aspects and evidence. Tech Reg Anesth Pain Manage. 2015;19:80–4. https://doi.org/10.1053/j.trap.2016.09.014.

4. Grassi A, Quaglia A, Canata GL, Zaffagnini S. An update on the grading of muscle injuries: a narrative review from clinical to comprehensive systems. Joints. 2016;4:39. https://doi.org/10.11138/JTS/2016.4.1.039.

5. Morelli KM, Brown LB, Warren GL. Effect of NSAIDs on recovery from acute skeletal muscle injury: a systematic review and meta-analysis. Am J Sports Med. 2018;46:224–33. https://doi.org/10.1177/0363546517697957.

6. Fernandes TL, Pedrinelli A, Hernandez AJ. Muscle injury – physiopathology, diagnosis, treatment and clinical presentation. Rev Bras Ortop. 2011;46:247. https://doi.org/10.1016/S2255-4971(15)30190-7.

7. Whalan M, Lovell R, McCunn R, Sampson JA. The incidence and burden of time loss injury in Australian men's sub-elite football (soccer): a single season prospective cohort study. J Sci Med Sport. 2019;22:42–7. https://doi.org/10.1016/j.jsams.2018.05.024.

8. Sánchez M, Albillos J, Angulo F, et al. Platelet-rich plasma in muscle and tendon healing. Oper Tech Orthop. 2012;22:16–24. https://doi.org/10.1053/j.oto.2011.11.003.

9. Mishra A, Woodall J, Vieira A. Treatment of tendon and muscle using platelet-rich plasma. Clin Sports Med. 2009;28:113–25. https://doi.org/10.1016/j.csm.2008.08.007.

10. Barlow Y, Willoughby J. Pathophysiology of soft tissue repair. Br Med Bull. 1992;48:698–711. https://doi.org/10.1093/OXFORDJOURNALS.BMB.A072572.

11. Mosca MJ, Rodeo SA. Platelet-rich plasma for muscle injuries: game over or time out? Curr Rev Musculoskelet Med. 2015;8:145–53. https://doi.org/10.1007/S12178-015-9259-X/TABLES/2.

12. Piuzzi NS, Dominici M, Long M, et al. Proceedings of the signature series symposium "cellular therapies for orthopaedics and musculoskeletal disease proven and unproven therapies—promise, facts and fantasy," International Society for Cellular Therapies. Cytotherapy. 2018;20:1381–400. https://doi.org/10.1016/j.jcyt.2018.09.001.

13. Stilhano RS, Martins L, Ingham SJM, et al. Gene and cell therapy for muscle regeneration. Curr Rev Musculoskelet Med. 2015;8:182–7. https://doi.org/10.1007/s12178-015-9268-9.

14. Jain NK, Gulati M. Platelet-rich plasma: a healing virtuoso. Blood Res. 2016;51:3. https://doi.org/10.5045/BR.2016.51.1.3.

15. Terada S, Ota S, Kobayashi M, et al. Use of an antifibrotic agent improves the effect of platelet-rich plasma on muscle healing after injury. J Bone Joint Surg Ser A. 2013;95:980–8. https://doi.org/10.2106/JBJS.L.00266.

16. Li H, Usas A, Poddar M, et al. Platelet-rich plasma promotes the proliferation of human muscle derived progenitor cells and maintains their stemness. PLoS One. 2013;8:e64923. https://doi.org/10.1371/journal.pone.0064923.

17. Dimauro I, Grasso L, Fittipaldi S, et al. Platelet-rich plasma and skeletal muscle healing: a molecular analysis of the early phases of the regeneration process in an experimental animal model. PLoS ONE. 2014;9:e102993. https://doi.org/10.1371/JOURNAL.PONE.0102993.

18. Tsai WC, Yu TY, Chang GJ, et al. Use of platelet-rich plasma plus suramin, an antifibrotic agent, to improve muscle healing after injuries. Am J Sports Med. 2021;49:3102–12. https://doi.org/10.1177/03635465211030295.

19. Li H, Hicks JJ, Wang L, et al. Customized platelet-rich plasma with transforming growth factor β1 neutralization antibody to reduce fibrosis in skeletal muscle. Biomaterials. 2016;87:147–56. https://doi.org/10.1016/J.BIOMATERIALS.2016.02.017.

20. Rossi LA, Molina Rómoli AR, Bertona Altieri BA, et al. Does platelet-rich plasma decrease time to return to sports in acute muscle tear? A randomized controlled trial. Knee Surg Sports Traumatol Arthrosc. 2017;25:3319–25. https://doi.org/10.1007/s00167-016-4129-7.

21. Hamid A, Ali M, Yusof A, et al. Platelet-rich plasma injections for the treatment of hamstring injuries: a randomized controlled trial. Am J Sports Med. 2014;42:2410–8. https://doi.org/10.1177/0363546514541540.

22. Bezuglov E, Maffulli N, Tokareva A, Achkasov E. Platelet-rich plasma in hamstring muscle injuries in professional soccer players: a pilot study. Muscles Ligaments Tendons J. 2019;9:112–8.

23. Zanon G, Combi F, Combi A, et al. Platelet-rich plasma in the treatment of acute hamstring injuries in professional football players. Joints. 2016;4:17–23.

24. Rettig AC, Meyer S, Bhadra AK. Platelet-rich plasma in addition to rehabilitation for acute hamstring injuries in NFL players: clinical effects and time to return to play. Orthop J Sports Med. 2013;1:2325967113494354.

25. Hamilton B, Tol JL, Almusa E, et al. Platelet-rich plasma does not enhance return to play in hamstring injuries: a randomised controlled trial. Br J Sports Med. 2015;49:943–50.

26. Reurink G, Goudswaard GJ, Moen MH, et al. Platelet-rich plasma injections in acute muscle injury. N Engl J Med. 2014;370:2546–7.

27. Grassi A, Napoli F, Romandini I, et al. Is platelet-rich plasma (PRP) effective in the treatment of acute muscle injuries? A systematic review and meta-analysis. Sports Med. 2018;48:971–89. https://doi.org/10.1007/S40279-018-0860-1/FIGURES/7.

Peter A. Everts, Ignacio Dallo, José Fábio Lana,
and Luga Podesta

## 18.1 Introduction

Orthobiology and regenerative medicine, nonsurgical interventional procedures, involve the use of autologous-prepared biologics to stimulate the body's natural healing processes. These biologics act as a scaffold for tissue repair, immunomodulation, painkilling, and tissue regeneration. The most well-known orthobiological treatment products include platelet-rich plasma (PRP), BMAC, and adipose tissue (AT) preparations. In this chapter we will focus on how to perform a bone marrow aspiration procedure to extract BMA to prepare a BMAC product for injection in patients with knee OA. This procedure can be safely performed by well-trained physicians at POC. The reader should be familiar with their national regulatory requirements to utilize BMAC, which is beyond the scope of this chapter.

### 18.1.1 Safety and Contraindications

Before performing a BMA procedure, patients should be well informed about the procedure, and it is essential to obtain an informed written consent. Patients need to be informed about potential risks, like infection, hematoma, and anemia [1]. After obtaining an informed consent, any medications, supplements, or activities of daily living that might have an impact in the BMA extraction procedure, bone marrow cell viability, or potential therapeutic effect need to be reviewed and discussed with the patient. It is important that patients avoid specific medications to maintain bone marrow cell viability and post BMAC treatment cellular function in the recipient environment. These medications include nonsteroidal anti-inflammatory drug (NSAIDs), corticosteroid injections, systemic and inhaled steroids, antibiotics (fluoroquinolone), anticoagulants, and statins [2–5]. Contraindications to perform a BMA procedure are severe anemia, active systemic or local infection at the BMA extraction and injection site, and active cancer.

P. A. Everts (✉)
Research and Education Division, Gulf Coast
Biologics, Fort Myers, FL, USA

OrthoRegen Group, Max-Planck University,
Indaiatuba, SP, Brazil
e-mail: peter@gulfcoastbiologics.com

I. Dallo
Department of Orthopaedic Surgery and Sports
Medicine, Sport Me Medical Center, Unit of
Biological Therapies and MSK Interventionism,
Seville, Spain

J. F. Lana
OrthoRegen Group, Max-Planck University,
Indaiatuba, SP, Brazil

Department of Orthopaedics, The Bone and Cartilage
Institute, Indaiatuba, SP, Brazil

L. Podesta
Bluetail Medical Group and Podesta Orthopedic
Sports Medicine, Naples, FL, USA


B. Kocaoglu et al. (eds.), *Musculoskeletal Injections Manual*,
https://doi.org/10.1007/978-3-031-52603-9_18

## 18.2 Bone Marrow Tissue

The bone is composed of cortical and trabecular bone, cartilage, and connective tissues. Spongy, or trabecular, bone is composed of a lattice of fine bone plates filled with hematopoietic marrow, fat-containing marrow, and arterial-venous sinusoidal blood vessels. Furthermore, it consists of bone cells at different developmental stages (including pre-osteoblasts, osteoblasts, and osteocytes), collagen fibrils, and calcium and phosphate deposits [6]. Bone marrow tissue is soft, similar to the peripheral blood. Two categories of bone marrow tissue exist, the red and yellow marrow. Depending on age, the red marrow is replaced by the yellow marrow. Friedenstein and colleagues reported first on the isolation of bone marrow-derived stem cells from bone marrow stroma in plastic culture dishes and identified mesenchymal stem cells as colony-forming unit fibroblasts (CFU-Fs) [7]. The bone marrow stroma is made up of a network of many different cells, like fibroblast-like cells, and includes a subpopulation of multipotent cells which are able to generate the mesenchyme, cells that are referred to as mesenchymal stem cells (MSCs) [8].

The red bone marrow is a rich source of bone marrow-derived cells and present in most skeletal system bones of the iliac crest, tibia, spine vertebrae, humerus, calcaneus, ribs, and near point of attachment of long bones of legs and arms. It has been estimated that more than 500 billion cells per day can be produced in the bone marrow, in particular erythrocytes, leukocytes, and platelets [9]. Orthobiological BMAC procedures focus on the extraction of marrow from the red bone marrow as it contains myeloid and lymphoid stem cells and MSCs.

### 18.2.1 Bone Marrow-Specific Regions

The bone marrow cavity can be partitioned into four regions: endosteal, sub-endosteal, central, and perisinusoidal regions, according to the model of Lambertsen and Weis, have been adopted and modified [10]. In general, the bone

marrow consists of a hematopoietic component (parenchyma) and a vascular component (stroma). The parenchyma includes hematopoietic progenitor and hematopoietic stem cells (HSCs), which are localized close to the endosteum and around the blood vessels. Bone marrow stroma cells, including endothelial cells, are recognized as multipotential non-hematopoietic progenitor cells, capable of differentiating into various tissues of mesenchymal origin, including osteoblasts, chondrocytes, tenocytes, endothelial cells, myocytes, fibroblasts, and adipocytes [11, 12].

### 18.2.2 Bone Marrow Niches

Bone marrow niches are three-dimensional microenvironments. In Table 18.1, an overview of the classical bone marrow niches is presented. It is assumed that they control genes and properties that define "stemness," including the control and balance between quiescence, self-renewal, proliferation, and differentiation of diverse cell types. Stem cell niches are defined as specific cellular and molecular microenvironments, regulating bone marrow HSC, MSC, and progenitor functions, consisting of autonomous signaling molecules and mechanisms and facilitates intercellular contact, and the interaction between stem cells and their neighboring extracellular matrix [13, 14]. Harvested bone marrow stem cells and subsequently injected into a totally different microenvironment can potentially differentiate into cell types of this new local environment [15]. Zhao et al. revealed the plasticity potential of bone marrow MSCs, as these cells were capable of de-differentiation into cells from other cell lineages [16]. Their finding has potentially great clinical implications for orthobiological treat-

**Table 18.1** Classical bone marrow niches

| |
|---|
| Arteriolar niche |
| Endosteal niche |
| Hematopoietic stem cell niche |
| Megakaryocyte niche |
| Mesenchymal stem cell niche |
| Perivascular niche |

ments, since autologous BMAC originates from their specific and original bone marrow niche but are frequently used in other pathoanatomic tissue types to treat various pathologies.

### 18.2.3  Bone Marrow Aspiration

Clinicians utilizing orthobiological, regenerative medicine applications have a growing interest in harvesting BMA to prepare a minimal manipulative BMAC, as this is a plentiful source of bone marrow stem cells and their progenitor cells, megakaryocytes, platelets, leukocytes, and other cells, easily accessible via a BMA harvesting procedure [17, 18].

## 18.3  BMAC Procedural Preparations

Patients should be informed to hydrate in the days before the procedure, and they should avoid eating solid food 3–5 h prior the BMA harvesting to avoid nausea. The extraction procedure can be performed in an office setting and should be executed using aseptic skin preparation techniques, including harvesting tissue site draping. It is recommended that the physician is wearing sterile gloves, a hair cover, and a facemask. Consider wearing a sterile gown. Under normal circumstances, patients can be monitored with a pulse oximeter that includes heart rate monitoring. Supplemental oxygen, blood pressure, automated external defibrillator, and crash cart supplies should be available.

### 18.3.1  BMA Harvesting Sites

It is important to choose a harvesting site that has the most MSCs that can be extracted, as they represent a small population of the total of bone marrow cells [19]. In humans, the most common anatomical location to obtain bone marrow with the highest potential for MSCs is the posterior superior iliac spine (PSIS), compared to the tibia,

**Table 18.2**  Anatomical location for BMA

Calcaneus
Iliac crest: anterior and PSIS
Proximal humerus
Sternum
Tibia: distal and proximal
Vertebral body

calcaneus [20–22], and other harvesting sites (Table 18.2) [23].

### 18.3.2  Imaging Options

A certain volume of BMA needs to be extracted to produce a BAMC. It is imperative to precisely locate the donor site, as most MSCs are located in endosteal and subendosteal areas [24, 25]. Safe trocar placement to penetrate the cortical bone is accomplished by using image guidance during aspiration procedure. Here, we focus on a BMA procedure from the PSIS sites, as it is the most frequently reported anatomical site for BMA.

#### 18.3.2.1  Ultrasound

When the PSIS is targeted, patients are positioned in the prone position. A pillow is placed under the waist to elevate the pelvis and avoiding lumbar lordosis and superficially positioning the PSIS. Sonographic assessment using a portable ultrasound system with either a high-frequency linear or 5–2-MHz low-frequency curvilinear transducer is positioned in a transverse plane over the hyperechoic L5 spinous process. The transducer is then translated laterally toward the physician until the hyperechoic iliac crest comes into view. The transducer can then be toggled to identify the broadest and flattest portion of the PSIS. Once identified, skin markings are made adjacent to the center of the transducer's width and length. Transecting lines drawn from the X and Y skin markings localize the center and needle target point of the PSIS, as shown if Fig. 18.1 [26]. With the ultrasound transducer in the transverse plane over the PSIS, the top (most superficial depth) and slope of the PSIS is noted for correct angulation of the trocar. This mark is

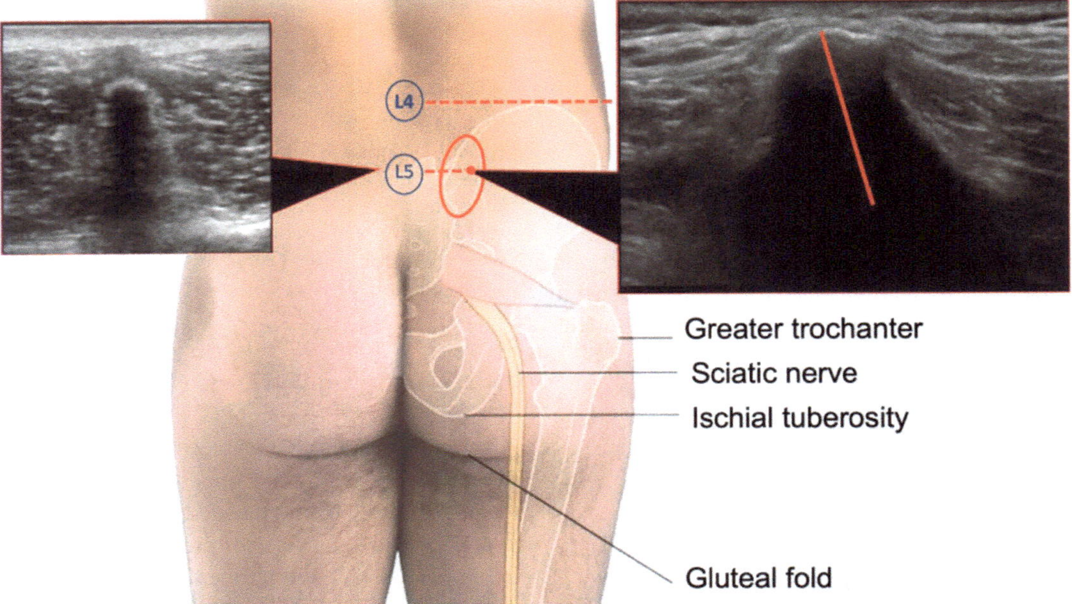



**Fig. 18.1** Surface anatomy of the PSIS and view with probe in transverse plane. (Courtesy of L. Podesta, MD)

maintained during the delivery of local anesthetics and antiseptic skin preparations prior to the introduction of the needle trocar.

### 18.3.2.2 Fluoroscopy

Fluoroscopic imaging, using ipsilateral or contralateral oblique beam angulations for viewing the PSIS site, is initiated. The perpendicular fluoroscopic approach requires a beam angle around 15° ipsilateral to the PSIS, entering laterally with angulation toward the sacroiliac joint. This angle will view the lateral ilium outer wall, and a needle is directed anteromedially. Fluoroscopic images support in positioning the tip of the trocar above the target area for entering the PSIS. The parallel fluoroscopic approach results in viewing down the PSIS table, at a 25° contralateral oblique beam position. This results in a classic view of the "teardrop," as shown in Fig. 18.2. Imaging can confirm the entry point into the PSIS table and visualize the angle through the cortex, allowing for safe trocar advancement through the cortical bone in the marrow cavity [27]. Using the parallel approach technique allows for a safe deeper marrow penetration.

### 18.3.3 Cortical Bone Penetration Options

Following ultrasound or fluoroscopic imaging, with visualization of the PSIS, a sharp trocar is used to penetrate the cortical bone, using either a manual force only technique, perpendicular and slightly lateral to the patient at 9–12 counterclockwise-clockwise rotations, or a mallet. Some physicians prefer to use a battery-powered drill to pass the cortical bone. However, at all times, regardless of the approach, avoid increased manipulation and tissue trauma using the sharp trocar, as this will increase the risk for neurovascular injury, bleeding, tearing of lateral gluteal muscle origins, and post-procedural pain.

### 18.3.4 Anesthetic Considerations

A local anesthetic is required to control pain. It is injected around the PSIS cortex and periosteal sleeve, making sure to "walk off" the PSIS in four directions (superiorly, medially, laterally, and inferiorly). Typically, 1% lidocaine is used,

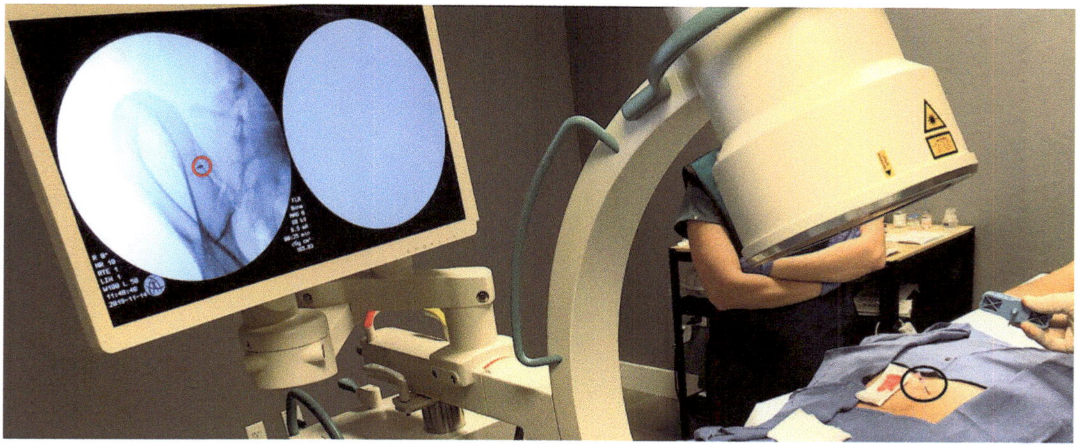

**Fig. 18.2** Fluoroscopic imaging. Fluoroscopy imaging of the PSIS, with the patient in prone on a fluoroscopic table. The parallel fluoroscopic approach results in viewing down the PSIS table, at a 25° contralateral oblique beam position. This results in a classic view of the "teardrop," referring to the outline of the medial and lateral borders, as shown in the monitor. The tip of a needle (black circle), in the numbed skin, is marking the entry site of the bone marrow trocar to be placed in the marrow cavity, while the physician is on the ipsilateral side of the fluoroscope, viewing the correct position on the monitor (red circle) (courtesy of G. Flanagan II, MD)

without epinephrine, up to 10 mL total per aspiration side. Alternatively, 0.25–0.5% ropivacaine can be used as well. A 22- to 27-gauge, 2- to 3.5-inch needle is typically used. Local anesthetic should be applied to the skin at the needle entry site, soft tissue trajectory of the needle, and the surface of the bone/periosteum. The needle used to anesthetize the bone should never be longer than the needle used to harvest the bone marrow. This helps to avoid needle length mismatch in reaching the bone for harvesting. Never inject anesthetic through the trocar or needle used to aspirate marrow, as this can have a deleterious effect on the viability of the bone marrow cells. After the anesthetic is administered, wait at least 3–5 min until an adequate level of anesthesia is achieved.

### 18.3.5 BMA Harvesting Needle Devices

Different bone marrow needle harvesting systems are available on the market, with distinctive different design characteristics, affecting the marrow harvesting dynamics and bone marrow cellularity (Fig. 18.3). Traditionally, physicians have been using the Jamshidi™ harvesting needle (Ranfac Corporation, Avon, MA). Recently, the Aspire Bone Marrow Harvesting System™ (EmCyte Corporation, Fort Myers, FL) and the Marrow Cellution Bone Marrow Aspiration Device™ (Ranfac Corporation, Avon, MA) have been introduced. A significant difference between the latter two harvesting systems is that the Aspire™ system has a separate introducer and an aspiration needle with a closed, blunt, tip for minimal peripheral blood aspiration from the marrow cavity. Aspirating BMA is only feasible through the three aspiration needle side orifices [28]. In Fig. 18.4, the Marrow Cellution device is developed for a low volume of BMA aspiration with direct patient injection, without concentration, whereas the BMA following Aspire™ harvesting is intended for centrifugation processing to produce a high-concentration MSC product [29].

### 18.3.6 Anticoagulation with Heparin Solution

Prior to a BMA procedure, it is highly recommended that all the bone marrow harvesting com-

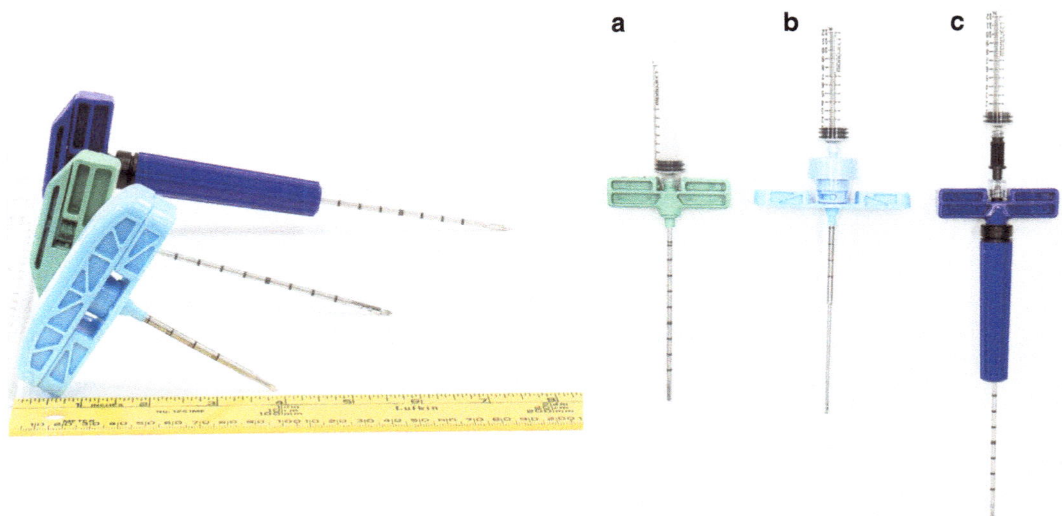


**Fig. 18.3** Three different BMA needle systems. (**a**) Jamshidi™ harvesting needle (Ranfac Corporation, Avon, MA); (**b**) Aspire Bone Marrow Harvesting System™ (EmCyte Corporation®, Fort Myers, FL); (**c**) Marrow Cellution Bone Marrow Aspiration Device™ (Ranfac Corporation, Avon, MA)

**Fig. 18.4** Detailed aspect of fenestrated blunt aspirating canula for BMA aspiration. The aspiration cannula has a blunt tip and triple core side ports for quiescent BMA and progenitor stem cell collection from the endosteal and sub-endosteal bone marrow niche, minimizing RBC contamination in the collected BMA volume (Aspire™ bone marrow harvesting system, EmCyte Corporation®, Fort Myers, FL)

ponents, collection syringes, and all the processing accessories that will be in contact with bone marrow are thoroughly heparin rinsed as bone marrow tissue has the potential for fast clotting during the extraction procedure [30]. Clot formation will prevent an adequate BMAC procedure to concentrate MSCs and other marrow constituents. Contingent on the desired amount of bone marrow, the total needed volume of heparin to be used, at a concentration of 1000 IU/mL, is calculated. Good laboratory practice indicates a ratio of 1 mL of 1000 IU heparin concentration to 9 mL of bone marrow. For example, when using a 10 mL collection syringe, add 1 mL of heparin to the syringe, and mix 9 mL of BMA thoroughly to ensure an adequate level of anticoagulation. Depending on the instructions for use and physicians' preference, the heparinized BMA can be filtered through a 200-micron, heparin rinsed, filter to eliminate potential particles, fibrin strands, and fat tissue.

## 18.3.7 Extraction/Collection Syringes

Hernigou et al. indicated to use small (10 mL) syringes for bone marrow collection, as this resulted in significantly higher BMA MSC and progenitor cell concentrations, when compared to a 50 mL syringe [17]. Furthermore, filling the

**Fig. 18.5** Aspect of BMAC and cellular densities following a two-spin procedure. After the second spin, the BMAC is resuspended and extracted from the concentration chamber of the BMAC device (**a**) (PureBMC® Supraphysiologic, EmCyte Corporation®, Fort Myers FL). (**b**) Magnification of the BMAC buffy coat layer. All BMAC cells are located in this multicellular stratum, according to their individual cellular densities. Note, the MSC density is close to the density of erythrocytes. Therefore, avoiding collecting a fraction of erythrocytes will decrease the capture rate of MSCs

extraction syringe to 10–20% of its total volume capacity resulted in an improved yield of MSCs. Lately, physicians tend to use 10 ml syringes, employing a fast and intermittent pull technique to collect small volumes from different intratrabecular depths and or cortical sites. Another advantage for using 10 ml syringes is that anticoagulation protocols can be better managed, since smaller syringes fill considerably faster than larger syringes.

### 18.3.8 BMAC Device Function

An effective BMAC injection is reliant on the performance of the BMA procedure, with minimal cellular trauma, while maximizing MSC cellular yields, avoiding peripheral RBC infiltration [28, 31]. Dedicated, national regulatory compliant and registered processing kits are commercially available for BMAC preparations and orthobiological treatment procedures. Kits may include a BMA concentration device and a BMA harvesting needle system. BMAC preparation protocols and techniques are streamlined and validated methods to concentrate marrow cells.

Double-spin centrifugation protocols, performed at POC, create a layered BMAC buffy coat stratum, based on different centrifugal forces that accomplish density cellular separation, because of the specific cellular gravity of the individual marrow components, as shown in Fig. 18.5.

## 18.4 BMC Injection Knee Osteoarthritis

### 18.4.1 Background

Osteoarthritis (OA) of the knee is a common progressive degenerative joint disease, with a high percentage of chronic pathoanatomic degenerative changes with loss of cartilage, subchondral bone changes, synovial inflammation, and various meniscal pathologies. Safe, nontraditional, and nonsurgical treatment plans for knee OA may include orthobiological therapies, using autologous preparations derived from whole blood (PRP), bone marrow (BMAC), and adipose tissue [32–34]. These orthobiologics play a beneficial role in reducing local inflammation and promote synovial and cartilage anabolism by

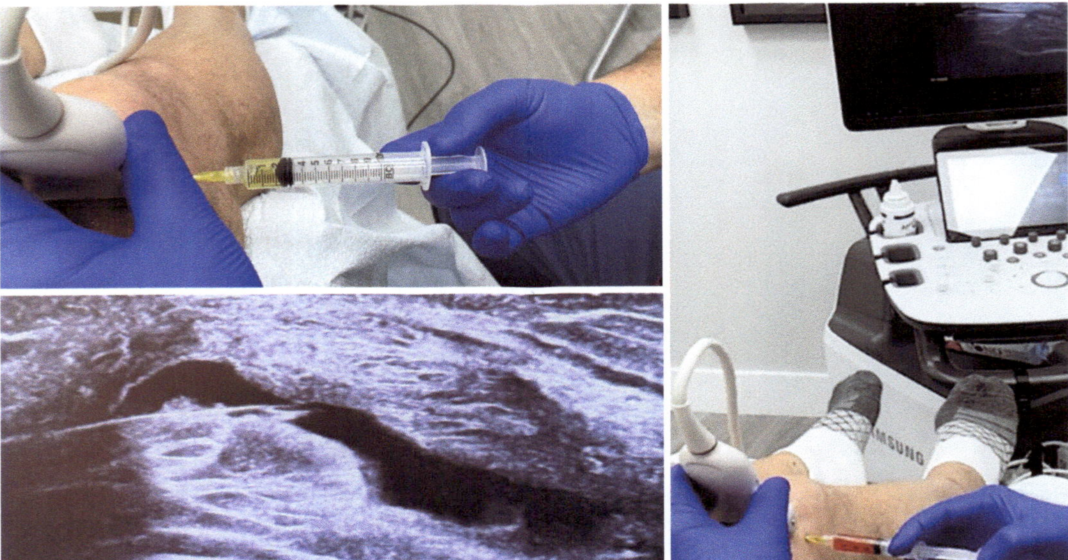

**Fig. 18.6** BMAC injection via the superolateral approach

releasing a series of proteins, like platelet-derived growth factors, HSCs, MSCs, and particular leukocytes and their specific phenotypes, to provide pain relief and functional improvement. The goal of BMAC treatment in knee OA is aiming at achieving restoration of all impaired articular components. The paracrine impact of the MSCs present in BMAC is essential in tissue repair and the regenerative ability by stimulating immuno-modulatory, anti-catabolic, anti-apoptotic, and chondrogenic effects [35]. Hence, BMAC has been used successfully to treat a variety of MSK pathologies, including patients with moderate to severe knee OA [36]. Favorable outcomes are thought to be associated with the presence of MSCs in the BMAC [37]. More specifically, higher BMAC MSC concentrations, measured as colony-forming unit fibroblast (CFU-F)®, indicate more positive patient reported outcomes [38–40].

### 18.4.2 Intra-articular Setup

(a) Anatomical structures: The knee joint is an encapsulated hinge-type synovial joint, consisting of four bones: distal femur, proximal tibia, proximal fibula, and patella, involving femorotibial, patella-femoral, and tibiofibular articulations.

(b) Patient positioning: Supine or seated for a superolateral injection approach is preferred for intra-articular knee injections, especially when an effusion is present [41]. The physician is standing/seated on the injection side of the affected knee (Fig. 18.6).

(c) Imaging: Ultrasound high-frequency linear array transducer; with transducer in in short axis to quadriceps tendon; needle position in-plane (long axis), lateral to medial approach.

(d) Needles: 22–18 gauge, 1.5–2.0 inch for effusion evacuation

   27–22 gauge, 1.5–2.0 inch for biological injection

(e) Injection volume: Dependent on the amount of BMAC prepared, ranging 2–10 mL total.

In a recent systematic review on BMAC for the treatment of knee OA (keeling), 8 studies met the inclusion criteria, including a total of 299 knees with a mean follow-up of 12.9 months. BMAC injections were effective in improving pain and patient-reported outcomes in patients with knee OA ($P < 0.01$ and $P < 0.05$, respectively) [42].

Before BMAC injection, the synovial fluid is evacuated via the superolateral approach. After evacuation of the effusion, the BMAC is injected. All manipulations are ultrasound guided.

## References


1. Bowen JE. Technical issues in harvesting and concentrating stem cells (bone marrow and adipose). PM R. 2015;7(4S):S8–18. https://doi.org/10.1016/j.pmrj.2015.01.025.

2. Wyles CC, Houdek MT, Wyles SP, Wagner ER, Behfar A, Sierra RJ. Differential cytotoxicity of corticosteroids on human mesenchymal stem cells. Clin Orthop Relat Res. 2015;473(3):1155–64. https://doi.org/10.1007/s11999-014-3925-y.

3. Stephenson AL, Wu W, Cortes D, Rochon PA. Tendon injury and fluoroquinolone use: a systematic review. Drug Saf. 2013;36(9):709–21. https://doi.org/10.1007/s40264-013-0089-8.

4. Izadpanah R, Schächtele DJ, Pfnür AB, Lin D, Slakey DP, Kadowitz PJ, et al. The impact of statins on biological characteristics of stem cells provides a novel explanation for their pleiotropic beneficial and adverse clinical effects. Am J Physiol Cell Physiol. 2015;309(8):C522–31. https://doi.org/10.1152/ajpcell.00406.2014.

5. van Esch RW, Kool MM, van As S. NSAIDs can have adverse effects on bone healing. Med Hypotheses. 2013;81(2):343–6. Available from: https://www.sciencedirect.com/science/article/pii/S0306987713001667.

6. Everts PA, Panero AJ. Basic science of autologous orthobiologics: part 2. Mesenchymal stem cells. Orthobiologics. 2023;34(1):25–47. Available from: https://www.sciencedirect.com/science/article/pii/S1047965122000730.

7. Charbord P. Bone marrow mesenchymal stem cells: historical overview and concepts. Hum Gene Ther. 2010;21(9):1045–56. https://doi.org/10.1089/hum.2010.115.

8. Everts PA, Flanagan G, Podesta L. Autologous orthobiologics. In: Mostoufi SA, George TK, Tria Jr AJ, editors. Clinical guide to musculoskeletal medicine: a multidisciplinary approach. Cham: Springer; 2022. p. 651–79. https://doi.org/10.1007/978-3-030-92042-5_62.

9. McDaniel JS, Antebi B, Pilia M, Hurtgen BJ, Belenkiy S, Necsoiu C, et al. Quantitative assessment of optimal bone marrow site for the isolation of porcine mesenchymal stem cells. Stem Cells Int. 2017;2017:1–10. Available from: https://www.hindawi.com/journals/sci/2017/1836960/.

10. Lambertsen RH, Weiss L. Studies on the organization and regeneration of bone marrow: origin, growth, and differentiation of endocloned hematopoietic colonies. Am J Anat. 1983;156(4):369–92. https://doi.org/10.1002/aja.1001660402.

11. Muguruma Y, Yahata T, Miyatake H, Sato T, Uno T, Itoh J, et al. Reconstitution of the functional human hematopoietic microenvironment derived from human mesenchymal stem cells in the murine bone marrow compartment. Blood. 2006;107(5):1878–87. Available from: https://ashpublications.org/blood/article/107/5/1878/133429/Reconstitution-of-the-functional-human.

12. Fliedner TM, Graessle D, Paulsen C, Reimers K. Structure and function of bone marrow hemopoiesis: mechanisms of response to ionizing radiation exposure. Cancer Biother Radiopharm. 2002;17(4):405–26. Available from: http://www.liebertpub.com/doi/10.1089/108497802760363204.

13. Spradling A, Drummond-Barbosa D, Kai T. Stem cells find their niche. Nature. 2001;414(6859):98–104. Available from: http://www.nature.com/articles/35102160.

14. Li L, Xie T. Stem cell niche: structure and function. Annu Rev Cell Dev Biol. 2005;21:605–31.

15. Wu P, Tarasenko YI, Gu Y, Huang LYM, Coggeshall RE, Yu Y. Region-specific generation of cholinergic neurons from fetal human neural stem cells grafted in adult rat. Nat Neurosci. 2002;5(12):1271–8. Available from: http://www.nature.com/articles/nn974.

16. Zhao LR, Duan WM, Reyes M, Keene CD, Verfaillie CM, Low WC. Human bone marrow stem cells exhibit neural phenotypes and ameliorate neurological deficits after grafting into the ischemic brain of rats. Exp Neurol. 2002;174(1):11–20. Available from: https://linkinghub.elsevier.com/retrieve/pii/S0014488601978537.

17. Hernigou P, Homma Y, Flouzat Lachaniette CH, Poignard A, Allain J, Chevallier N, et al. Benefits of small volume and small syringe for bone marrow aspirations of mesenchymal stem cells. Int Orthop. 2013;37(11):2279–87. Available from: http://link.springer.com/10.1007/s00264-013-2017-z.

18. Oliver K, Awan T, Bayes M. Single-versus multiple-site harvesting techniques for bone marrow concentrate: evaluation of aspirate quality and pain. Orthop J Sports Med 2017;5(8):2325967117724398. https://doi.org/10.1177/2325967117724398.

19. Pittenger MF. Multilineage potential of adult human mesenchymal stem cells. Science. 1999;284(5411):143–7. https://doi.org/10.1126/science.284.5411.143.

20. Ricci WM, Spiguel A, McAndrew C, Gardner M. What's new in orthopaedic trauma. JBJS. 2013;95(14):7170. Available from: https://journals.lww.com/jbjsjournal/Fulltext/2013/07170/What_s_New_in_Orthopaedic_Trauma.12.aspx.

21. Friedlis MF, Centeno CJ. Performing a better bone marrow aspiration. Phys Med Rehabil Clin N Am. 2016;27(4):919–39. Available from: https://linkinghub.elsevier.com/retrieve/pii/S104796511630050X.

22. Marx RE, Tursun R. A qualitative and quantitative analysis of autologous human multipotent adult stem cells derived from three anatomic areas by marrow aspiration: tibia, anterior ilium, and posterior ilium. Int J Oral Maxillofac Implants. 2013;28(5):290–4. https://doi.org/10.11607/jomi.te10.

23. Piuzzi NS, Hussain ZB, Chahla J, Cinque ME, Moatshe G, Mantripragada VP, et al. Variability in the preparation, reporting, and use of bone marrow aspirate concentrate in musculoskeletal disorders: a systematic review of the clinical orthopaedic literature. J Bone Jt Surg. 2018;100(6):517–25. Available from: http://Insights.ovid.com/crossref?an=00004623-201803210-00010.

24. Cordeiro-Spinetti E, Taichman RS, Balduino A. The bone marrow endosteal niche: how far from the surface? The Bone Marrow Endosteal Niche. J Cell Biochem. 2015;116(1):6–11. https://doi.org/10.1002/jcb.24952.

25. Siclari VA, Zhu J, Akiyama K, Liu F, Zhang X, Chandra A, et al. Mesenchymal progenitors residing close to the bone surface are functionally distinct from those in the central bone marrow. Bone. 2013;53(2):575–86. Available from: https://linkinghub.elsevier.com/retrieve/pii/S8756328212014317.

26. Hirahara AM, Panero A, Andersen WJ. An MRI analysis of the pelvis to determine the ideal method for ultrasound-guided bone marrow aspiration from the iliac crest. Am J Orthop. 2018;47(5):38. https://doi.org/10.12788/ajo.2018.0038.

27. Hernigou J, Alves A, Homma Y, Guissou I, Hernigou P. Anatomy of the ilium for bone marrow aspiration: map of sectors and implication for safe trocar placement. Int Orthop. 2014;38(12):2585–90. https://doi.org/10.1007/s00264-014-2353-7.

28. Everts PA, Ferrell J, Mahoney CB, Ii GF, de Roman MI, Paul R, et al. A comparative quantification in cellularity of bone marrow aspirated with two new harvesting devices, and the non-equivalent difference between a centrifugated bone marrow concentrate and a bone marrow aspirate as biological injectates, using a bi-lateral patient model. J Stem Cell Res Ther. 2020;10:1–10.

29. Dallo I, Chahla J, Mitchell JJ, Pascual-Garrido C, Feagin JA, LaPrade RF. Biologic approaches for the treatment of partial tears of the anterior cruciate ligament: a current concepts review. Orthop J Sports Med. 2017;5(1):232596711668172. https://doi.org/10.1177/2325967116681724.

30. Everts P, Flanagan G II, Rothenberg J, Mautner K. The rationale of autologously prepared bone marrow aspirate concentrate for use in regenerative medicine applications. In: Regenerative medicine. London: IntechOpen; 2020. Available from: https://www.intechopen.com/online-first/the-rationale-of-autologously-prepared-bone-marrow-aspirate-concentrate-for-use-in-regenerative-media.

31. Mautner K, Jerome MA, Easley K, Nanos K, Everts PA. Laboratory quantification of bone marrow concentrate components in unilateral versus bilateral posterior superior iliac spine aspiration. J Stem Cell Res Ther. 2020;466:9.

32. Vora A, Borg-Stein J, Nguyen RT. Regenerative injection therapy for osteoarthritis: fundamental concepts and evidence-based review. PM R. 2012;4(5S):S104–9. https://doi.org/10.1016/j.pmrj.2012.02.005.

33. Centeno CJ, Al-Sayegh H, Freeman MD, Smith J, Murrell WD, Bubnov R. A multi-center analysis of adverse events among two thousand, three hundred and seventy two adult patients undergoing adult autologous stem cell therapy for orthopaedic conditions. Int Orthop. 2016;40(8):1755–65. https://doi.org/10.1007/s00264-016-3162-y.

34. Mautner K, Bowers R, Easley K, Fausel Z, Robinson R. Functional outcomes following microfragmented adipose tissue versus bone marrow aspirate concentrate injections for symptomatic knee osteoarthritis. Stem Cells Transl Med. 2019;8(11):1149–56. https://doi.org/10.1002/sctm.18-0285.

35. Ossendorff R, Walter S, Schildberg F, Khoury M, Salzmann G. Controversies in regenerative medicine: should knee joint osteoarthritis be treated with mesenchymal stromal cells? Eur Cell Mater. 2022;43:89–111. Available from: https://www.ecmjournal.org/papers/vol43/pdf/v043a09.pdf.

36. Centeno CJ, Al-Sayegh H, Bashir J, Goodyear S, Freeman MD. A dose response analysis of a specific bone marrow concentrate treatment protocol for knee osteoarthritis. BMC Musculoskelet Disord. 2015;16(1):258. https://doi.org/10.1186/s12891-015-0714-z.

37. Muschler GF, Midura RJ. Connective tissue progenitors: practical concepts for clinical applications. Clin Orthop Relat Res. 2002;395:66–80. Available from: https://journals.lww.com/clinorthop/Fulltext/2002/02000/Connective_Tissue_Progenitors__Practical_Concepts.8.aspx.

38. El-Kadiry AEH, Lumbao C, Salame N, Rafei M, Shammaa R. Bone marrow aspirate concentrate versus platelet-rich plasma for treating knee osteoarthritis: a one-year non-randomized retrospective comparative study. BMC Musculoskelet Disord. 2022;23(1):23. https://doi.org/10.1186/s12891-021-04910-5.

39. Dulic O, Rasovic P, Lalic I, Kecojevic V, Gavrilovic G, Abazovic D, et al. Bone marrow aspirate concentrate versus platelet rich plasma or hyaluronic acid for the treatment of knee osteoarthritis. Medicina. 2021;57(11):1193. Available from: https://www.mdpi.com/1648-9144/57/11/1193.

40. Kon E, Boffa A, Andriolo L, Di Martino A, Di Matteo B, Magarelli N, et al. Subchondral and intra-articular injections of bone marrow concentrate are a safe and effective treatment for knee osteoarthritis: a pro-

spective, multi-center pilot study. Knee Surg Sports Traumatol Arthrosc. 2021;29(12):4232–40. https://doi.org/10.1007/s00167-021-06530-x.

41. Hermans J, Bierma-Zeinstra SMA, Bos PK, Verhaar JAN, Reijman M. The most accurate approach for intra-articular needle placement in the knee joint: a systematic review. Semin Arthritis Rheum. 2011;41(2):106–15. Available from: https://www.sciencedirect.com/science/article/pii/S0049017211000679.

42. Keeling LE, Belk JW, Kraeutler MJ, Kallner AC, Lindsay A, McCarty EC, et al. Bone marrow aspirate concentrate for the treatment of knee osteoarthritis: a systematic review. Am J Sports Med. 2022;50(8):2315–23. https://doi.org/10.1177/03635465211018837.

# Fat-Derived Orthobiologics for Knee OA



Peter A. Everts, Raphael Barnabe, Luga Podesta, and Rowan Paul

## 19.1 Introduction

Over the past decades, PRP- and mesenchymal stem cell (MSC)-based autologous products have been broadly applied in orthobiology and regenerative medicine applications. MSC-based biologics can be prepared from bone marrow and adipose tissue, and they are ideal cell types for the treatment of damaged tissues given their restorative and pro-regenerative properties [1]. MSCs cells can be found in particular in most of the vascularized tissues, being a subtype of pericytes with pro-regenerative properties, and in particular in bone marrow and adipose tissue (AT) [2, 3]. Autologous AT is a heterogeneous biological source of various cellular tissue components, and it contains one of the best MSC populations [4]. AT is abundantly present in humans and relatively easy to harvest through a mini-liposuction procedure, at point-of-care [5]. In this chapter we focus on the harvesting AT via a mini-liposuction procedure (MLS) and the preparation of several fat-derived products.

### 19.1.1 Safety and Contraindications

Before performing a fat-derived orthobiological treatment, patients should be well informed about the procedure and MLS. It is essential to obtain an informed written consent. Patients need to be informed about potential risk factors and complications [6]. Reported complications include ecchymosis, edema, surgical site infection, seroma, and venous thromboembolism. Contraindications to perform an MLS and a fat-derived procedure are ongoing infections, severe diabetes, and active cancer.

## 19.2 Adipose Tissue Background

Adipose tissue, otherwise known as body fat, is loose connective tissue that extends throughout the body and is mainly composed of adipocytes and acts as a central metabolic organ in the regulation of whole-body energy homeostasis.

P. A. Everts (✉)
Research and Education Division, Gulf Coast Biologics, Fort Myers, FL, USA

OrthoRegen Group, Max-Planck University, Indaiatuba, SP, Brazil
e-mail: peter@gulfcoastbiologics.com

R. Barnabe
Instituto de Pesquisa Clinica Anna Vitoria Lana, Indaiatuba, SP, Brazil
e-mail: dr.rafael@birm.com.br

L. Podesta
Bluetail Medical Group and Podesta Orthopedic Sports Medicine, Naples, FL, USA

R. Paul
Department of Community and Family Medicine, Dartmouth Geisel School of Medicine, Hanover, NH, USA
e-mail: Rowan_Paul@alumni.brown.edu

© The Author(s), under exclusive license to Springer Nature Switzerland AG 2024
B. Kocaoglu et al. (eds.), *Musculoskeletal Injections Manual*,
https://doi.org/10.1007/978-3-031-52603-9_19

Adipocytes have secretory capacities, as they release a variety of effector cells like endocrine hormones, exosomes, lipids, inflammatory cytokines, and peptide hormones, impacting local and systemic metabolic responses [7]. Autologous AT-prepared products consist of a heterogeneous source of various cellular tissue components that can be used as a cell-based, disease modifying, therapies in orthobiological and other regenerative medicine procedures. Furthermore, concentrated adipose tissue provides clinicians with a physiological 3D multicellular scaffold, including adipose-derived stem cells (ASCs) and stromal cells. These features are of high interest when treating MSK disorders [8, 9].

### 19.2.1 Adipose Tissue Structure

Adipose tissue is a highly vascularized connective tissue, abundantly present throughout the human body. White AT (WAT) is responsible for energy storage and plays a pivotal physiological role in maintaining metabolic homeostasis in the body by releasing release of several adipocytokines, growth factors, and cytokines that may act in an endocrine or paracrine fashion [10]. Brown AT (BAT) plays a significant role in thermogenesis and have the ability to diffuse energy by producing heat to ensure body temperature regulation, rather than storing it as triglycerides [11].

### 19.2.2 Adipose-Derived Tissue Stem Cells

ASCs are prevalent in adipose connective tissue and surrounding blood vessels [12]. The first characterization of adult mesenchymal stem cells in lipoaspirates, was performed by Zuk in 2002 [13], followed by several others, confirming that adipose mesenchymal stem cells have a high proliferation capacity and multilineage cell differentiation potential, capable of differentiating into adipogenic, chondrogenic, myogenic, osteogenic, and neurogenic cells [14]. Furthermore,

ASCs contribute in the reduction of proinflammatory cytokines, chemokines, and cellular apoptosis [15, 16]. Plentiful literature has demonstrated the anti-inflammatory and adaptive properties of ASCs based on environmental conditions, with anti-inflammatory effects on chondrocytes and synoviocytes [17]. These AT characteristics have great potential in clinical orthobiological tissue repair applications [18], to counteract inflammatory processes, but also to promote regenerative processes in the joint synovium and chondral surfaces [19]. These non-lipid-laden stromal cells can be isolated from suction-aspirated adipose tissue by either enzymatic collagenase or mechanical emulsification.

### 19.2.3 Adipose Harvesting Technique Considerations

Subcutaneous AT harvesting is performed using a mini-liposuction technique. The procedure is executed by means of aspiration cannulas, introduced through small skin incisions, assisted by controlled suction. Its basic principles have been elaborated by Illouz, who was the first to introduce the modern, safe, and widespread method of liposuction with a blunt-tipped cannula and subcutaneous infiltration to facilitate adipose breakdown and controlled fat aspiration [20]. The procedure preserves neurovascular structures with minimal patient discomfort. In 1985, Klein introduced the tumescent liposuction technique, using a specific adipose infiltration fluid to optimize harvesting of AT (Table 19.1) [21].

Later, Coleman introduced a new three-step technique to decrease trauma to adipose tissue following liposuction, employing manual low negative pressure lipo-aspiration. Harvested fat is transferred to a dedicated (concentration) processing device to purify the AT, using minimal manipulative processing steps [22]. Properly executed lipo-aspiration procedures, followed by AT processing, will not affect ASC viability and functionality, while yielding significant concentrations of AD-tSVF cells [23].

**Table 19.1** Classical formulation of Klein's tumescent solution (80)

| Ingredient | Concentration | Volume | Activity |
| --- | --- | --- | --- |
| Sodium chloride | 0.9% | 1L | Diluent |
| Lidocaine | 1% | 50 mL | Anesthesia |
| Epinephrine | 1:1000 | 1 mL | Vasoconstriction |
| Sodium bicarbonate | 8.4% | 10 mL | Acidity compensation |

## 19.3 Preparation of Minimally Manipulated Products from Adipose Tissue

There are various protocols and preparation methods available to produce different AT-derived minimally manipulated orthobiological adipose products. Overall, the objective is to have access to ASCs and other reparative adipose-derived cells that are present in the stroma portion of AT [12, 24]. Furthermore, AT provides clinicians with a physiological 3D multicellular scaffold, including AT extracellular matrix, ASCs, and other SVF constituents, that can be utilized in tissue grafting like tendon tears. In Table 19.2, more detailed ATC preparation steps are presented.

### 19.3.1 Adipose Tissue Concentrate

Similar to the preparation of PRP and BMAC, cellular density gradient centrifugation can also be used to create a viable biological AT concentrate specimen (ATC) (Table 19.2) [3]. Centrifugation techniques have proven to be an effective means to produce a clean and viable ATC. Both, single- and double-spin centrifugation procedures have been used to separate the residual infranatant extracellular fluid, residual pro-inflammatory oil, and adipose tissue debris from the final ATC product, based on the principle of density separation, as demonstrated in Fig. 19.1. Thereafter, the ATC is aspirated from the concentration tube. Subsequently, the ATC can be subjected to micro fat homogenization techniques to produce an intact stromal vascular environment, which subsequently can be applied to tissue sites.

**Table 19.2** Detailed representation of the preparation steps to produce ATC

*A. Tumescent fluid preparation:* a sterile NaCl solution consisting of anesthetics (lidocaine for pain relieve and epinephrine for blood vessels to constrict to minimize RBC contamination in fat tissue during harvesting)

*B. Tumescent injection:* via small skin cuts, a thin blunt injector needle is injected in the target adipose harvesting area

*C. Waiting time:* reports indicate to wait at least 20 minutes before starting the fat harvesting procedure. This time is needed for the fluid to cause the injected area to swell and stiffen, supporting in easy in fat removal

*D. Adipose harvesting:* with a dedicated harvester cannula, fat tissue is harvested using liposuction, by applying manually negative pressure to a collection syringe fat is removed from the area that was injected with tumescent fluid

*E. Racking and decanting:* syringes filled with harvested fat is placed in a rack, with plunger in upward direction. After the adipose harvesting, leave all syringes in the in the rack for 10-15 minutes, with the Luer end of the syringe capped. Decant the supernatant (tumescent fluid) by removing the cap, until adipose tissue starts to block the Luer

*F. Transfer the decanted fat* into a disposable processing device and place it in a dedicated centrifuge to concentrate the AT specimen

*G. Centrifugation protocol:* density layer separation by centrifugation, producing ATC. Follow the instruction for use of the preparation device to extract ATC

*H. Mechanical emulsification:* a method to emulsify the ATC, by moving the two syringes back and forward through a restraining device to size the ATC, making it suitable for tissue injection

### 19.3.2 Micro-fragmented Adipose Tissue (MFAT)

After AT harvesting, the fat is processed with the devoted device [25]. In this closed-loop device, the AT is rinsed with saline solution to remove

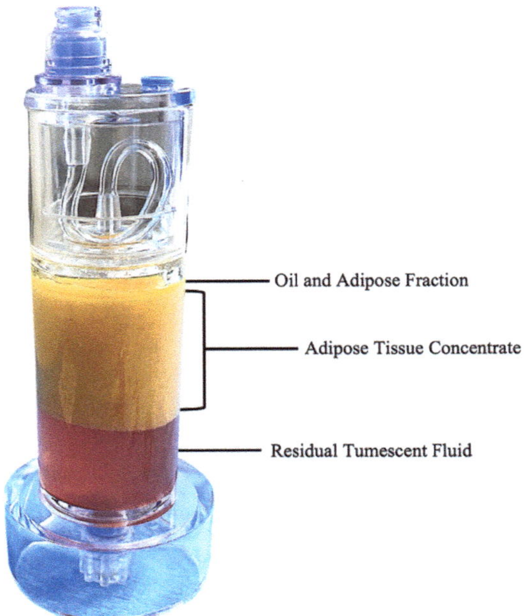


**Fig. 19.1** Centrifugated density separation of adipose tissue. After a single centrifugation spin, the AT is clean and concentrated. Typically, with adipose centrifugation, the first, top, layer consists mainly of inflammatory oil and adipose tissue cell residues. The middle layer is comprised of ATC, and the third bottom layer contains residual tumescent fluid and some RBCs (Progenikine® Adipose Concentration System, EmCyte Corporation®, Fort Myers, FL, with permission)

cellular debris, oil, erythrocytes, and other contaminants. During this washing process, the fat is mechanically sized to isolate ASCs, to produce micro-fragmented adipose tissue graft (MFAT) [26], without the use of enzymes. AT is gently micro-fractured and washed, leaving oil and blood residues behind in the process. The final adipose product is then collected and ready to be delivered to treatment sites.

Additional emulsification steps can lead to the preparation of microfat or nanofat, differing in the adipose tissue cluster size.

1. Macrofat: fat grafts are harvested larger than 2.4 mm; indicated for filling a wide range of tissue sites.
2. Microfat: fat grafts are harvested using a cannula with a hole diameter ranging from 1.2 to 2.4 mm, an emulsifier with a hole diameter of 1.2 mm, and a parcel diameter of less than 1.2 mm.

Nanofat: fat grafts extracted using a 1.2- to 2.4-mm-diameter cannula, 400–600 μm emulsifier, and 400–600-μm parcel diameter.

### 19.3.3 Stromal Vascular Fraction (SVF)

The use of AT in orthobiological applications is based on the separation of the stromal fraction (SVF) contained in adipose tissue, allowing for access to ASCs and other AT constituents [24]. This heterogeneous mixture of cells which includes all the cell population excluding adipocytes can be created by different AT disruption techniques, like enzymatic digestion, or mechanical emulsification preparation methods. The overall processing of adipose harvesting and preparations is illustrated in Fig. 19.2. AT-derived orthobiologics comprise of ASCs, acting as the principal therapeutic effector cells [27], in combination with precursor and mature endothelial cells, pericytes, lymphocytes, pre-adipocytes, macrophages, and mature adipocytes [3, 4], as shown in Table 19.3. Noteworthy, ASCs meet the four criteria for MSCs as defined by the International Society for Cellular Therapy (ISCT) [4].

### 19.3.4 SVF by Enzymatic Digestion

Enzymatic digestion techniques use enzymes (mainly collagenase) to isolate ASCs and stromal cells by digesting the AT peptide collagen bonds, while destructing extracellular structures. Centrifugation techniques are used to separate the floating adipocytes from the pelleted SVF [28]. Collagenase-based SVF protocols produce an adipose-derived cellular SVF, which is directly applied, without the need for in vitro cell expansion. Additionally, erythrocytes are usually routinely removed during this enzymatic processing protocol. This is a significant dissimilarity compared with BMACs, which contain a significant number of erythro-

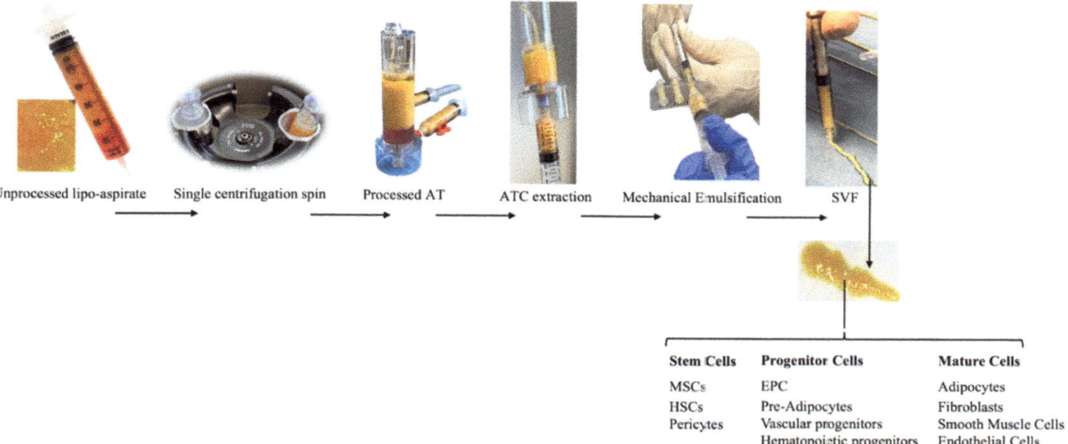


**Fig. 19.2** Mechanical SVF preparation method. Adipose tissue is harvested following lipo-aspiration and subsequently processed in a centrifuge for density cellular layering. After centrifugation, the oil and tumescent fluid are removed from the concentration device. Thereafter, ATC is extracted and collected in a syringe for microfragmentation of the ATC prior application. (Progenikine® Adipose Concentration System, and Adicen™ Emulsifier device, EmCyte Corporation®, Fort Myers, FL, with permission). *AT* adipose tissue, *ATC* adipose tissue concentrate, *SVF* stromal vascular fraction, *MSC* mesenchymal stem cells, *HSCs* hematopoietic stem cells, *EPC* endothelial progenitor cell

**Table 19.3** SVF cellular distribution

| Cell type | Percentage range in SVF | Cells |
|---|---|---|
| Stromal | 15–30 | ASC |
| | | Pre-adipocytes |
| | | Fibroblasts |
| Hematopoietic origin | 34–45 | Platelets |
| | | Neutrophil |
| | | Lymphocyte |
| | | Monocyte/macrophage |
| | | Erythrocyte |
| HSCs and progenitor | 1–15 | |
| Pericytes | 3–5 | |
| Endothelial cells | | |
| Smooth muscle cells | 5–15 | |

*ASC* adipose mesenchymal stem cells, *HSC* hematopoietic stem cells

cytes [27]. In the current regulatory landscape, enzymatic-prepared SVF products are listed as "more than minimal manipulative preparations" and are therefore in many countries not allowed to be used as an orthobiological treatment product.

## 19.4 SVF by Mechanical Disruption

There are various fat-sizing devices available on the market that can generate an emulsified AT treatment specimen to produce adipose-derived mechanical SVF. This method signifies to the cellular population in adipose SVF in a bioactive scaffold or extracellular matrix [29]. In this case adipose tissue is mechanically sized by a mechanical disruption achieved by different commercially available systems, among which mechanical emulsification. Depending on the mechanical method, the yield of SVF can differ, with some methods being more efficient in terms of cell purification (Table 19.3) [30]. An advantage of mechanical SVF is that it is composed of both cellular and native structural fat fragments, providing a bioactive cellular adipose tissue matrix [31]. Compared to enzymatic digestion, mechanical SVF preparations are not a 100% concentrated cellular product, a distinct difference compared to enzymatic SVF. Noteworthy, mechanical SVF is therefore more likely to be approved by national regulatory administrations.

A graphical schematic presentation shows the processing steps from lipoaspirate collection to mechanical SVF (Fig. 19.2).

## 19.5 Adipose Tissue Orthobiological Properties

Several studies have indicated the potential for ASCs to demonstrate paracrine activity and exhibit differentiation potential toward different cell lineages (adipogenic, osteogenic, chondrogenic, and myogenic lineages), while providing immunosuppressive and angiogenetic properties [32, 33].

### 19.5.1 Immunomodulation

Cells contained in AT preparations are proficient in secreting anti-inflammatory and immunosuppressive factors that can exert immunomodulatory effect [32]. In particular, ASCs regulate the immune system via direct cell-cell communication and indirectly through the secretion of soluble mediators, growth factors, and extravascular vesicles [34, 35], regulating the activities of T cells, B cells, and macrophages [36]. Pro-inflammatory and anti-inflammatory molecules

that are regulators and influencers in adipose immunomodulatory activities are adipokines, antioxidative, pro-angiogenic, anti-apoptotic, growth factors (like, VEGF, FGF, TGF), and specific interleukins (IL-6, IL-7) [37]. Interestingly, several studies compared the immunomodulatory abilities of ASCs and BMSCs and found similar effects when used in chronic inflammatory conditions [38, 39].

### 19.5.2 Angiogenesis

AT and its SVF cellular variations secrete angiogenic factors such as angiopoietin-2, VEGF, and several adipokines (e.g., leptin and adiponectin), capable of modulating angiogenesis and vascular structures [40]. The plasticity of AD-MSCs have demonstrated to enhance (neo)angiogenetic processes, through endothelial cell activities, which is essential in the treatment of tissue repair [41]. These specialized adipose characteristics were also well demonstrated by Miranville et al., revealing the differentiation of AD-MSCs into endothelial cells, ultimately contributing to angiogenesis [42]. Additionally, AD-MSCs stimulate angiogenesis through paracrine activities, with SVF releasing various pro-angiogenic factors like growth factors and macrophages [43, 44].

## 19.6 Adipose-Derived Products in Knee Osteoarthritis

Sharma et al. described the by-products of adipose tissue with regenerative potential, including expanded ASCs, microfat, nanofat, SVF, and other AT-derived products [45]. In MFAT processing, the lipoaspirate is refined to a cluster of 0.2–0.8 mm, while the supportive vascular stromal niche remains intact and ASCs stay in their natural habitat prior to intra articular knee injections [46]. In a recent systematic review by Aletto et al., a total of 24 clinical trials were analyzed using AT-SVF products [47]. The application significantly increased clinical outcomes in all studies. VAS and WOMAC were the most used scores to analyze clinical outcomes. Additionally, Koh et al. showed that, through second-look arthroscopy, clinical improvements persisted for more than 2 years and that 87.5% of elderly patients improved or maintained cartilage healing status at 2 years postoperatively [48]. Importantly, they also concluded that AT-SVF injections are safe intraarticular knee injection procedures, with good clinical and imaging outcomes in the early follow-up period (6–24 months). The safety findings were recently confirmed by Jeyaram and associates [49]. The authors also stated that given the presence of a heterogeneous population of cells in AT-SVF, there is no need for the culture expansion of cells, which further decreases the risk of culture-induced chromosomal aberrations.

More studies are definitely needed to establish standardized protocols to prepare AT-SVF injectates as minimal invasive treatment modalities for the management of knee OA and other orthobiological applications.

# References


1. Muthu S, Jeyaraman M, Jain R, Gulati A, Jeyaraman N, Prajwal GS, et al. Accentuating the sources of mesenchymal stem cells as cellular therapy for osteoarthritis knees—a panoramic review. Stem Cell Investig. 2021;8:71189. Available from: https://sci.amegroups.org/article/view/71189.
2. Everts P, Flanagan G II, Rothenberg J, Mautner K. The rationale of autologously prepared bone marrow aspirate concentrate for use in regenerative medicine applications. In: Regenerative medicine. London: IntechOpen; 2020. Available from: https://www.intechopen.com/online-first/the-rationale-of-autologously-prepared-bone-marrow-aspirate-concentrate-for-use-in-regenerative-media.
3. Everts PA, Flanagan G, Podesta L. Autologous orthobiologics. In: Mostoufi SA, George TK, Tria Jr AJ, editors. Clinical guide to musculoskeletal medicine: a multidisciplinary approach. Cham: Springer; 2022. p. 651–79. https://doi.org/10.1007/978-3-030-92042-5_62.
4. Bourin P, Bunnell BA, Casteilla L, Dominici M, Katz AJ, March KL, et al. Stromal cells from the adipose tissue-derived stromal vascular fraction and culture expanded adipose tissue-derived stromal/stem cells: a joint statement of the International Federation for Adipose Therapeutics and Science (IFATS) and the International Society for Cellular Therapy (ISCT). Cytotherapy. 2013;15(6):641–8. Available from: https://linkinghub.elsevier.com/retrieve/pii/S1465324913003873.
5. Everts PA, Panero AJ. Basic science of autologous orthobiologics part 2. Mesenchymal stem cells. Orthobiologics. 2023;34(1):25–47. Available from: https://www.sciencedirect.com/science/article/pii/S1047965122000730.
6. Wu S, Coombs DM, Gurunian R. Liposuction: concepts, safety, and techniques in body-contouring surgery. Cleve Clin J Med. 2020;87(6):367–75. https://doi.org/10.3949/ccjm.87a.19097.
7. Argentati C, Morena F, Bazzucchi M, Armentano I, Emiliani C, Martino S. Adipose stem cell translational applications: from bench-to-bedside. Int J Mol Sci. 2018;19(11):3475. Available from: http://www.mdpi.com/1422-0067/19/11/3475.
8. Lim MH, Ong WK, Sugii S. The current landscape of adipose-derived stem cells in clinical applications. Expert Rev Mol Med. 2014;16:e8. Available from: https://www.cambridge.org/core/article/current-landscape-of-adiposederived-stem-cells-in-clinical-applications/FA953E3B268FB0386B2763F7A2F84716.
9. Docheva D, Müller SA, Majewski M, Evans CH. Biologics for tendon repair. Adv Drug Deliv Rev. 2015;84:222–39. Available from: https://linkinghub.elsevier.com/retrieve/pii/S0169409X14002786.
10. Karastergiou K, Mohamed-Ali V. The autocrine and paracrine roles of adipokines. Mol Cell Endocrinol. 2010;318(1–2):69–78. https://doi.org/10.1016/j.mce.2009.11.011.
11. Cinti S. Reversible physiological transdifferentiation in the adipose organ. Proc Nutr Soc. 2009;68(4):340–9. https://doi.org/10.1017/S0029665109990140.
12. Brown JC, Shang H, Li Y, Yang N, Patel N, Katz AJ. Isolation of adipose-derived stromal vascular fraction cells using a novel point-of-care device: cell characterization and review of the literature. Tissue Eng Part C Methods. 2017;23(3):125–35. https://doi.org/10.1089/ten.tec.2016.0377.
13. Zuk PA, Zhu M, Ashjian P, Ugarte DAD, Huang JI, Mizuno H, et al. Human adipose tissue is a source of multipotent stem cells. Mol Biol Cell. 2002;13:17.
14. Yang HJ, Kim KJ, Kim MK, Lee SJ, Ryu YH, Seo BF, et al. The stem cell potential and multipotency of human adipose tissue-derived stem cells vary by cell donor and are different from those of other types of stem cells. Cells Tissues Organs. 2014;199(5–6):373–83. Available from: https://www.karger.com/Article/FullText/369969.
15. Manferdini C, Maumus M, Gabusi E, Piacentini A, Filardo G, Peyrafitte J, et al. Adipose-derived mesenchymal stem cells exert antiinflammatory effects on chondrocytes and synoviocytes from osteoarthritis patients through prostaglandin E2. Arthritis Rheum. 2013;65(5):1271–81.
16. Yun S, Ku SK, Kwon YS. Adipose-derived mesenchymal stem cells and platelet-rich plasma synergisti-

cally ameliorate the surgical-induced osteoarthritis in Beagle dogs. J Orthop Surg. 2016;11(1):9. https://doi.org/10.1186/s13018-016-0342-9.

17. Ortiz-Virumbrales M, Menta R, Pérez LM, Lucchesi O, Mancheño-Corvo P, Avivar-Valderas Á, et al. Human adipose mesenchymal stem cells modulate myeloid cells toward an anti-inflammatory and reparative phenotype: role of IL-6 and PGE2. Stem Cell Res Ther. 2020;11(1):462. https://doi.org/10.1186/s13287-020-01975-2.

18. Pak J, Chang JJ, Lee JH, Lee SH. Safety reporting on implantation of autologous adipose tissue-derived stem cells with platelet-rich plasma into human articular joints. BMC Musculoskelet Disord. 2013;14(1):337. https://doi.org/10.1186/1471-2474-14-337.

19. Bora P, Majumdar AS. Adipose tissue-derived stromal vascular fraction in regenerative medicine: a brief review on biology and translation. Stem Cell Res Ther. 2017;8(1):145. https://doi.org/10.1186/s13287-017-0598-y.

20. Illouz Y. Body contouring by lipolysis: a 5-year experience with over 3000 cases. Plast Reconstr Surg. 1983;72(5):591–7. https://doi.org/10.1097/00006534-198311000-00001.

21. Klein JA. The tumescent technique: anesthesia and modified liposuction technique. Dermatol Clin. 1990;8(3):425–37. Available from: https://www.sciencedirect.com/science/article/pii/S0733863518304741.

22. Coleman SR. Structural fat grafting: more than a permanent filler. Plast Reconstr Surg. 2006;118(Suppl):108S–20S. Available from: http://journals.lww.com/00006534-200609011-00015.

23. Gentile P, Calabrese C, De Angelis B, Pizzicannella J, Kothari A, Garcovich S. Impact of the different preparation methods to obtain human adipose-derived stromal vascular fraction cells (AD-SVFs) and human adipose-derived mesenchymal stem cells (AD-MSCs): enzymatic digestion versus mechanical centrifugation. Int J Mol Sci. 2019;20(21):5471. Available from: https://www.mdpi.com/1422-0067/20/21/5471.

24. Gimble JM, Katz AJ, Bunnell BA. Adipose-derived stem cells for regenerative medicine. Circ Res. 2007;100(9):1249–60. https://doi.org/10.1161/01.RES.0000265074.83288.09.

25. Giori A, Tremolada C, Vailati R, Navone SE, Marfia G, Caplan AI. Recovery of function in anal incontinence after micro-fragmented fat graft (Lipogems®) injection: two years follow up of the first 5 cases. CellR4. 2015;3(2):e1544.

26. Tremolada C, Colombo V, Ventura C. Adipose tissue and mesenchymal stem cells: state of the art and Lipogems® technology development. Curr Stem Cell Rep. 2016;2(3):304–12. https://doi.org/10.1007/s40778-016-0053-5.

27. Varma MJO, Breuls RGM, Schouten TE, Jurgens WJFM, Bontkes HJ, Schuurhuis GJ, et al. Phenotypical and functional characterization of freshly isolated adipose tissue-derived stem cells.

28. Sugii S, Kida Y, Berggren WT, Evans RM. Feeder-dependent and feeder-independent iPS cell derivation from human and mouse adipose stem cells. Nat Protoc. 2011;6(3):346–58.

29. Alexander RW. Understanding mechanical emulsification (nanofat) versus enzymatic isolation of tissue stromal vascular fraction (tSVF) cells from adipose tissue: potential uses in biocellular regenerative medicine. J Prolother. 2016;8:e947–60.

30. Vasilyev VS, Borovikova AA, Vasilyev SA, Khramtsova NI, Plaksin SA, Kamyshinsky RA, et al. Features and biological properties of different adipose tissue based products. Milli-, micro-, emulsified (nano-) fat, SVF, and AD-multipotent mesenchymal stem cells. In: Plastic and aesthetic regenerative surgery and fat grafting: clinical application and operative techniques. Cham: Springer; 2022. p. 91–107.

31. Tonnard P, Verpaele A, Peeters G, Hamdi M, Cornelissen M, Declercq H. Nanofat grafting: basic research and clinical applications. Plast Reconstr Surg. 2013;132(4):1017–26. Available from: http://journals.lww.com/00006534-201310000-00053.

32. Ong WK, Chakraborty S, Sugii S. Adipose tissue: understanding the heterogeneity of stem cells for regenerative medicine. Biomolecules. 2021;11(7):918. Available from: https://www.mdpi.com/2218-273X/11/7/918.

33. Wang K, Yu LY, Jiang LY, Wang HB, Wang CY, Luo Y. The paracrine effects of adipose-derived stem cells on neovascularization and biocompatibility of a macroencapsulation device. Acta Biomater. 2015;15:65–76. Available from: https://www.sciencedirect.com/science/article/pii/S1742706114005984.

34. Fang Y, Zhang Y, Zhou J, Cao K. Adipose-derived mesenchymal stem cell exosomes: a novel pathway for tissues repair. Cell Tissue Bank. 2019;20(2):153–61. https://doi.org/10.1007/s10561-019-09761-y.

35. Braga OGSA, Goncalves Reis R, Jorge Carvalho Sousa N, Gimble J, Salgado A, Reis R, et al. Adipose tissue derived stem cells secretome: soluble factors and their roles in regenerative medicine. Curr Stem Cell Res Ther. 2010;5(2):103–10. Available from: http://www.eurekaselect.com/openurl/content.php?genre=article&issn=1574-888X&volume=5&issue=2&spage=103.

36. Naik S, Larsen SB, Cowley CJ, Fuchs E. Two to Tango: dialog between immunity and stem cells in health and disease. Cell. 2018;175(4):908–20. Available from: https://linkinghub.elsevier.com/retrieve/pii/S0092867418311723.

37. Mitchell R, Mellows B, Sheard J, Antonioli M, Kretz O, Chambers D, et al. Secretome of adipose-derived mesenchymal stem cells promotes skeletal muscle regeneration through synergistic action of extracellular vesicle cargo and soluble proteins. Stem Cell Res Ther. 2019;10(1):116. https://doi.org/10.1186/s13287-019-1213-1.

38. Hao T, Chen J, Zhi S, Zhang Q, Chen G, Yu F. Comparison of bone marrow-vs. adipose tissue-derived mesenchymal stem cells for attenuating liver fibrosis. Exp Ther Med. 2017;14:5956–64. https://doi.org/10.3892/etm.2017.5333.

39. Pendleton C, Li Q, Chesler DA, Yuan K, Guerrero-Cazares H, Quinones-Hinojosa A. Mesenchymal stem cells derived from adipose tissue vs bone marrow. PLoS ONE. 2013;8(3):e58198. https://doi.org/10.1371/journal.pone.0058198.

40. Sorop O, Olver TD, van de Wouw J, Heinonen I, van Duin RW, Duncker DJ, et al. The microcirculation: a key player in obesity-associated cardiovascular disease. Cardiovasc Res. 2017;113(9):1035–45. Available from: http://academic.oup.com/cardiovascres/article/113/9/1035/3803704/The-microcirculation-a-key-player-in.

41. Planat-Benard V, Silvestre JS, Cousin B, André M, Nibbelink M, Tamarat R, et al. Plasticity of human adipose lineage cells toward endothelial cells: physiological and therapeutic perspectives. Circulation. 2004;109(5):656–63. https://doi.org/10.1161/01.CIR.0000114522.38265.61.

42. Miranville A, Heeschen C, Sengenès C, Curat CA, Busse R, Bouloumié A. Improvement of postnatal neovascularization by human adipose tissue-derived stem cells. Circulation. 2004;110(3):349–55. https://doi.org/10.1161/01.CIR.0000135466.16823.D0.

43. Koh YJ, Koh BI, Kim H, Joo HJ, Jin HK, Jeon J, et al. Stromal vascular fraction from adipose tissue forms profound vascular network through the dynamic reassembly of blood endothelial cells. Arterioscler Thromb Vasc Biol. 2011;31(5):1141–50. https://doi.org/10.1161/ATVBAHA.110.218206.

44. Rehman J, Traktuev D, Li J, Merfeld-Clauss S, Temm-Grove CJ, Bovenkerk JE, et al. Secretion of angiogenic and antiapoptotic factors by human adipose stromal cells. Circulation. 2004;109(10):1292–8. https://doi.org/10.1161/01.CIR.0000121425.42966.F1.

45. Sharma S, Muthu S, Jeyaraman M, Ranjan R, Jha SK. Translational products of adipose tissue-derived mesenchymal stem cells: bench to bedside applications. World J Stem Cells. 2021;13(10):1360.

46. Bianchi F, Maioli M, Leonardi E, Olivi E, Pasquinelli G, Valente S, et al. A new nonenzymatic method and device to obtain a fat tissue derivative highly enriched in pericyte-like elements by mild mechanical forces from human lipoaspirates. Cell Transplant. 2013;22(11):2063–77. https://doi.org/10.3727/096368912X657855.

47. Aletto C, Oliva F, Maffulli N. Knee intra-articular administration of stromal vascular fraction obtained from adipose tissue: a systematic review. J Clin Orthop Trauma. 2022;25:101773. Available from: https://linkinghub.elsevier.com/retrieve/pii/S0976566222000091.

48. Koh YG, Choi YJ, Kwon SK, Kim YS, Yeo JE. Clinical results and second-look arthroscopic findings after treatment with adipose-derived stem cells for knee osteoarthritis. Knee Surg Sports Traumatol Arthrosc. 2015;23:1308–16.

49. Jeyaraman M, Maffulli N, Gupta A. Stromal vascular fraction in osteoarthritis of the knee. Biomedicine. 2023;11(5):1460. Available from: https://www.mdpi.com/2227-9059/11/5/1460.

Tahsin Beyzadeoglu and Onur Cetin


## 20.1 Background

Musculoskeletal disorders, including osteoarthritis, tendinopathies, ligament injuries, and muscular injuries, pose significant challenges to patients and healthcare systems worldwide. Traditional treatment options for these conditions often focus on symptom management and may not address the underlying pathological processes or promote tissue regeneration. As a result, there is a growing interest in regenerative therapies that harness the body's innate healing mechanisms. In the last decades, local injections of autologous blood preparations begin to be a solid option in the conservative treatment of the musculoskeletal injuries and degenerative changes.

Autologous conditioned serum (ACS) therapy has emerged as a promising regenerative treatment modality and is one of the methods of autologous blood preparations which produce anti-inflammatory cytokines and growth factors from the patient's own blood to facilitate tissue repair and regeneration [1–3]. The rationale behind ACS therapy lies in its ability to modulate inflammation, promote tissue repair, and stimulate the regenerative capacity of the affected site. It offers a potentially safer and more targeted approach compared to traditional interventions such as corticosteroid injections or non-steroidal anti-inflammatory drugs (NSAIDs) [4]. By utilizing the patient's own blood, ACS therapy reduces the risk of adverse reactions and immune responses. The World Anti-Doping Agency (WADA) approved the ACS-Orthokine therapy for the use of intraarticular or soft tissue injections with other autologous blood products such as PRP.

## 20.2 Mechanisms of Action

Autologous conditioned serum (ACS) therapy harnesses the regenerative potential of the patient's own blood components to promote tissue healing and repair. The therapeutic effects of ACS therapy are mediated through various mechanisms, including the modulation of inflammation, growth factor, and cytokine regulation.

### 20.2.1 Inflammatory Modulation

One of the key mechanisms of ACS therapy is the modulation of inflammatory processes. ACS contains anti-inflammatory cytokines, particularly


T. Beyzadeoglu (✉)
Faculty of Health Sciences, Halic University,
Istanbul, Turkey

Orthopedics and Traumatology, Beyzadeoglu Clinic,
Istanbul, Turkey
e-mail: tbeyzade@superonline.com

O. Cetin
Orthopedics and Traumatology, Medar Atasehir
Hospital, Istanbul, Turkey

interleukin-1 receptor antagonist (IL-1Ra), which competitively inhibits the binding of interleukin-1 (IL-1) to its receptors 1 and 2 [1–3]. IL-1 is a potent pro-inflammatory cytokine involved in the pathogenesis of various musculoskeletal disorders. By blocking the IL-1 signaling pathway, ACS therapy reduces inflammation and mitigates its detrimental effects on tissues. In addition to IL-1Ra, ACS also contains other anti-inflammatory cytokines, such as IL-4, IL-10, and IL-13, which further contribute to the downregulation of inflammatory responses. They have been shown to exert anti-inflammatory effects by increasing the synthesis of IL-1Ra and reducing the pro-inflammatory cytokine production. IL-1 pro-inflammatory effects can go passive on the tissue if the IL-1Ra folds it by 10 to 1000 [4, 5].

Cytokine production with recombinant DNA techniques is technically challenging. A limited number of them are approved for clinical treatment, which are IL-1Ra, IL-2, IFN, IGF-1, EPO, PDGF BB, G-CSF, GM-CSF, and BMP2+7. Because of strong potency, these factors need to be used with care in the safe dosages. However, IL-1Ra is free of side effects and can be applied in high doses.

### 20.2.2 Growth Factors and Cytokines

The discovery of the cytokines made us understand the biological and inflamatuar effect of osteoarthritis and gives us a chance to create point shot treatments. There has been found many cytokines, and the nomenclature is heterogenic such as group of interleukins (IL) named in order of their discovery, other group named according their first described functions such as TNF (tumor necrosis factor), GCSF (granulocyte colony-stimulating factor), and another group named of their cellular origin such as monocyte-derived monokine. IL and interferons (IFN) have immunomodulator effects having the potential to be specifically treat the joint and connective tissue diseases.

IL-1, TNF$\alpha$, and IL-6 cytokines are playing an important role in acute phase response and in tissue healing; this response will result in a good outcome by activating endothelial cells and leukocytes if it is transient. But in chronic inflammation such as chronic osteoarthritis, tendinitis, or autoimmune disease, cytokines such as IL-1, IL-17, IL-18, and TNF$\alpha$ can be the major role players for the detrimental effects on tissues and impair the healing, degrade the matrix, and erode the articular surface with the enhanced synthesis of collagenases (MMP-1, MMP-8, MMP-9, MMP-13) and aggrecans (ADAMTS4, ADAMTS5) [6]. ACS therapy harnesses the regenerative potential of various growth factors and cytokines present in autologous serum. These bioactive molecules play crucial roles in tissue healing and repair processes.

## 20.3 Clinical Applications, Indication, and Dosage of Autologous Conditioned Serum Therapy

Autologous conditioned serum (ACS) therapy has demonstrated promising clinical applications in various musculoskeletal disorders. The regenerative and anti-inflammatory properties of ACS make it a potential treatment option for conditions where tissue healing and repair are impaired. The following are some of the clinical applications of ACS therapy.

### 20.3.1 Osteoarthritis

Osteoarthritis (OA) is a degenerative joint disease characterized by cartilage degradation, inflammation, and pain. ACS therapy has shown promising results in alleviating symptoms and improving joint function in patients with OA. Several clinical studies have reported reductions in pain, improved physical function following autologous anti-inflammatory therapies [6, 7]. A randomized controlled study compared and demonstrated the superiority of ACS injection over Hyaluronan and saline injections on the patients with grade 2–3 knee osteoarthritis. A study by Hashemi et al. reviewed 60 patients with knee osteoarthritis and compare the ACS and hyaluronan within two groups. After 6-month follow-up, ACS group showed a significantly better results that hyaluronan group [8]. Fotouhi et al. studied ACS along with the PRP and SVF (stromal vascular fraction). ACS injections on knee osteoarthritis gives an excellent result with major decrease in WOMAC score [9].

## 20.3.2 Tendinopathies

Tendinopathies, including Achilles tendinopathy and lateral epicondylitis, are characterized by degenerative changes in tendons, resulting in pain and functional impairment. ACS therapy has shown promise in the management of tendinopathies by promoting tendon healing and reducing pain. Majewski et al. conducted a study on rat models to demonstrate Achilles tendon healing after dissection and re-suturing with ACS vs saline injections. At 4 weeks, ACS group showed a better tendon strength biomechanically, but at 8 weeks, both groups have similar results. In conclusion, ACS may not accelerate the healing, but tendon may gain a better quality when inflammation is reduced and growth factors are supplied [10]. A retrospective study by Godek et al. reported an excellent or good results on rotator cuff tendinopathy (72%), Achilles tendon tendinopathy (75%), and tennis elbow enthesopathy (88.2%) [11].

## 20.3.3 Ligament Injuries

ACS therapy has been investigated as a potential adjunctive treatment in ligament injuries to enhance the healing process and improve outcomes. Another hypothesis that IL-1 thought to be responsible is the bone lysis of the tibial tunnel after ACL surgery. Darabos et al. showed the postoperative intraarticular ACS injection reduces the tibial tunnel widening compared to placebo, but the clinical results about the laxity are open to debate [11].

## 20.3.4 Other Musculoskeletal Conditions

A study conducted by Wright-Carpenter et al. showed that muscle injuries can be treated with ACS, injecting into the affected muscle. It is compared to the Actovegin®/Traumeel® injections in athletes. Return to full training time is reduced by 5.7 days (16.6 vs 22.3 days) in the ACS group [12]. A mouse model study with a severe contusion on the gastrocnemius muscle showed 30 vs 48 hours accelerated tissue recovery with three injections of ACS compared to saline injections. IL-1a may inhibit the myogenic terminal differentiation, and IL-1Ra in ACS may block it to regenerate the muscle much faster [13].

## 20.3.5 Epidural Peri-radicular Injection for Back Pain

Radicular pain can be the cause of chronic inflammation of spinal nerve roots. Becker et al. demonstrated that ACS injection via epidural peri-radicular approach is superior to 5 mg and 10 mg triamcinolone groups after 6 months [14]. Becker et al. reported a study on patients with discopathies. Regardless of the size of the hernia, excellent and good results demonstrated with cervical discopathies with a 56% rate and lumbar discopathies with a 62% rate. However with the lumbar stenosis group, unsatisfactory results are much higher [15].

## 20.4  Application of ACS

Injection frequency and dosage with indication and application sites were given in Table 20.1. There are several products to prepare ACS at the market. In Orthokine® (Orthogen GmbH, Düsseldorf, Germany) system, 10 mL of patient's venous blood is taken and stored in a tube with glass spheres and incubated 24 h at 37 °C. Including IL-1Ra, anti-inflammatory cytokines which produced by leukocytes are found in

**Table 20.1** Injection frequency and dosage with indication and application site

| Application area | Total number of doses | Frequency in a week | Volume of doses (mL) |
|---|---|---|---|
| Hip joint | 4–6 | 2–3 | 2–4 |
| Knee joint | 6 | 2–3 | 2–4 |
| Ankle joint | 4 | 2–3 | 2–4 |
| Shoulder joint | 4 | 2–3 | 2–4 |
| Small joints | 4 | 2–3 | 0.5–2.5 |
| Tendinopathies | 4 | 1–2 | 1–4 |
| Muscle injuries | 5–6 | 3 | 2.5–5 |
| Post-surgery of tendon repairs (4–6 weeks later) | 4 | 1 | 0.5–2.5 |
| Cervical spine | 3–4 | 1–3 | 3–4 |
| Lumbar spine | 4–6 | 1–3 | 4 |

a richer amount in the serum. After the incubation, serum is centrifuged, and ACS is extracted and ready for use or can be stored at $\leq 18\ °C$, up to 12 months in the deep freezer.

A recent product, Idria S+® (Biological Innovations, Switzerland) prepares ACS from 30 mL of venous blood without any incubation and with only centrifugation claiming to get more IL-1Ra through small but highly surfaced glass beads and positive-charged ions.

## 20.5 Side Effects and Contraindications

There have been no serious side effects attributed to the ACS therapy. Side effect rate is nearly 1.3% and has the same frequency with the placebo injections like heat, swelling, and pain at the application area. It is advised to inject ACS with a 0.2-μm filter.

## 20.6 Future Perspectives and Conclusion

Autologous conditioned serum (ACS) therapy has shown promising results in the management of various musculoskeletal disorders. However, there are still several areas that warrant further investigation and hold potential for future advancements. Future research should focus on

optimizing the treatment protocols for ACS therapy. This includes determining the optimal concentration of growth factors and cytokines in ACS, refining the processing techniques to maximize their bioavailability, and identifying the most effective dosing regimens. By fine-tuning these parameters, clinicians can enhance the therapeutic efficacy of ACS therapy and improve patient outcomes.

Although there is growing clinical evidence supporting the use of ACS therapy, further research is needed to expand the evidence-base. Large-scale randomized controlled trials, long-term follow-up studies, and comparative effectiveness research can provide more robust evidence on the efficacy, safety, and cost-effectiveness of ACS therapy in various musculoskeletal conditions. Additionally, the exploration of novel applications, such as in cartilage repair or spinal disorders, can further expand the clinical utility of ACS therapy. With ongoing advancements and collaborative efforts, ACS therapy has the potential to revolutionize the management of musculoskeletal conditions, offering improved outcomes and quality of life for patients.

## References


1. Dinarello CA, Thompson RC. Blocking IL-1: interleukin 1 receptor antagonist in vivo and in vitro. Immunol Today. 1991;12(11):404–10.
2. Dinarello CA. Interleukin-1 and interleukin-1 antagonism. Blood. 1991;77(8):1627–52.
3. Granowitz EV, Clark B, Mancilla J, Dinarello C. Interleukin-1 receptor antagonist competitively inhibits the binding of interleukin-1 to the type II interleukin-1 receptor. J Biol Chem. 1991;266(22):14147–50.
4. Firestein G, Berger A, Tracey D, Chosay J, Chapman D, Paine M, et al. IL-1 receptor antagonist protein production and gene expression in rheumatoid arthritis and osteoarthritis synovium. J Immunol. 1992;149(3):1054–62.
5. Arend WP, Welgus H, Thompson RC, Eisenberg S. Biological properties of recombinant human monocyte-derived interleukin 1 receptor antagonist. J Clin Invest. 1990;85(5):1694–7.
6. Van Buul GM, Koevoet WL, Kops N, Bos PK, Verhaar JA, Weinans H, et al. Platelet-rich plasma releasate inhibits inflammatory processes in osteoarthritic chondrocytes. Am J Sports Med. 2011;39(11):2362–70.

7. Kon E, Mandelbaum B, Buda R, Filardo G, Delcogliano M, Timoncini A. Platelet-rich plasma versus hyaluronic acid viscosupplementation as treatments for cartilage pathology: from early degeneration to osteoarthritis. Arthroscopy. 2011;27:1490–501.

8. Hashemi M, Taheri M, Adlkhoo H, Dadkhah P, Abbasian MR. Comparison of the effect of intra-articular injection of autologous (orthokine) interleukin-1 receptor antagonist (il-1ra) and hyaluronic acid in pain control of knee osteoarthritis. Novelty Biomed. 2019;7(4):210–7.

9. Fotouhi A, Maleki A, Dolati S, Aghebati-Maleki A, Aghebati-Maleki L. Platelet rich plasma, stromal vascular fraction and autologous conditioned serum in treatment of knee osteoarthritis. Biomed Pharmacother. 2018;104:652–60.

10. Majewski M, Ochsner PE, Liu F, Flückiger R, Evans CH. Accelerated healing of the rat Achilles tendon in response to autologous conditioned serum. Am J Sports Med. 2009;37(11):2117–25.

11. Godek P, Szajkowski S, Golicki D. Evaluation of the effectiveness of orthokine therapy: retrospective analysis of 1000 cases. Ortop Traumatol Rehabil. 2020;22(2):107–19.

12. Darabos N, Haspl M, Moser C, Darabos A, Bartolek D, Groenemeyer D. Intraarticular application of autologous conditioned serum (ACS) reduces bone tunnel widening after ACL reconstructive surgery in a randomized controlled trial. Knee Surg Sports Traumatol Arthrosc. 2011;19:36–46.

13. Heisterbach PE, Todorov A, Flückiger R, Evans CH, Majewski M. Effect of BMP-12, TGF-β1 and autologous conditioned serum on growth factor expression in Achilles tendon healing. Knee Surg Sports Traumatol Arthrosc. 2012;20(10):1907–14.

14. Wright-Carpenter T, Klein P, Schäferhoff P, Appell H, Mir L, Wehling P. Treatment of muscle injuries by local administration of autologous conditioned serum: a pilot study on sportsmen with muscle strains. Int J Sports Med. 2004;25(8):588–93.

15. Becker C, Heidersdorf S, Drewlo S, de Rodriguez SZ, Krämer J, Willburger RE. Efficacy of epidural perineural injections with autologous conditioned serum for lumbar radicular compression: an investigator-initiated, prospective, double-blind, reference-controlled study. Spine. 2007;32(17):1803–8.

# Alpha-2-Macroglobulin Concentrate as Orthobiologic in Osteoarthritis

**21**

Peter A. Everts, Luga Podesta, José Fábio Lana, Gayan Poovendran, Gabriel Silva Santos, and Stephany Cares Huber

## 21.1 Introduction

Osteoarthritis (OA) is a degenerative and debilitating joint disease and is one of the most prevalent diseases in the United States [1]. OA is characterized by a painful inflammatory disease, causing progressive articular cartilage destruction, limiting patients in their activities. An increase in prevalence and incidence in osteoarthritis can be attributed to many factors, with age and obesity being the most frequent factors [2]. Currently, many orthobiologic treatments options are available to treat various osteoarthritic pathologies (Fig. 21.1). Nonsurgical treatment options are still limited, other than the use of autologous biologics like PRP and bone aspirate marrow concentrate (BMAC), or similar derived products [3]. It has been suggested that A2M, a serum protease inhibitor protein, inhibits the many endogenous and exogenous proteinases presenting in the pathogenesis of OA [4], despite the fact that only few clinical studies report on the application of A2M in OA pathologies [5]. Unfortunately, robust clinical trials are lacking regarding the A2M treatment efficacy and mid- to long-term results.

### 21.1.1 Safety and Contraindications

A2M proteins are naturally occurring macromolecules and are present in a low concentration in the circulation or in the synovial fluid of joints. In order to be an effective orthobiologic product, it has been suggested to use concentrated A2M as an orthobiologic injectate [6]. Zhu et al. postulated that A2M is an autologous proteinase inhibitor with no autoimmune rejection potential and only one active ingredient, while inhibiting various inflammatory factors and degenerative proteinase [7]. They mentioned no negative effects of A2M injections, or adverse events following the preparation method, which in part is very similar to the preparation of PRP.


P. A. Everts (✉)
Research and Education Division, Gulf Coast Biologics, Fort Myers, FL, USA

OrthoRegen Group, Max-Planck University, Indaiatuba, SP, Brazil
e-mail: peter@gulfcoastbiologics.com

L. Podesta
Bluetail Medical Group and Podesta Orthopedic Sports Medicine, Naples, FL, USA

J. F. Lana
OrthoRegen Group, Max-Planck University, Indaiatuba, SP, Brazil

G. Poovendran
Pro-Institute Poovendran Regenerative Orthopedics, Davie, FL, USA

G. S. Santos · S. C. Huber
Brazilian Institute of Regenerative Medicine (BIRM), Indaiatuba, Brazil



© The Author(s), under exclusive license to Springer Nature Switzerland AG 2024
B. Kocaoglu et al. (eds.), *Musculoskeletal Injections Manual*,
https://doi.org/10.1007/978-3-031-52603-9_21

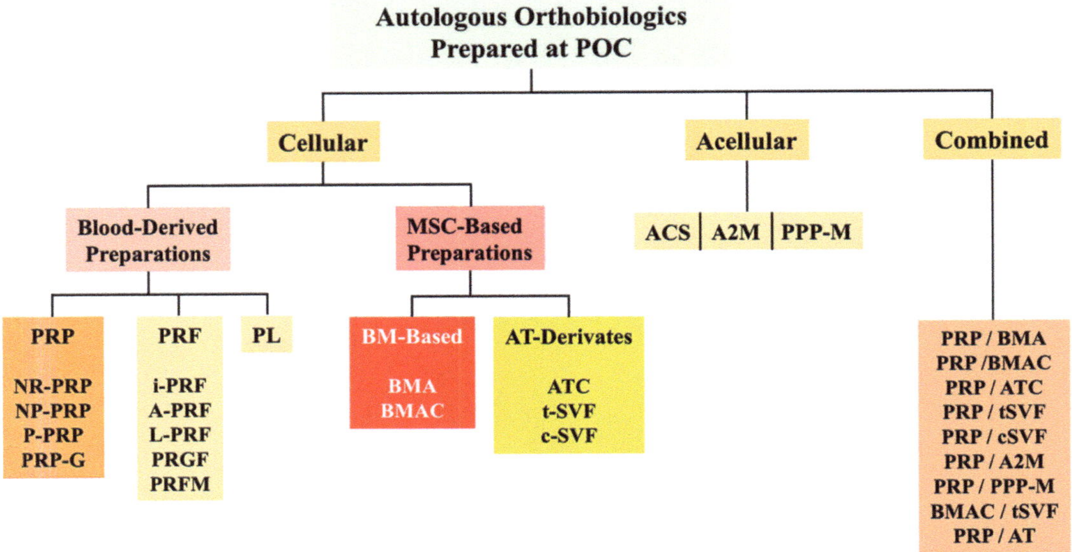

**Fig. 21.1** At point-of-care autologous orthobiologic preparations. Many orthobiologic products are available to treat a plethora of musculoskeletal pathologies. In this figure an overview is presented only from autologous prepared products, prepared at point-of-care, and a combination of this products. *POC* point-of-care, *MSC* mesenchymal stem cell, *ACS* autologous conditioned serum, *A2M* alpha-2-marcoglobulin, *PPP-M* platelet-poor plasma matrix, *PRP* platelet-rich plasma, *PRF* platelet-rich fibrin, *PL* platelet lysate, *BM* bone marrow, *AT* adipose tissue, *NR-PRP* neutrophil-rich PRP, *NP-PRP* neutrophil-poor PRP, *P-PRP* pure PRP, *PRP-G* PRP gel, *BMA* bone marrow aspirate, *BMAC* bone marrow aspirate concentrate, *ATC* adipose tissue complex, *t-SVF* tissue stromal vascular fraction, *c-SVF* cellular stromal vascular fraction, *AT* autologous thrombin

## 21.2 General Background A2M Protein Complex

Human A2M is a homo-tetrameric glycoprotein with four identical subunits each with a molecular mass of 180 kDa; consequently the entire A2M weight is more than 720 kDa. A2M is one of the largest plasma proteins in the human body, produced in the liver and present in plasma [8]. A2M proteins occur primarily in serum at 1.5 mg/ml in healthy patients, and at approximately a tenth of that concentration in healthy synovial fluid (SF) [6]. A2M acts as protease inhibitor, transporting hormones and enzymes; displays effector and inhibitor functions in the development of the lymphatic system; and hinders components of the complement system and hemostasis system. More specifically, A2M treatments are hypothesized to act as a treatment modality due to inhibitory properties, suppressing proteinates that are known to be harmful to cartilage.

## 21.2.1 Pharmacological Aspects A2M

A2M proteins interact with a broad range of proteolytic endo-proteinases that break peptide bonds of nonterminal amino acids, by acting as an inhibitor of active catalytic proteins A2M protein complexes have a broad inhibition range, mainly because of their unique mechanisms of action [9]. By forming a tetrameric cage around active proteases, A2M inhibits protease-substrate interaction physically, a mechanism referred to as the protease "snap-trap" mechanism [10]. This leads to the fact that proteases that are "trapped" by A2M are prevented from cleaving large substrate molecules (e.g., collagen), while small peptides (which escape the A2M cage) will be digested as normal. By cleaving the bait region with a protease, A2M becomes activated (a-A2M) and undergoes a conformational change, thereby trapping active proteases in its cage. Furthermore, along with sterically capturing proteases, a-A2M also exposes a reactive thioester that forms covalent A2M/protease complexes with small primary amines [11, 12]. However, during conformational rearrangements that take place during the transition into a-A2M, the receptor binding domains (RBDs) are exposed onto the surface of the protease-a-A2M complex during conformational rearrangements, wearying. In turn, this causes a-A2M complexes to be flagged by low-density lipoprotein receptor-related protein-1 for uptake by cells [13]. In Fig. 21.2, a graphical representation of the various A2M biological activation steps is presented.

## 21.2.2 Growth Factors, Cytokines, and A2M Application

It has been implied that the presence of A2M in SF acts to reduce concentrations of various proteinases harmful to cartilage, potentially attenuating OA symptoms and encouraging chondrogenesis. Studies on biochemical interaction revealed that many growth factors and cytokines bind to A2M-binding proteins. In particular, growth factors such as transforming growth factor (TGF)-β1, fibroblast growth factor (FGF)-2, macrophage activation factors (MACFs), and

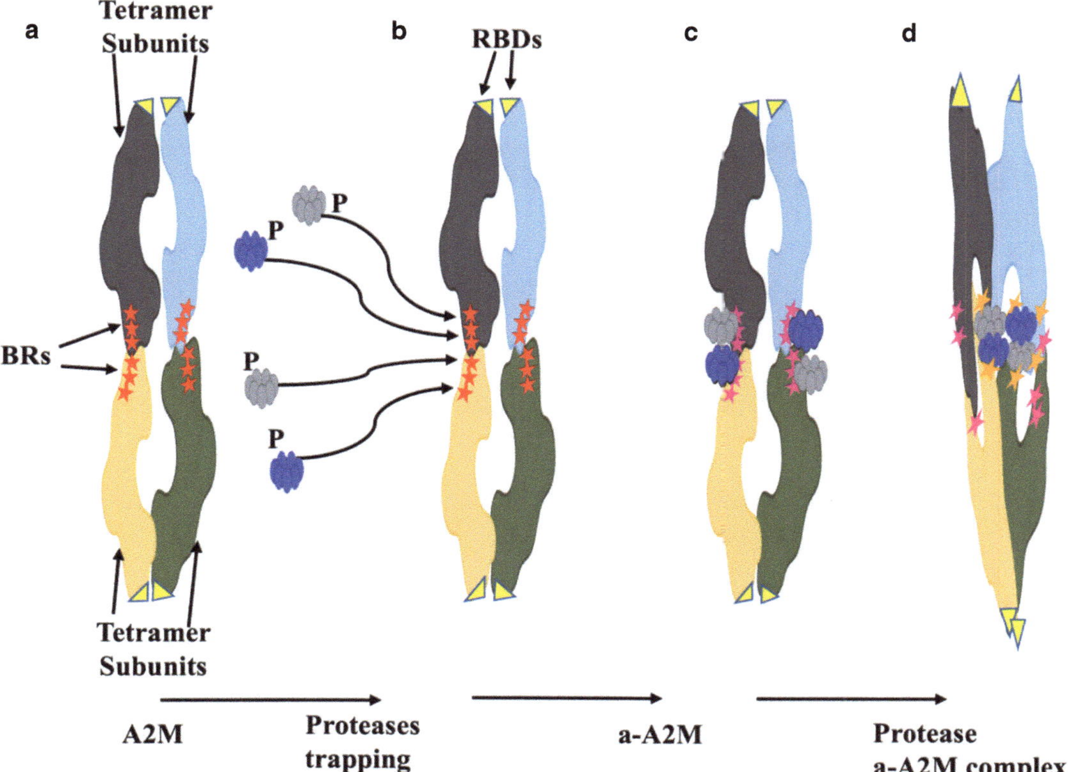



**Fig. 21.2** Illustration of the A2M interactions with proteases. The four monomer subunits of the nonactivated A2M tetramer, each 180 kDa in molecular weight, each with their own bait regions (BRs) (red stars), are shown in (**a**). Furthermore, each monomer contains receptor-binding domains (RBDs). In (**b**), active proteolytic endoproteinases (P) will cleave the BR, resulting in protease trapping and A2M activation (a-A2M), as shown in (**c**), followed by conformational rearrangement of the protein and subsequent development of covalently bonding between A2M and the protease residues (orange stars), setting up for an a-A2M-protease complex and change in RBDs, as they are now open to the A2M protein surface, represented in (**d**)

tumor necrosis factor-alpha (TNF-α) have a high binding affinity for a-A2M and reach a binding equilibrium within 15 min after binding to a-A2M [14–17]. Additionally, mild evidence hints toward the potential of A2M to inhibit activated matrix metalloprotease-13 (MMP-13) in vivo [18]. Other experimental and in vitro studies demonstrated that A2M can capture MMPs by the creation of A2M/MMP complexes [19] in the presence of their natural occurring tissue inhibitors of matrix metalloproteinase (TIMPs), even when there is a surplus of active MMPs [18]. Extracellular proteases known as a disintegrin and metalloproteinase with thrombospondin motifs (ADAMTS) are known for their roles in extra cellular matrix (ECM) degeneration and OA [20]. Furthermore, reports indicate that ADAMTS-1, ADAMTS-4, ADAMTS-5, ADAMTS-7, and ADAMTS-12 were inhibited by A2M, in a dose-dependent manner [20]. Cuellar et al. showed that A2M can additionally reduce cytokine-induced upregulation of collagenases in chondrocytes via trapping IL-1β and TNF-α [21].

## 21.3 Modulating Inflammation and Cartilage Degradation in OA

In OA, chronic inflammation contributes to pain mediated by TNF-α, interleukin (IL)-1β and IL-6, and other cytokines and chemokines [22]. Also, they downregulate ECM protein production, while inducing the up regulation of MMPs and ADAMTS leading to the degradation collagen [23, 24]. Subsequently, the degradative products of cartilage catabolism will stimulate the production of inflammatory proteases, in addition to cytokines, which further contributes to increases in inflammatory proteases, specifically elastase and cathepsins [25]. It has been shown that A2M decreases cartilage catabolism, inhibiting the protease activity of MMPs, ADAMTS, TNF-α, and IL-1β and other mediators of the pathologic cartilage catabolism process and ECM breakdown [6].

### 21.3.1 The Protective Effects of A2M in Synovial Fluid (SF)

Specific studies like Western blot analysis, protein mass spectrometry, ELISA, and immunohistochemistry have indicated that A2M is an element of joint synovial fluid (SF). However, the A2M concentration in SF is much lower than in serum, even in patients with knee OA [7]. Wang et al. demonstrated that the A2M concentration in SF is too low to adequately inhibit catabolic proteinases to counteract on intra-articular inflammation [26]. In this way, the concept behind employing c-A2M is based on the idea that its molecular structure enables it to perform a unique mechanistic function in binding inflammatory mediators in a specific way. OA inflammation and degradation might be limited by providing supplemental intra-articular c-A2M injections to facilitate chondrogenic and chondroprotective effects.

## 21.4 Procedural Steps Producing A2M as Orthobiologic Injectate

A2M preparation methods involve a unit of fresh autologous peripheral blood, a whole blood density separation procedure, and the use of ultrafiltration devices. PRP preparation methods will not be discussed here and are presented previously in this book.

### 21.4.1 Ultrafiltration Process

Microfiltration and ultrafiltration techniques have been used widely used for decades as blood-plasma contact devices in blood apheresis [27]. Later, these techniques were used to concentrate, increase, the hematocrit of diluted blood, in particular in infant cardiac surgery [28]. Thereafter, modified ultrafiltration techniques were successfully developed to alleviate inflammatory responses caused by cytokines [29]. Ultrafiltration techniques are sterile, convective processes, wherein PPP is either mechanically or manually, transported through a microporous membrane. Plasma proteins with a molecular mass, molecular weight, less than the pore size of are filtered out of the device. Ultrafiltration devices are made of high-performance medical-grade polymers, elastomers. Devices should offer an excellent biocompatibility and viscoelasticity, with weak intermolecular forces. Furthermore, they should present a consistent performance with a minimal residual volume in the device after processing. Ultra-filtrating plasma water will occur at a rate proportional to a transmembrane pressure gradient of the ultrafiltration device.

## 21.4.2 Clinical A2M Preparation Techniques

The preparation of concentrated A2M can be executed as an office-based preparation procedure, and requires approximately 60 minutes, depending on the PRP preparation technology used, plasma concentration method, and total whole blood volume needed to prepare the desired amount of concentrated plasma volume for the therapeutic application [30]. A2M therapy involves a unit of fresh peripheral blood that is processed by single or double spin PRP preparation devices, and ultimately PPP ultrafiltration techniques to eliminate plasma water, small proteins, proteases, and chemokines, while increasing the larger plasma protein concentrations (albumen, fibrinogen, and A2M).

A predetermined unit of whole blood is centrifugated, producing PRP, PPP, and a concentrated layer of erythrocytes. After the PRP density separation procedure [31], the PPP fraction is not discarded, but properly isolated and collected, using aseptic techniques. The PPP fraction is exposed to plasma ultrafiltration, employing dedicated disposable ultrafiltration systems, using either mechanical or manual preparation techniques to transport the PPP through the synthetic fibers of the ultrafiltration device. Ultrafiltration techniques will result in concentrating, yielding, large plasma proteins, like A2M and other proteins. The final c-A2M treatment vial is, ideally, image guided delivered to pathoanatomic tissues.

The initial technique for concentrating autologous plasma, producing A2M concentrate, with the intent to prepare an orthobiologic treatment product capable of regulating endogenous catabolic cytokines and inhibiting the activities of serine proteases and MMPs in OA, is a method described by Cuellar et al. They described the use of Autologous Platelet Integrated Concentration system (APIC™, Cytonics, Jupiter FL, USA), to prepare an orthobiologic injectate, as shown in Fig. 21.3 [32].

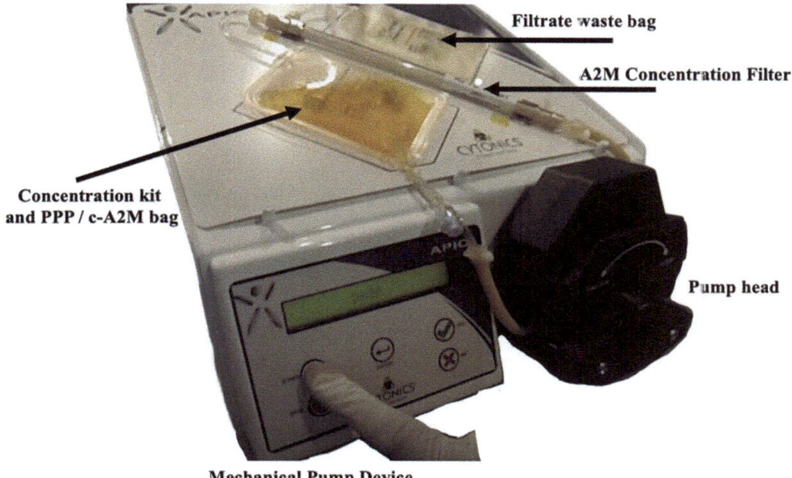


**Fig. 21.3** The Cytonics APIC™ A2M mechanical preparation setup. The Cytonics setup includes a concentration kit with a plasma concentration filter pump tubing and collection bags. The PPP is mechanically pumped through the filter to eliminate water and other plasma constituents that can pass the filter poor size. The manufacturer indicates that the filtration process only takes 20 min, apart from the whole blood centrifugation steps

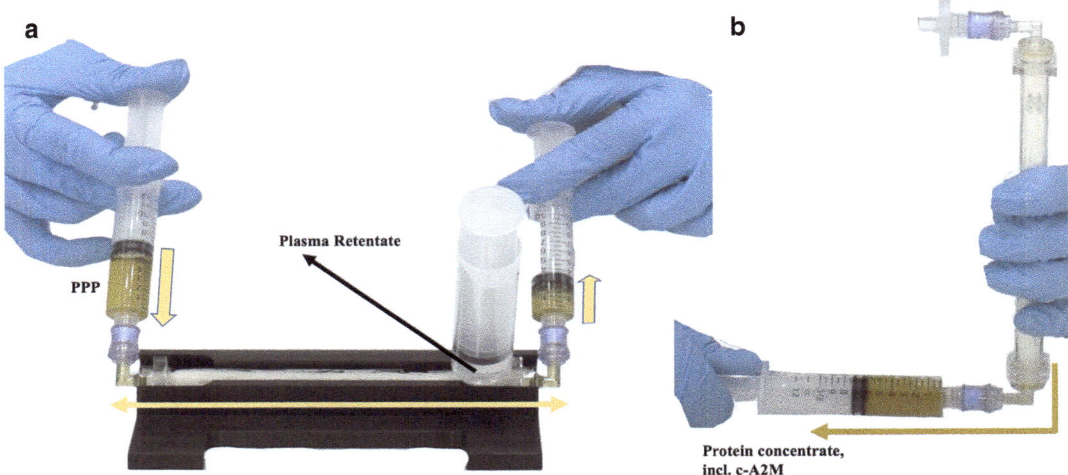


**Fig. 21.4** The EmCyte Corporation® CORE™ Ultrafiltration System for PPP protein concentrtion. In (**a**), the CORE™ ultrafiltration device is placed in a black table manifold for easy processing. The plasma in the syringes is injected back and forth through the device. The plasma retentate, or ultrafiltrate volume, is collected in an effluent syringe, attached to the effluent port. When the desired volume of protein concentrate (including A2M proteins) is reached, the protein concentrate can be extracted, while emptying the filter with an air filter attached, as shown in (**b**)

Thereafter, more ultrafiltration devices have entered the orthobiologic market to concentrate PPP following a PRP procedure to prepare c-A2M, along with other large plasma proteins. Figure 21.4 illustrates the latest ultrafiltration device, using a simple, pumpless manual force syringe technique for the ultrafiltration process to produce protein concentrates, that includes A2M (Core™ ultrafiltration device, EmCyte Corp, Fort Myers FL).

## 21.5 Clinical Data

There are several distinct clinical applications of employing autologous concentrated A2M, including the treatment of painful intra-articular and extra-articular joints, following mild to moderate OA of the knee, hip, and shoulder, as well as spinal discogenic pain [33, 34].

Human clinical trials addressing the efficacy of concentrated A2M are scarce. According to Clinical Trials.gov, some studies are underway and yet to be published [3]. However, several animal studies have reported positive outcomes [22, 35]. Klein et al., in 75 patients randomized controlled clinical trial, patients had symptomatic knee osteoarthritis with Kellgren-Lawrence grade 2 or 3, revealed that c-A2M was not significantly better than regular PRP, assessing visual analog scale scores, Western Ontario and McMaster Universities Osteoarthritis Index (WOMAC), Knee Injury and Osteoarthritis Outcome Score (KOOS), and Lysholm and Tegner scores [36].

Nevertheless, large clinical studies have not (yet) been performed and are needed to assess its true potential benefits for OA, including comparative studies using accepted and well-published orthobiologic products like PRP and mesenchymal stem cell-based tissue concentrates [37].

## 21.6 Future Directions

Early data from experimental and small clinical case series are encouraging, indicating the potential of c-A2M supplementation in a variety of musculoskeletal disorders. Larger, prospective, and comparative clinical trials need to be designed to further support and enhance the current available preliminary results.

Based on available scientific data, new research and scientific steps should be directed towards further understanding the activities of proteolytic endo-proteinases and application of c-A2M mechanisms, effective concentrations, and activation about a better comprehension of chronic nociceptive inflammatory tissues, modulating the local tissue immune microenvironment, related to cartilage pathologies. This includes further characterization of anti-proteinase proteins to specifically target mechanisms responsible for joint pathophysiology, disease progression, and hopefully an ability for tissue repair.

Lastly, A2M is easily isolated as a major contributor of PRP therapies. The combination of validated and effective PRP therapies with concentrated PPP, including c-A2M, represents an exciting scientific opportunity to investigate potential combined anti-inflammatory, pain modulatory, and tissue repair properties in a single injectate. Developing new safe and effective biotherapy treatment modalities has the potential to increase the number of patient and pathology specific treatment options. This will enable physicians to deliver more multicellular orthobiologics to chronic and degenerated musculoskeletal pathologies [38].

# References


1. Andia I, Atilano L, Maffulli N. Moving toward targeting the right phenotype with the right platelet-rich plasma (PRP) formulation for knee osteoarthritis. Ther Adv Musculoskelet Dis. 2021;13:1759720X2110043. https://doi.org/10.1177/1759720X211004336.
2. Wallace IJ, Worthington S, Felson DT, Jurmain RD, Wren KT, Maijanen H, et al. Knee osteoarthritis has doubled in prevalence since the mid-20th century. Proc Natl Acad Sci. 2017;114(35):9332–6.
3. Patel S, Jindal K, Dhillon M. The future of injectable orthobiologic substances for knee osteoarthritis: options beyond platelet-rich plasma. J Musculoskelet Surg Res. 2020;4(4):173. Available from: https://journalmsr.com/the-future-of-injectable-orthobiologic-substances-for-knee-osteoarthritis-options-beyond-platelet-rich-plasma/.
4. Cuellar JM. Intradiscal injection of an autologous alpha-2-macroglobulin (A2M) concentrate alleviates back pain in FAC-positive patients. Orthop Rheumatol Open Access J. 2017;4(2):42–6. Available from: https://juniperpublishers.com/oroaj/OROAJ.MS.ID.555634.php.
5. Thompson K, Klein D, Campbell K, Gonzlez-Lomas G, Alaia M, Strauss E, et al. The effectiveness of alpha-2-macroglobulin injections for osteoarthritis of the knee. Arthrosc J Arthrosc Relat Surg. 2021;37(1):e80. Available from: https://linkinghub.elsevier.com/retrieve/pii/S0749806320312238.
6. Wang S, Wei X, Zhou J, Zhang J, Li K, Chen Q, et al. Identification of $\alpha_2$-macroglobulin as a master inhibitor of cartilage-degrading factors that attenuates the progression of posttraumatic osteoarthritis: $\alpha_2$ M attenuates posttraumatic OA progression. Arthritis Rheumatol. 2014;66(7):1843–53. https://doi.org/10.1002/art.38576.
7. Zhu M, Zhao B, Wei L, Wang S. Alpha-2-macroglobulin, a native and powerful proteinase inhibitor, prevents cartilage degeneration disease by inhibiting majority of catabolic enzymes and cytokines. Curr Mol Biol Rep. 2021;7(1):1–7. https://doi.org/10.1007/s40610-020-00142-z.
8. Vandooren J, Itoh Y. Alpha-2-macroglobulin in inflammation, immunity and infections. Front Immunol. 2021;12:803244. https://doi.org/10.3389/fimmu.2021.803244/full.
9. Armstrong PB, Quigley JP. $\alpha$2-macroglobulin: an evolutionarily conserved arm of the innate immune system. Dev Comp Immunol. 1999;23(4-5):375–90.
10. Marrero A, Duquerroy S, Trapani S, Goulas T, Guevara T, Andersen GR, et al. The crystal structure of human $\alpha$2-macroglobulin reveals a unique molecular cage. Angew Chem Int Ed. 2012;51(14):3340–4.
11. Salvesen GS, Barrett AJ. Covalent binding of proteinases in their reaction with $\alpha$2-macroglobulin. Biochem J. 1980;187(3):695–701.
12. Goulas T, Garcia-Ferrer I, Marrero A, Marino-Puertas L, Duquerroy S, Gomis-Rüth FX. Structural and functional insight into pan-endopeptidase inhibition by $\alpha$2-macroglobulins. Biol Chem. 2017;398(9):975–94.
13. Kristensen T, Moestrup SK, Gliemann J, Bendtsen L, Sand O, Sottrup-Jensen L. Evidence that the newly cloned low-density-lipoprotein receptor related protein (LRP) is the $\alpha$2-macroglobulin receptor. FEBS Lett. 1990;276(1-2):151–5.
14. McDaniel MC, Laudico R, Papermaster BW. Association of macrophage-activation factor from a human cultured lymphoid cell line with albumin and $\alpha$2-macroglobulin. Clin Immunol Immunopathol. 1976;5(1):91–104.
15. Dennis P, Saksela O, Harpel P, Rifkin D. $\alpha$2-macroglobulin is a binding protein for basic fibroblast growth factor. J Biol Chem. 1989;264(13):7210–6.
16. Huang SS, O'Grady P, Huang J. Human transforming growth factor beta. alpha 2-macroglobulin complex is a latent form of transforming growth factor beta. J Biol Chem. 1988;263(3):1535–41.
17. Danielpour D, Sporn MB. Differential inhibition of transforming growth factor beta 1 and beta 2 activity by alpha 2-macroglobulin. J Biol Chem. 1990;265(12):6973–7.

18. Wan R, Hu J, Zhou Q, Wang J, Liu P, Wei Y. Application of co-expressed genes to articular cartilage: new hope for the treatment of osteoarthritis. Mol Med Rep. 2012;6(1):16–8.

19. Tchetverikov I. Matrix metalloproteinases-3, -8, -9 as markers of disease activity and joint damage progression in early rheumatoid arthritis. Ann Rheum Dis. 2003;62(11):1094–9. https://doi.org/10.1136/ard.62.11.1094.

20. Luan Y, Kong L, Howell DR, Ilalov K, Fajardo M, Bai XH, et al. Inhibition of ADAMTS-7 and ADAMTS-12 degradation of cartilage oligomeric matrix protein by alpha-2-macroglobulin. Osteoarthr Cartil. 2008;16(11):1413–20. Available from: https://linkinghub.elsevier.com/retrieve/pii/S1063458408000940.

21. Cuéllar JM, Cuéllar VG, Scuderi GJ. α2-macroglobulin. Phys Med Rehabil Clin N Am. 2016;27(4):909–18. Available from: https://linkinghub.elsevier.com/retrieve/pii/S1047965116300493.

22. Lefebvre V, Peeters-Joris C, Vaes G. Modulation by interleukin 1 and tumor necrosis factor α of production of collagenase, tissue inhibitor of metalloproteinases and collagen types in differentiated and dedifferentiated articular chondrocytes. Biochim Biophys Acta BBA Mol Cell Res. 1990;1052(3):366–78. Available from: https://www.sciencedirect.com/science/article/pii/0167488990901454.

23. Saklatvala J. Tumour necrosis factor α stimulates resorption and inhibits synthesis of proteoglycan in cartilage. Nature. 1986;322(6079):547–9. https://doi.org/10.1038/322547a0.

24. Saal JA, Saal JS. Nonoperative treatment of herniated lumbar intervertebral disc with radiculopathy: an outcome study. Spine. 1989;14(4):431–7. Available from: https://journals.lww.com/spinejournal/Fulltext/1989/04000/Nonoperative_Treatment_of_Herniated_Lumbar.18.aspx.

25. Miller RE, Lu Y, Tortorella MD, Malfait AM. Genetically engineered mouse models reveal the importance of proteases as osteoarthritis drug targets. Curr Rheumatol Rep. 2013;15:1–15.

26. Sun C, Cao C, Zhao T, Guo H, Fleming BC, Owens B, et al. A2M inhibits inflammatory mediators of chondrocytes by blocking IL-1β/NF-κB pathway. J Orthop Res. 2023;41(1):241–8.

27. Senthilkumar S, Rajesh S, Mohan D, Soundararajan P. Preparation, characterization, and performance evaluation of poly(ether-imide) incorporated cellulose acetate ultrafiltration membrane for hemodialysis. Sep Sci Technol. 2013;48(1):66–75. https://doi.org/10.1080/01496395.2012.674603.

28. Lawson DS, Smigla GR, Shearer IR, Ing R, Schulman S, Kern F, et al. A clinical comparison of two commercially available pediatric hemoconcentrators. J Extra Corpor Technol. 2004;36(1):66–8.

29. Suzuki H, Oshima N, Watari T. Effect of modified ultrafiltration on cytokines and hemoconcentration in dogs undergoing cardiopulmonary bypass. J Vet Med Sci. 2020;82(11):1589–93. Available from: https://www.jstage.jst.go.jp/article/jvms/82/11/82_20-0143/_article.

30. Everts PA, Mazzola T, Mautner K, Randelli PS, Podesta L. Modifying orthobiological PRP therapies are imperative for the advancement of treatment outcomes in musculoskeletal pathologies. Biomedicine. 2022;10(11):2933. Available from: https://www.mdpi.com/2227-9059/10/11/2933.

31. Everts P, Onishi K, Jayaram P, Lana JF, Mautner K. Platelet-rich plasma: new performance understandings and therapeutic considerations in 2020. Int J Mol Sci. 2020;21(20):7794. Available from: https://www.mdpi.com/1422-0067/21/20/7794.

32. Cuéllar JM, Cuéllar VG, Scuderi GJ. α2-macroglobulin: autologous protease inhibition technology. Phys Med Rehabil Clin. 2016;27(4):909–18.

33. Oshita H, Sandy JD, Suzuki K, Akaike A, Bai Y, Sasaki T, et al. Mature bovine articular cartilage contains abundant aggrecan that is C-terminally truncated at Ala719-Ala720, a site which is readily cleaved by m-calpain. Biochem J. 2004;382(1):253–9.

34. Homandberg GA, Wen C, Hui F. Cartilage damaging activities of fibronectin fragments derived from cartilage and synovial fluid. Osteoarthr Cartil. 1998;6(4):231–44.

35. Kanbe K, Takemura T, Takeuchi K, Chen Q, Takagishi K, Inoue K. Synovectomy reduces stromal-cell-derived factor-1 (SDF-1) which is involved in the destruction of cartilage in osteoarthritis and rheumatoid arthritis. J Bone Joint Surg Br. 2004;86(2):296–300.

36. Klein D, Bloom D, Campbell K, Gonzalez-Lomas G, Alaia M, Strauss E, et al. Alpha-2-macroglobulin not significantly better than regular PRP for knee arthritis symptoms. Orthop J Sports Med. 2020;8(7_suppl6):2325967120S0045. https://doi.org/10.1177/2325967120S00454.

37. Everts PA, Panero AJ. Basic science of autologous orthobiologics: part 2. Mesenchymal stem cells. Orthobiologics. 2023;34(1):25–47. Available from: https://www.sciencedirect.com/science/article/pii/S1047965122000730.

38. Lana JF, Purita J, Everts PA, De Mendonça Neto PAT, De Moraes Ferreira Jorge D, Mosaner T, et al. Platelet-rich plasma power-mix gel (ppm)—an orthobiologic optimization protocol rich in growth factors and fibrin. Gels. 2023;9(7):553. Available from: https://www.mdpi.com/2310-2861/9/7/553.

# Part IV

# Injections of Anatomical Regions and Diseases

# Injections of Anatomical Regions and Diseases: Shoulder

Mocini Fabrizio, Candura Dario,
Proietti Lorenzo, Ciolli Gianluca,
Brancaccio Vincenzo, and Cerciello Simone

## 22.1 Sternoclavicular Joint

### 22.1.1 Anatomy and Biomechanics

The sternoclavicular joint (SCJ) is the only joint between the axial skeleton and the upper limb. It is a saddle-shaped diarthrodial joint with the two articular surfaces covered with hyaline cartilage. The joint is concave in the vertical axis and convex in the anteroposterior axis. The articulating surfaces are separated by an articular disc. The joint is stabilized medially by the anterior and posterior sternoclavicular ligaments and laterally by the interclavicular and costoclavicular ligaments. The anterior and posterior sternoclavicular ligaments fix the clavicle to the sternum and provide anterior and posterior stability to the joint. The interclavicular ligament connects the two clavicles together, while the costoclavicular ligament connects the clavicle to the first rib. Vascular supply to the SCJ comes from the suprascapular artery and internal thoracic artery. Nerve supply of the joint is from the medial suprascapular nerve and the nerve to the subclavius. The normal SCJ achieves 35° of movement both in the coronal and horizontal plane during shoulder abduction along with 45° of rotation. The muscular stabilizers which also move the SC joint include the sternocleidomastoid, pectoralis major, deltoid, trapezius, and rhomboids. These muscles work together to stabilize the joint during movements of the upper limb [1].

### 22.1.2 Pathologies and Indication for Injections

The SCJ can be affected by many pathologic conditions such as dislocation and subluxation (anterior and less frequently posterior), synovitis-acne-pustulosis-hyperostosis-osteitis (SAPHO), avascular necrosis (Friedrich disease), rheumatic arthritis, and crystalline arthropathies. Besides these, the most common pathologic condition of the SCJ is osteoarthritis (OA), which is asymptomatic in most cases [2]. Cadaveric studies in patients aged over 60 years have shown OA to be present in over 50% of subjects. Pain caused by OA of the SCJ is typically increased with abduction or forward flexion beyond 90°. The treatment is usually conservative with NSAIDs and/or steroid injections [3]. Surgery is recommended for symptomatic patients unresponsive to conservative treatment. Surgical options are

M. Fabrizio (✉) · P. Lorenzo
Casa di Cura Villa Betania GIOMI, Rome, Italy

C. Dario · B. Vincenzo
Fondazione Universitaria Policlinico Agostino Gemelli, Roma, Italy

C. Gianluca · C. Simone
Casa di Cura Villa Betania GIOMI, Rome, Italy

Fondazione Universitaria Policlinico Agostino Gemelli, Roma, Italy


B. Kocaoglu et al. (eds.), *Musculoskeletal Injections Manual*,
https://doi.org/10.1007/978-3-031-52603-9_22

arthroscopic or open resection arthroplasty. SCJ injections can be also performed as diagnostic purpose: the patient's pain is relieved following the injection; it can be concluded that the joint is the source of the pain.

### 22.1.3 Appliances

The type of injector, needle, and syringe used depends on the preference of the healthcare provider and the specific pathologic conditions. Usually, a 25- to 27-G, 25- to 38-mm needle is used, and the SCJ is injected with 1–3 mL of injectate to avoid overdistension.

### 22.1.4 Agents

Corticosteroids, local anesthetics, visco-supplements, and orthobiologic agents.

### 22.1.5 Injection Technique

The patient is placed in supine position with the head turned to the opposite side and the top of the bed elevated at 30–40°. In most cases, the clavicle is more prominent than the sternum, so a step is easily palpable as a landmark of the SCJ. In the case of difficulty in identifying the SCJ due to anatomical abnormalities, the shoulder can be passively abducted and externally rotated in order to increase the joint space. Sterile technique is applied, and structures are marked. A small amount of local anesthetic may be injected to numb the skin and deeper tissues around the injection site. Once the SCJ is localized, a 25- to 27-G 25- to 38-mm needle is introduced with an anterior-to-posterior approach, and the SCJ is injected with usually 1–3 mL of injectate to avoid overdistension (Fig. 22.1). An overall accuracy of 78% for palpation-guided SCJ injections in cadavers has been demonstrated [4]. To facilitate visualization of the SCJ and increase accuracy, injections can be performed under CT or, more easily, ultrasound guidance. The SCJ is best visualized using a high-frequency (>10 MHz) linear transducer placed long axis to the clavicle (short axis to the joint) with the center of the joint in the center of the transducer. Color Doppler can be helpful to identify the carotid artery which is located deep to the SCJ [5]. The needle is then removed, and a sterile dressing is applied to the injection site.

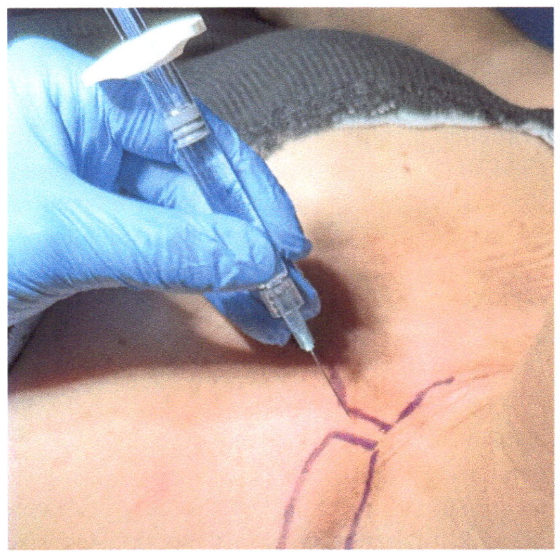
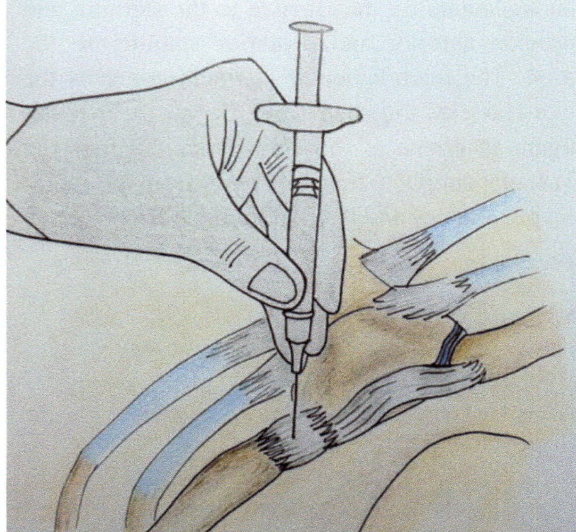

**Fig. 22.1** Sternoclavicular joint injection: patient placed in supine position with the head turned to the opposite side and the top of the bed at 30–40°, the joint is identified by palpation, and 1–3 mL are injected with a 25- to 27-G 25- to 38-mm with an anterior-to-posterior approach

### 22.1.6 Aftercare

The patient is observed for a while. In the first 24–48 h, information is given about the symptoms that may worsen. Rest is recommended for at least 24–48 h after the injection. Avoiding activities that may stress the joint, such as heavy lifting, is recommended. Applying ice to the injection site for 20 min at a time, several times a day, for the first 48 h can help reduce pain and swelling. Pain medication such as paracetamol or NSAIDs can help relieving pain. It is obviously important to watch out for side effects in the injection site such as redness, swelling, or drainage that could indicate an infection. Depending on the underlying condition that led to the sternoclavicular joint injection, physical therapy may be recommended to help improve joint mobility and strength.

## 22.2  Acromioclavicular Joint

### 22.2.1 Anatomy and Biomechanics

The acromioclavicular joint (ACJ) is a small diarthrodial joint that connects the distal clavicle to the acromion process of the scapula. The two articular extremities are separated by a fibrocartilaginous disc, and the joint is surrounded by a synovial membrane. The acromioclavicular ligaments provide horizontal stability to the joint, whereas the coracoclavicular ligaments (trapezoid and conoid) provide vertical stability. Additional stabilization is provided by coracoacromial ligament and the attachments of the deltoid and trapezius muscles [6]. The biomechanics of the ACJ is complex and involves multiple structures: during shoulder abduction, the scapula rotates upward and outward, and the clavicle moves upward and rotates posteriorly at the AC joint. Moreover, the ACJ transfers forces from the arm to the trunk during activities such as throwing or pushing, allowing the arm to generate more force [7].

### 22.2.2 Pathologies and Indication for Injections

The main indications for injection of the ACJ are rheumatic arthritis, osteolysis of the distal clavicle, and mainly primary or post-traumatic OA. Patients with chronic ACJ pain are usually older than 40 years and complain a focal pain located over the superior aspect of the ACJ. The pain is insidious and exacerbated with direct palpation of ACJ or with cross-body adduction of the shoulder. Patients may also describe a grinding sensation. Conservative treatment with physical therapy, NSAIDs, and intraarticular injections represents the first therapeutic approach in most cases [8]. Surgery is recommended for symptomatic patients unresponsive to conservative measures and include arthroscopic and open distal clavicle excision, but the current scientific literature has not shown significant differences between conservative and surgical treatment [9]. ACJ injections can be both diagnostic and therapeutic, resulting in significant pain relief for affected patients. The duration can last weeks to months, and it is variable among patients. Steroid injections can be repeated, but multiple injections should be avoided because they can result in degradation of joint cartilage over time.

### 22.2.3 Appliances

The type of injector, needle, and syringe used depends on the preference of the healthcare provider and the specific condition being treated. Usually, a 22- to 27-G 25- to 38-mm curved or straight needle is used for the injection of a volume of 5–10 mL.

### 22.2.4 Agents

Corticosteroids, local anesthetics, viscosupplements, and orthobiologic agents.

## 22.2.5 Injection Technique

The patient is placed in supine or seated position with the arm at the side. To identify the ACJ, palpate the clavicle distally until a small depression just lateral to termination of the bone. The ACJ has a very variable anatomy, further worsened by the OA changes [10]. After aseptic preparation of the skin, a local anesthetic can be used to numb the skin and underlying tissues to reduce discomfort during the injection. Insert a 22- to 27-G 25- to 38-mm needle from the superior-anterior approach with the needle perpendicular to the joint (Fig. 22.2). The needle should not meet any resistance. Once in the joint, if fluid is present in the joint and creates overdistension, the provider may use the needle to withdraw the fluid for testing or to relieve pressure in the joint. Any solution must be injected slowly. Accuracy of blind needle placement for ACJ is very low, since it was found to be about 40% [11]. Fluoroscopy or ultrasound, if available, can significantly improve the accuracy of ACJ injections up to 100% [12]. A high-frequency (>10 MHz) linear-array transducer is placed short axis to the clavicle and moved laterally until the ACJ; then the needle is introduced and described before but now under direct visualization, and the ACJ is injected under with the desired amount of injectate (usually <10 mL) avoiding excess overdistension of the joint.

## 22.2.6 Aftercare

The patient is observed for a while. In the first 24–48 h, information is given about the symptoms that may worsen. Rest is recommended for at least 24–48 h after the injection. Avoid activities that may stress the joint, such as heavy lifting. Applying ice to the injection site for 20 min at a time, several times a day, for the first 48 h can help reducing pain and swelling. Pain medication such as paracetamol or NSAIDs can help relieve pain. It is obviously important to watch out for side effects in the injection site such as redness, swelling, or drainage that could indicate an infection. Depending on the underlying condition that led to the ACJ injection, physical therapy may be recommended to help improve joint mobility and strength.

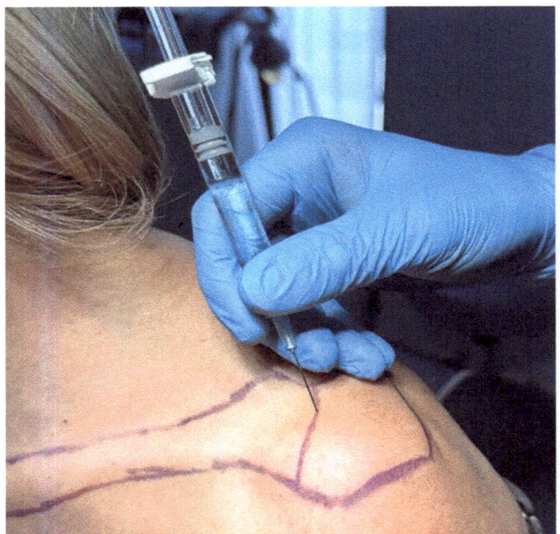

**Fig. 22.2** Acromioclavicular joint injection: patient placed in seated position, the joint is identified by palpation, and the fluid is injected with a 22- to 27-G 25- to 38-mm needle from the superior-anterior approach perpendicular to the joint

## 22.3 Subacromial Bursa

### 22.3.1 Anatomy and Biomechanics

The subacromial bursa (SB) is a virtual synovial space that is located deep to the deltoid muscle, acromion, and the coracoacromial (CA) ligament and superficial to the supraspinatus tendon, rotator interval (RI), greater tuberosity, and intertubercular groove [13]. The bursa is composed of subacromial and subdeltoid portions, with a subcoracoid extension in some cases. The subdeltoid bursa lies between the deltoid muscle and the acromion, while the subacromial bursa proper lies between the rotator cuff tendons and the acromion. The anterior portion of the bursa under the CA arch has a significant role in outlet impingement [14]. The SB is about 1 mm in thickness in normal conditions, and it is covered superficially by a layer of peri bursal fat. During movement of the shoulder joint, the rotator cuff tendons slide back and forth beneath the acromion. The subacromial bursa acts as a lubricated cushion, reducing friction and preventing damage to the rotator cuff tendons. It also helps to distribute the force of movement evenly across the joint.

### 22.3.2 Pathologies and Indication for Injections

SB bursopathy is a very common pathologic condition that can be a primary or secondary cause of shoulder pain. The pathologic conditions most often associated are subacromial impingement and rotator cuff tears. The use of subacromial injections of corticosteroid and local anesthetic represents an inexpensive and efficacious way both to diagnose and treat symptomatic rotator cuff disease and subacromial impingement. Although corticosteroids will not heal the rotator cuff tear, they can reduce the associated inflammation of the bursa and the tendons, reducing pain and allowing patients their activities of daily living (ADL) [15]. Subacromial infiltrations may also play a role in calcific tendinopathy. Although not validated in the literature, many authors agree to the use of subacromial infiltrations with cortisone in the acute phase of calcific tendinopathy together with NSAIDs and physiotherapy for the maintenance of ROM [16]. Finally, subacromial corticosteroid injections have also been described for the treatment of adhesive capsulitis with results comparable to the glenohumeral joint injections [17]. In addition, hyaluronic acid (HA) injections have also been described in the attempt of a conservative approach of cuff and bursal disorders. They have proven to be a valuable safe alternative to other conservative methods for the treatment of rotator cuff disorders [18].

### 22.3.3 Appliances

The type of injector, needle, and syringe used depends on the preference of the healthcare provider and the specific condition being treated. Usually, a 21- to 25-G 38-mm needle is used for the injection of a volume of 5–10 mL.

### 22.3.4 Agents

Corticosteroids, local anesthetics, viscosupplements, and orthobiologic agents.

### 22.3.5 Injection Technique

The most common sites of injection are the standard posterior and anterolateral (lateral) arthroscopy portals.

***Posterior Approach*** The patient is placed in seated or prone position with the affected arm in a relaxed position. After palpatory identification of the posterolateral corner of the acromion, insert the needle about 2 cm inferiorly and 1 cm medially (soft spot), and direct it anteriorly aiming upward at a 10° angle (Fig. 22.3). Gently pull back on the plunger as you advance to rule out intravascular placement. Slowly inject and withdraw the needle.

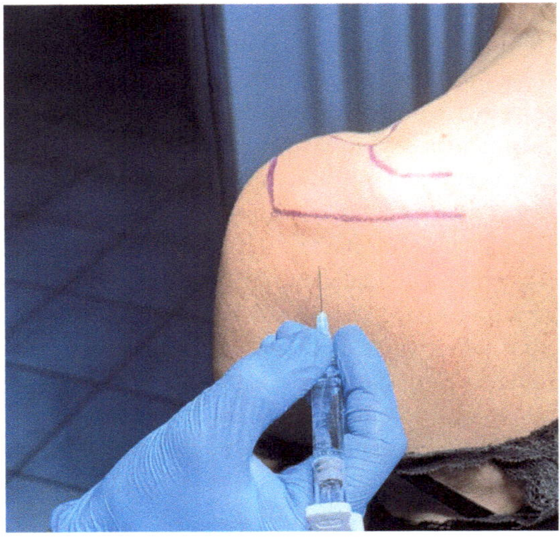

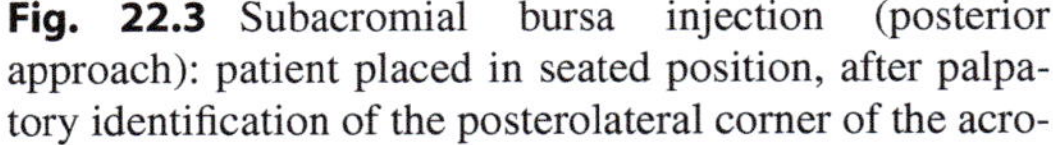

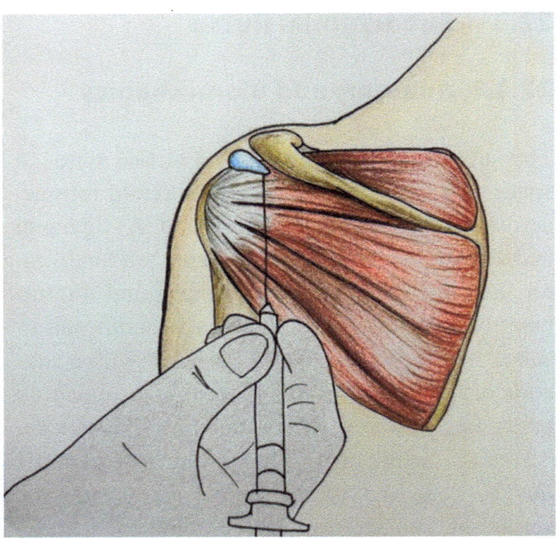

**Fig. 22.3** Subacromial bursa injection (posterior approach): patient placed in seated position, after palpatory identification of the posterolateral corner of the acromion, a 21- to 25-G 38-mm needle is inserted about 2 cm inferiorly and 1 cm medially and directed anteriorly upward at a 10° angle

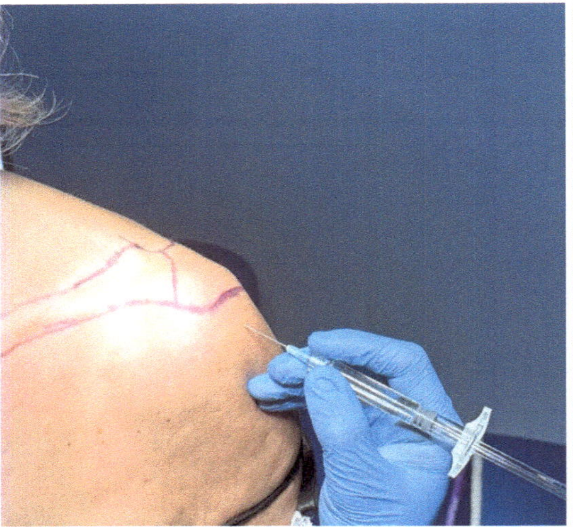

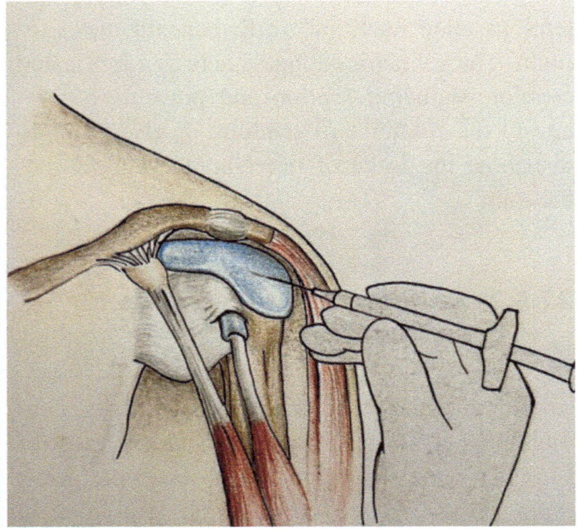

**Fig. 22.4** Subacromial bursa injection (lateral approach): patient placed in seated position, after palpatory identification of the lateral border of the acromion, a 21- to 25-G 38-mm needle is inserted about 2 cm below the border and 1 cm posterior to the anterior edge of the acromion

***Lateral Approach*** The patient is placed in seated or lateral decubitus on the contralateral side. After palpatory identification of the lateral border of the acromion, insert the needle about 2 cm below the border and 1 cm posterior to the anterior edge of the acromion (Fig. 22.4). If the needle or the injections meet resistance, the needle may be in contact with the acromion or within the supraspinatus tendon. Partially withdraw the needle, and then readvance in a slightly different direction so as not to encounter any resistance.

The provider may withdraw any fluid that may be present in the bursa for diagnostic purposes or to relieve pressure within the joint. Once the needle is properly positioned in the subacromial bursa, the fluid is gently injected into the bursa.

The accuracy of the blind subacromial injections is about 80%, and no difference has been demonstrated between the two approaches [14]. To improve the visualization of the anatomical structures and increase the accuracy, the procedure can be performed under ultrasound guidance (Fig. 22.5). Being a superficial structure, the SB is best visualized using a high-frequency (>10 MHz) linear-array transducer placed long axis to the supraspinatus tendon fibers. The needle is introduced under direct visualization of the ultrasound image to ensure proper placement within the SB, and the desired amount of drug is injected.

### 22.3.6 Aftercare

The patient is observed for a while. In the first 24–48 h, information is given about the symptoms that may worsen. Rest is recommended for at least 24–48 h after the injection. Avoid activities that involve heavy lifting or overhead movements for a few days to a few weeks following the injection. Gentle range of motion exercises are recommended to help prevent stiffness in the shoulder joint. Applying ice to the injection site for 20 min at a time, several times a day, for the first 48 h can help reduce pain and swelling. Pain medication such as paracetamol or NSAIDs can help relieve pain. It is obviously important to watch out for side effects in the injection site such as redness, swelling, or drainage that could indicate an infection. Depending on the underlying condition that led to the SB injection, physical therapy may be recommended to help improve joint mobility and strength.

## 22.4 Glenohumeral Joint

### 22.4.1 Anatomy and Biomechanics

The glenohumeral joint (GHJ) is a diarthrodial multiaxial joint with a ball-and-socket configuration. It is composed of the hyaline-covered spherical humeral head and the flat hyaline-covered glenoid fossa of the scapula with a fibrocartilaginous labrum that increases the congruence and the stability of the joint. The joint is surrounded by a fibrous capsule that is reinforced by ligaments that attach the humerus to the scapula. Moreover, there are several joint recesses associated including the axillary and the subscapular recesses and the LHBBT sheath [19, 20].

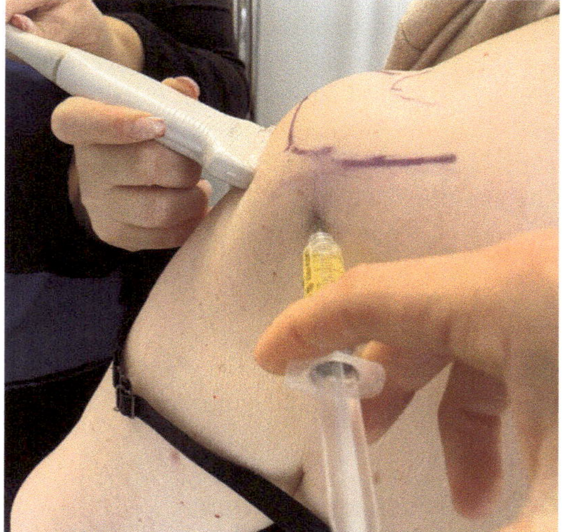

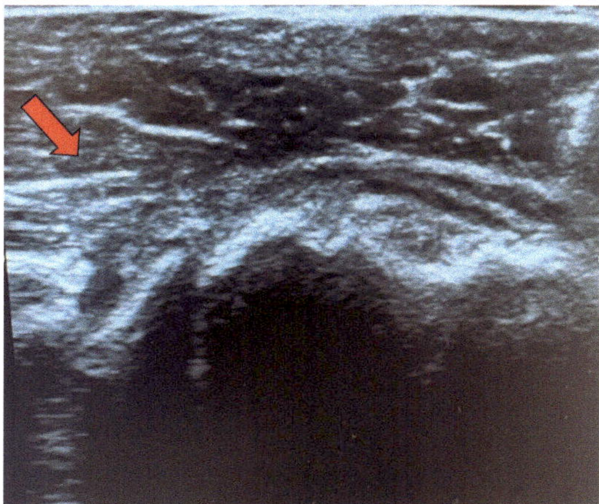

**Fig. 22.5** To increase the accuracy, shoulder injections can be performed under ultrasound guidance. The needle (arrow) is introduced under direct visualization of the ultrasound image to ensure proper placement, and the desired amount of drug is injected

The biomechanics of the GHJ is complex: the joint is inherently unstable due to the shallow socket of the glenoid fossa. This anatomic configuration allows a wide ROM (flexion, extension, abduction, adduction, internal and external rotation, and circumduction). To provide stability, the joint relies on a complex network of ligaments, muscles, and tendons, including the rotator cuff muscles, which work to keep the humeral head centered in the glenoid fossa. The forces acting on the GHJ during movement can be quite high, especially during activities such as throwing or lifting heavy objects. The joint is designed to distribute these forces across the joint surfaces and surrounding tissues, including the labrum, articular cartilage, and rotator cuff muscles [21, 22].

### 22.4.2 Pathologies and Indication for Injections

The GHJ can be affected by several post-traumatic or idiopathic pathologic conditions. The most common are rotator cuff tear (partial articular or complete), labral injury, inflammatory diseases such as rheumatoid arthritis and adhesive capsulitis, OA, and cuff tear arthropathy.

Conservative treatment often represents the first line of treatment, and it consists of physical therapy, oral NSAIDs, and intra-articular GHJ injections [23, 24]. Depending on the type and severity of pathological condition, the type of patient treated, these therapies can have a good effect. A good efficacy of GHJ intra-articular infiltrations of corticosteroids or orthobiologic agents has been demonstrated in improving symptoms in adhesive capsulitis and in partial rotator cuff tears at mid- and long-term follow-up [25–28]. Even in shoulder OA and rotator cuff arthropathy, intraarticular GHJ injections with HA or orthobiologic agents can give excellent results in the improvement of pain symptoms at short- and mid-term [29, 30]. When conservative therapy does not produce the desired effect on GHJ pathologies, surgical therapy (arthroscopic or open) may be indicated depending on the pathological conditions.

### 22.4.3 Appliances

The type of injector, needle, and syringe used depends on the preference of the healthcare provider and the specific condition being treated. Usually, a 21- to 25-G needle is used. The needle should be long enough to reach the joint space, but not so long that it risks damaging surrounding structures. A needle length of 4–7 mm is typically used, although shorter or longer needles may be used depending on the patient's anatomy. The amount of fluid injected is usually 10–20 mL. Larger volumes (20–40 ml) can be used in hydro-dilatation procedures in the case of adhesive capsulitis [31, 32].

### 22.4.4 Agents

Corticosteroids, local anesthetics, visco-supplements, and orthobiologic agents.

### 22.4.5 Injection Technique

The glenohumeral joint can be injected from an anterior or posterior approach. The accuracy rate of GHJ injections in cadaveric setting was demonstrated to be between 50% and 96% [33] with the anterior approach being considered slightly more accurate than the posterior [34]. As for the previously described anatomic regions of the shoulder, also in this case, the procedure can be performed under fluoroscopically or ultrasound guidance to increase the accuracy up to 100% [35].

***Anterior Approach*** The patient is placed in supine or seated position with the arm affected in a slight external rotation. A 23- to 25-G 50-mm needle is introduced just medial to the head of the humerus and 1 cm lateral to the coracoid process and directed posteriorly and slightly superiorly

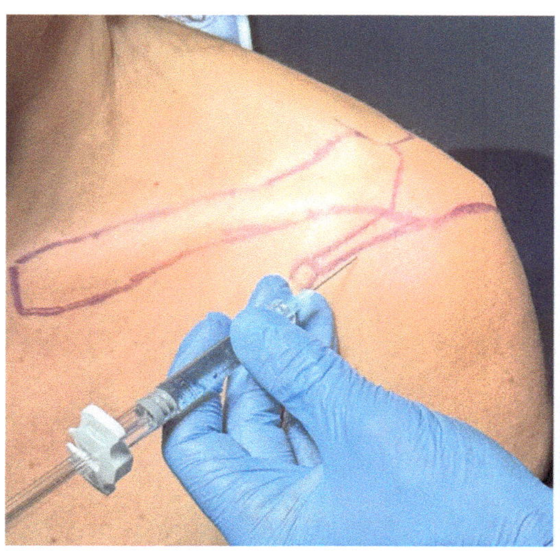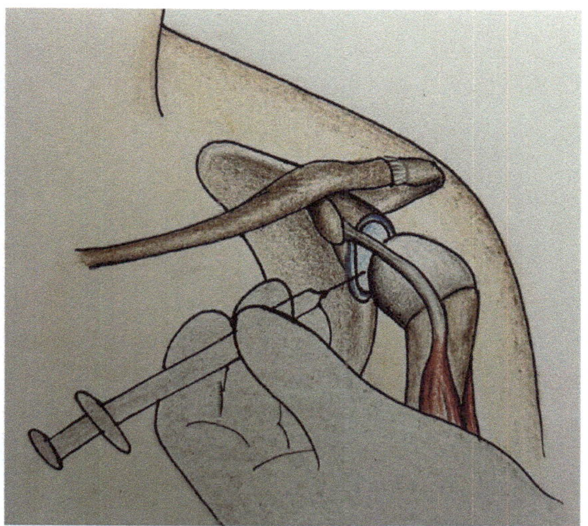

**Fig. 22.6** Glenohumeral joint injection (anterior approach): patient placed in seated position with the arm in a slight external rotation. A 23- to 25-G 50-mm needle is introduced just medial to the head of the humerus and 1 cm lateral to the coracoid process and directed posteriorly and slightly superiorly and laterally into the joint space through the rotator interval

and laterally into the joint space through the rotator interval (Fig. 22.6). If available, a high-frequency (>10 MHz) linear-array transducer is placed in the anatomic axial plane over the rotator interval, short axis to the long head of the biceps tendon. The needle is introduced under direct visualization into the GHJ and the desired amount of drug (usually 5–10 mL) in injected.

***Posterior Approach*** the patient is placed in seated, prone, or lateral position with the arm affected at the side. After palpatory identification of the posterolateral corner of the acromion, a 22- to 25-G 50- to 70-mm needle is introduced about 2 cm inferiorly and 1 cm medially (soft spot), and direct it anteriorly toward the coracoid process that can be easily palpated. It is important to avoid puncture of the glenoid labrum because it could be very painful for the patient (Fig. 22.7). If available, a 7.5- to 14-MHz linear array transducer is positioned long axis to the fibers of the infraspinatus in the anatomic axial oblique plane to the GHJ. Passive shoulder motion may be used to make the joint space more conspicuous. The needle is introduced under direct visualization into the GHJ, and the desired amount of drug (usually 5–10 mL) is injected.

In the case of adhesive capsulitis, a higher volume of injectate can be utilized to produce capsular distension, both from posterior or anterior approach. Volumes of up to 20 mL have been reported with good clinical outcomes [31, 32]. In this case, a bigger needle (21-G or larger) is used to allow easier capsular distension. Care must also be taken to avoid overdistension of the joint.

## 22.4.6 Aftercare

The patient is observed for a while. In the first 24–48 h, information is given about the symptoms that may worsen. Rest is recommended for at least 24–48 h after the injection. Avoid activities that involve heavy lifting or overhead movements for a few days to a few weeks following the injection. Gentle range of motion exercises and stretches are recommended to help prevent stiffness. Applying ice to the injection site for 20 min at a time, several times a day, for the first

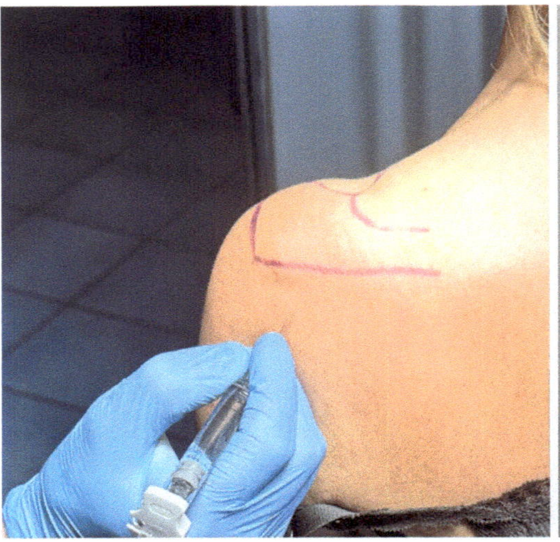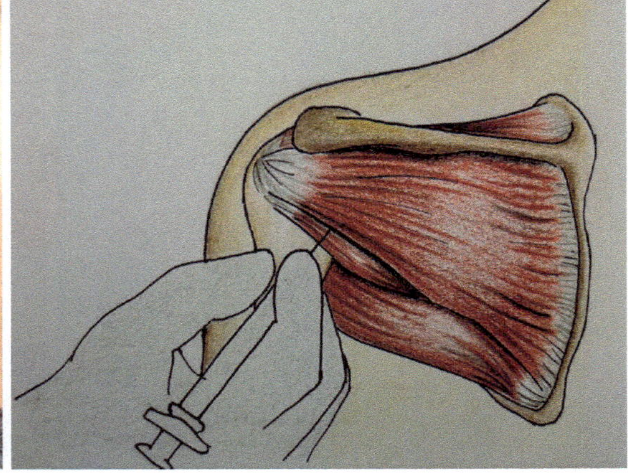

**Fig. 22.7** Glenohumeral joint injection (posterior approach): patient placed in seated position, after palpatory identification of the posterolateral corner of the acromion, a 22- to 25-G 50- to 70-mm needle is introduced about 2 cm inferiorly and 1 cm medially and direct it anteriorly toward the coracoid process

48 h can help reduce pain and swelling. Pain medication such as paracetamol or NSAIDs can help relieve pain. It is obviously important to watch out for side effects in the injection site such as redness, swelling, or drainage that could indicate an infection. Depending on the underlying condition that led to the GHJ injection, physical therapy may be recommended to help improve joint mobility and strength.

## References


1. van Tongel A, MacDonald P, Leiter J, Pouliart N, Peeler J. A cadaveric study of the structural anatomy of the sternoclavicular joint. Clin Anat. 2012;25(7):903–10.
2. Higginbotham TO, Kuhn JE. Atraumatic disorders of the sternoclavicular joint. J Am Acad Orthop Surg. 2005;13(2):138–45.
3. Robinson CM, Jenkins PJ, Markham PE, Beggs I. Disorders of the sternoclavicular joint. J Bone Joint Surg Br. 2008;90(6):685–96.
4. Weinberg AM, Pichler W, Grechenig S, Tesch NP, Heidari N, Grechenig W. Frequency of successful intra-articular puncture of the sternoclavicular joint: a cadaver study. Scand J Rheumatol. 2009;38(5):396–8.
5. Pourcho AM, Sellon JL, Smith J. Sonographically guided sternoclavicular joint injection: description of technique and validation. J Ultrasound Med. 2015;34(2):325–31.
6. Ha AS, Petscavage-Thomas JM, Tagoylo GH. Acromioclavicular joint: the other joint in the shoulder. AJR Am J Roentgenol. 2014;202(2):375–85.
7. Bontempo NA, Mazzocca AD. Biomechanics and treatment of acromioclavicular and sternoclavicular joint injuries. Br J Sports Med. 2010;44(5):361–9.
8. Merrigan B, Varacallo M. Acromioclavicular joint injection. In: StatPearls. Treasure Island: StatPearls Publishing; 2022. Available from: http://www.ncbi.nlm.nih.gov/books/NBK547727/.
9. Soler F, Mocini F, Djemeto DT, Cattaneo S, Saccomanno MF, Milano G. No differences between conservative and surgical management of acromioclavicular joint osteoarthritis: a scoping review. Knee Surg Sports Traumatol Arthrosc. 2021;29(7):2194–201.
10. Mall NA, Foley E, Chalmers PN, Cole BJ, Romeo AA, Bach BR. Degenerative joint disease of the acromioclavicular joint: a review. Am J Sports Med. 2013;41(11):2684–92.
11. Partington PF, Broome GH. Diagnostic injection around the shoulder: hit and miss? A cadaveric study of injection accuracy. J Shoulder Elb Surg. 1998;7(2):147–50.
12. Borbas P, Kraus T, Clement H, Grechenig S, Weinberg AM, Heidari N. The influence of ultrasound guidance in the rate of success of acromioclavicular joint injection: an experimental study on human cadavers. J Shoulder Elb Surg. 2012;21(12):1694–7.
13. Beals TC, Harryman DT, Lazarus MD. Useful boundaries of the subacromial bursa. Arthroscopy. 1998;14(5):465–70.
14. Hanchard NCA, Lenza M, Handoll HHG, Takwoingi Y. Physical tests for shoulder impingements and local lesions of bursa, tendon or labrum that may accom-

pany impingement. Cochrane Database Syst Rev. 2013;4:CD007427.
15. Akgün K, Birtane M, Akarirmak U. Is local subacromial corticosteroid injection beneficial in subacromial impingement syndrome? Clin Rheumatol. 2004;23(6):496–500.
16. Hurt G, Baker CL. Calcific tendinitis of the shoulder. Orthop Clin North Am. 2003;34(4):567–75.
17. Oh JH, Oh CH, Choi JA, Kim SH, Kim JH, Yoon JP. Comparison of glenohumeral and subacromial steroid injection in primary frozen shoulder: a prospective, randomized short-term comparison study. J Shoulder Elb Surg. 2011;20(7):1034–40.
18. Bansal S, Raja BS, Niraula BB, Regmi A, Choudhury AK, Sharma D, et al. Efficacy of hyaluronic acid in rotator cuff pathology compared to other available treatment modalities: a systematic review and meta-analysis. J Orthop Rep. 2023;2(3):100157.
19. Halder AM, Itoi E, An KN. Anatomy and biomechanics of the shoulder. Orthop Clin North Am. 2000;31(2):159–76.
20. Culham E, Peat M. Functional anatomy of the shoulder complex. J Orthop Sports Phys Ther. 1993;18(1):342–50.
21. Escamilla RF, Andrews JR. Shoulder muscle recruitment patterns and related biomechanics during upper extremity sports. Sports Med. 2009;39(7):569–90.
22. Fleisig GS, Barrentine SW, Escamilla RF, Andrews JR. Biomechanics of overhand throwing with implications for injuries. Sports Med. 1996;21(6):421–37.
23. McKee MD, Litchfield R, Hall JA, Wester T, Jones J, Harrison AJ. NASHA hyaluronic acid for the treatment of shoulder osteoarthritis: a prospective, single-arm clinical trial. Med Devices. 2019;12:227–34.
24. Schneider A, Burr R, Garbis N, Salazar D. Platelet-rich plasma and the shoulder: clinical indications and outcomes. Curr Rev Musculoskelet Med. 2018;11(4):593–7.
25. Giovannetti de Sanctis E, Franceschetti E, De Dona F, Palumbo A, Paciotti M, Franceschi F. The efficacy of injections for partial rotator cuff tears: a systematic review. J Clin Med. 2020;10(1):51.
26. Steuri R, Sattelmayer M, Elsig S, Kolly C, Tal A, Taeymans J, et al. Effectiveness of conservative interventions including exercise, manual therapy and medical management in adults with shoulder impingement: a systematic review and meta-analysis of RCTs. Br J Sports Med. 2017;51(18):1340–7.
27. Harna B, Gupta V, Arya S, Jeyaraman N, Rajendran RL, Jeyaraman M, et al. Current role of intra-articular injections of platelet-rich plasma in adhesive capsulitis of shoulder: a systematic review. Bioengineering. 2022;10(1):21.
28. Mocini F, Monteleone AS, Piazza P, Cardona V, Vismara V, Messinese P, et al. The role of adipose derived stem cells in the treatment of rotator cuff tears: from basic science to clinical application. Orthop Rev. 2020;12(Suppl 1):8682.
29. Zhang B, Thayaparan A, Horner N, Bedi A, Alolabi B, Khan M. Outcomes of hyaluronic acid injections for glenohumeral osteoarthritis: a systematic review and meta-analysis. J Shoulder Elb Surg. 2019;28(3):596–606.
30. Rossi LA, Piuzzi NS, Shapiro SA. Glenohumeral osteoarthritis: the role for orthobiologic therapies: platelet-rich plasma and cell therapies. JBJS Rev. 2020;8(2):e0075.
31. Lädermann A, Piotton S, Abrassart S, Mazzolari A, Ibrahim M, Stirling P. Hydrodilatation with corticosteroids is the most effective conservative management for frozen shoulder. Knee Surg Sports Traumatol Arthrosc. 2021;29(8):2553–63.
32. Park KD, Nam HS, Lee JK, Kim YJ, Park Y. Treatment effects of ultrasound-guided capsular distension with hyaluronic acid in adhesive capsulitis of the shoulder. Arch Phys Med Rehabil. 2013;94(2):264–70.
33. Sethi PM, El Attrache N. Accuracy of intra-articular injection of the glenohumeral joint: a cadaveric study. Orthopedics. 2006;29(2):149–52.
34. Sethi PM, Kingston S, Elattrache N. Accuracy of anterior intra-articular injection of the glenohumeral joint. Arthroscopy. 2005;21(1):77–80.
35. Patel DN, Nayyar S, Hasan S, Khatib O, Sidash S, Jazrawi LM. Comparison of ultrasound-guided versus blind glenohumeral injections: a cadaveric study. J Shoulder Elb Surg. 2012;21(12):1664–8.

# Injections of Anatomical Regions and Diseases: Elbow

Eduard Alentorn-Geli and Jorge Ramírez Haua

## 23.1 Elbow

### 23.1.1 Anatomy, Diagnosis, and Indications

The elbow is a complex joint with a number of specific features. Surgical procedures of the elbow not uncommonly produce stiffness, heterotopic ossification, or infection due to its close proximity with the skin. Therefore, the goal is to avoid surgery whenever possible, or to improve symptoms with a less invasive treatment strategy. Injections of the elbow can easily help in both goals. Several agents can be used for a number of conditions. Most commonly used medications are corticosteroids, hyaluronic acid, anesthetics (diagnostic injections), or orthobiologics. The field of orthobiologics includes platelet-rich plasma (PRP) or stem cells. Nowadays, this field is still growing, but there is enough evidence to conclude that orthobiologics in the elbow has

adequate evidence for efficacy in some conditions and is also a promising field. Injections can be applied after palpation, under ultrasound, or under fluoroscopy. It is obvious that image-guided injections will be more precise and may help better improve the patients' symptoms. The goal of this chapter is to report the existing evidence regarding injections of different medications for some of the most common elbow disorders.

## 23.1.2 Type of Injections and Agents

Injections can be classified regarding its way of application or the agent applied. The former refers to the use of image guidance, i.e., ultrasound-guided injections or fluoroscopy-guided injections. The latter refers to the type of medication given, typically corticosteroids, local anesthetic, hyaluronic acid, collagen-tenocytes, or orthobiologics (platelet-rich plasma and stem cells).

Numerous studies have determined that the use of ultrasound increases the injection accuracy compared to "blinded" conditions [1]. On occasions, the use of fluoroscopy is needed in order to assure intraosseous or intraarticular location of the needle in certain joins. Corticosteroid injections cannot be generally recommended as a predictable treatment option. Despite some patients may get benefit from this treatment modality when applied

E. Alentorn-Geli (✉)
Instituto Cugat, Barcelona, Spain

Fundación García Cugat, Barcelona, Spain

Mutualidad de Futbolistas, Delegación Cataluña, Barcelona, Spain
e-mail: ealentorn@institutocugat.com

J. R. Haua
Mutualidad de Futbolistas, Delegación Cataluña, Barcelona, Spain

Haua Sports Medicine, Granollers, Barcelona, Spain


B. Kocaoglu et al. (eds.), *Musculoskeletal Injections Manual*,
https://doi.org/10.1007/978-3-031-52603-9_23

in an inflamed bursa or in the peritendinous area, its repetitive use is not recommended due to potentially severe modification of tissue quality, particularly if applied intratendinously. Infiltration of corticosteroids inside the tendon decreases cellular activity and collagen formation and produces disorganization and even necrosis of collagen fibers and, therefore, is not recommended [2–5]. Interestingly, the application of PRP improved tissue quality after corticosteroid injections [5]. This is because PRP has been shown to increase tenocyte proliferation in vitro [6–9] and is in fact accepted as a safe and efficacious way of supporting tendon and ligament healing [10]. Moreover, PRP has demonstrated better outcomes for tendinopathies at long term compared to corticosteroid injections, despite similar or even superior outcomes at the short term (first 4–8 weeks) with corticosteroid injections [11]. In addition, care should be taken with corticosteroids to avoid subcutaneous atrophy and local depigmentation. In general, this type of injection is useful at providing a temporary relief that may help other concomitant therapies (i.e., physical therapy) to be more effective. Even in cases of improvement, lack of preventive measures easily can predispose to recurrence, as they do not improve tissue quality. In general, corticosteroid injections are not recommended as a continuous therapy for elbow tendinopathies, but as a single-use, second-line treatment option. Hyaluronic acid injections have also been tried in elbow tendinopathies with good outcomes [12–14].

### 23.1.3 Technique/Tricks/Pitfalls

The preparation of an injection is universal. Despite infection after injections are very rare, the maximum care must be taken in order to minimize its incidence. Thus, sterile conditions must be assured. The area to be infiltrated is first cleaned for macroscopic dirtiness. Then, iodine or chlorhexidine solution is applied to the injection side. Sterile gloves are strongly recom-

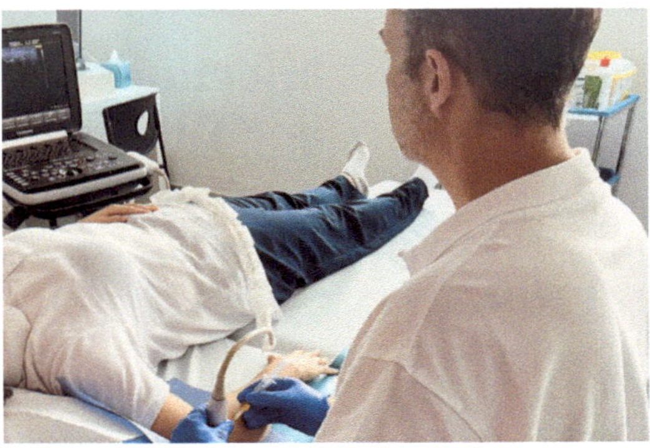

**Fig. 23.1** General, panoramic view of an elbow ultrasound-guided injection. The ultrasonographist comfortably seats next to the patient with the screen in front so that there is no need to twist the head to look at the screen and injection site

mended, but the use of a fenestrated drape is controversial because coverage of the examined joint may prevent adequate visualization of landmarks and optimal orientation. The probe of the ultrasound has to be adequately isolated using a sterile cover, as well. The ultrasonographist has to be in a comfortable position that facilitates both the performance of the injection and an adequate visualization of the screen without cervical spine rotation (Fig. 23.1). During ultrasound-guided injections, it is highly recommended to identify the infiltration side from both the axial and longitudinal views right before entering the needle. This will increase the precision of the injection.

### 23.1.4 Aftercare

After the injection, the patient is observed for 15 min for the development of any early symptoms. Depending on the injected agent, if no contraindication is present, icing for 15 min may be helpful. Then, the patient is informed about relative rest and low activity in first 24–48 h. And patient should be warned about red flags for complications such as redness, swelling, or drainage from the injection zone.

## 23.2 Radial Tuberosity (Biceps Tendon Insertion)

### 23.2.1 Anatomy, Diagnosis, and Indications

Most common distal biceps tendon disorders include bicipitoradial bursitis, symptomatic distal biceps tendinosis including high-grade intrasubstance tears, partial tear of the distal biceps, or complete tears. The insertion of the distal biceps is a complex area in terms of anatomical morphology and close proximity of neurovascular structures. Therefore, the first consideration when dealing with this condition is the need for accurate infiltrations. In this regard, the use of image-guided techniques is especially relevant. Sellon et al. conducted a cadaveric study where peritendinous (18 cases) and intratendinous (15 cases) distal biceps tendon infiltrations were conducted under ultrasound guidance by a single surgeon [15].

### 23.2.2 Technique/Tricks/Pitfalls

The authors performed the injection with the patient lying supine in a semi-supinated forearm position, infiltrating from distal to proximal under the long axis view of the ultrasound with the transducer in the antecubital fossa. This is the standard position for an ultrasound-guided injection through an anterior approach to the distal biceps tendon (Figs. 23.2 and 23.3). Despite the study lacks a control, comparative group, only 1 of the 33 cases was not injected in the desired location. Therefore, they concluded that ultrasound-guided distal biceps tendon injections were very accurate. Similarly, van der Vis et al. observed that manual injections around the distal biceps tendon (without ultrasound guidance) were not accurate enough (21% inaccuracy) [16]. The authors performed the injections through either an anterior or lateral approach and were conducted by upper limb specialist, general orthopedic surgeons, and orthopedic surgery residents. Both studies used a cadaver model and infiltrated latex and acrylic dye.

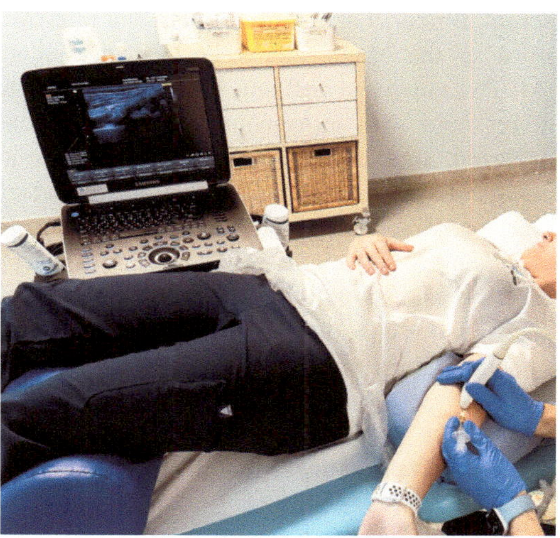

**Fig. 23.2** General overview of the patient's positioning, ultrasonographist position, and anterior approach to the distal biceps tendon injection. The patient is lying supine with the elbow in extension and the forearm supinated

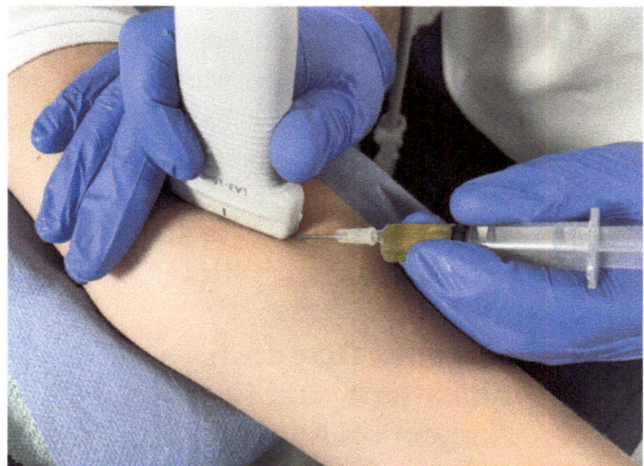

**Fig. 23.3** Detail of an ultrasound-guided anterior approach to the distal biceps. The transducer is first positioned to identify the antecubital bursa and the distal biceps. Then, the needle is advance following the long axis of the transducer until it is seen on the screen

### 23.2.3 Type of Injections and Agents

Regarding the type of medication, Barker et al. reported the outcomes of PRP infiltration for distal biceps tendinopathy [17]. The authors used ultrasound guidance and infiltrated the PRP intratendinously but using a preparation rich in leukocytes without platelet activation and with a concentration of platelets up to five times the serum. Barker et al. reported significant improvement in the Mayo Elbow Performance Score (MEPS) and visual analogue scale (VAS) for pain at rest and with movement. No complications occurred. Similarly, Sanli et al. reported the outcomes of a single injection of PRP for refractory biceps tendinopathy [18]. They found significant improvement in pain (at rest and during activity), function, and strength. Both studies had no comparative control group, which is a major limitation to obtain robust and meaningful conclusions.

### 23.2.4 Aftercare

After the injection, the patient is observed for 15 min for development of any early symptoms. Depending on the injected agent, if no contraindication is present, icing for 15 min may be helpful. Then, the patient is informed about relative rest and low activity in first 24–48 h, especially avoiding elbow flexion and supination movements. And patient should be warned about red flags for complications such as redness, swelling, or drainage from the injection zone.

## 23.3    Lateral Epicondylitis

### 23.3.1 Anatomy, Diagnosis, Indications, and Type of Injection Agents

Lateral epicondylitis is the most common and most studied elbow disorder [8, 10, 19]. This condition is caused by repetitive microtrauma or overload causing tendon fibers tearing and degeneration [20]. Given the previously mentioned evidence, it is not surprising that corticosteroid injections are not helpful for lateral epicondylitis over time. Gosens et al. observed better pain relief at short-term compared to PRP, but inferior outcomes at 2 years [11]. Other authors have also observed this finding [21–24]. In a systematic review and meta-analysis of studies comparing corticosteroid and PRP for lateral epicondylitis, Li et al. observed a superiority of the former at 4 and 8 weeks, but inferior outcomes compared to

PRP at 24 weeks [25]. In a very nice descriptive review by Kwapisz et al., a total of 15 studies investigated the effects of PRP injection for lateral epicondylitis [8]. Eleven of the 15 studies compared PRP to steroid injections, and 9 (81%) of them found a superiority of PRP over corticosteroid injections. In most of the studies, the corticosteroid injection group improved at the beginning (very short term) but worsened over time, whereas the PRP group maintained their improvement at the end of the study period. The periods for which PRP was still better than corticosteroid injections were at 3 months [21, 26], 6 months [23, 27], 1 year [24, 28], and 2 years [11]. Besides a favorable comparison of PRP over corticosteroid injections in the literature, the efficacy of PRP for the treatment of lateral epicondylitis has been systematically reproduced in various other systematic reviews and meta-analyses. Chen et al. published a couple of systematic review and meta-analysis involving 37 studies (1031 participants) in 1 article [10] and 16 level I studies (581 participants) in another article [19]. The authors found significant improvement with PRP for pain and function at the long-term. Niemiec et al. reported another systematic review and meta-analysis where the effectiveness of PRP to achieve a minimal clinically important difference for the treatment of lateral epicondylitis was evaluated [29]. The authors found that all scores (VAS, DASH, patient-rated tennis elbow evaluation, Mayo Clinic Performance Index) exceeded the respective minimal clinically important difference in almost all-time intervals evaluated. This finding occurred for both types of PRP preparation, leukocyte-rich and leukocyte-poor PRP [29]. One complication described for PRP injections has been temporary increase in pain [30]. However, this complication was most commonly observed in the leukocyte-rich as opposed to leukocyte-poor PRP injections [30]. This is most likely explained by the appearance of a leukocyte-mediated inflammatory response in leukocyte-rich PRP preparations. Interestingly, Kim et al. conducted a systematic review and meta-analysis comparing the outcomes of PRP infiltration with surgical treatment for lateral epicondylitis [31]. The investigation included 5 studies, involving 154 patients treated with PRP infiltrations and 186 patients undergoing surgical treatment. The authors did not find significant differences for any of the outcomes for any of the follow-up periods: VAS at 2 months, 6 months, and 12 months, and patient-related tennis elbow evaluation score at 12 weeks, 24 weeks, and 52 weeks.

Other products like autologous blood injection do not seem to provide any benefit over PRP [8]. On the other hand, bone marrow aspirate concentrate has provided promising results, but further research is needed in this field [32]. This is also the case for direct autologous tenocyte injection, a two-step surgical procedure where tenocytes are injected after harvesting and culturing [33]. Lastly, a comparative study to investigate the effectiveness of hyaluronic acid and saline (control group) injections was conducted by Zinger et al. [14]. The authors observed a significant improve in pain and function mostly at 3 months, lasting until 12 months despite a decrease in the improvement. However, the follow-up rate of the control group was less than 50% and could not be used as a comparative group. A comparison between hyaluronic acid and corticosteroid injections was conducted by Yalcin and Kayaalp [34]. Both treatment options yielded significant improvement in pain and function, but lasted for a short period. In general, hyaluronic acid can be considered a safe and effective treatment modality [12], but further comparative studies are needed before it can be considered a first-line injection agent.

As a result of the existing evidence, corticosteroid injection are no longer recommended except in cases that temporary, short-term relief is needed, provided not too many injections are accumulated over time. In terms of injection agents, PRP can be now considered the gold standard for treating lateral epicondylitis according to the conclusions from many systematic review and meta-analyses [8, 35]. The first line of treatment for lateral epicondylitis is physical therapy and activity modification. This is generally recommended before infiltration is performed because many patients may get full recover without any further more aggressive treatments.

When infiltration is needed, the use of ultrasound guidance is recommended to increase precision and accuracy.

## 23.3.2 Technique/Tricks/Pitfalls

The patient is placed in the supine position with the arm at the side with the hand in the belly or with 60° of abduction and internal rotation of the shoulder so that the forearm is on the bed or a resting table (Figs. 23.1 and 23.4). Lying supine is always preferred in any infiltration because the patient may get dizzy. The physician prepares the elbow using the standard protocol. The ultrasound is first used to identify the location where the tendon is mostly affected, and the transducer positioned in the long axis (Fig. 23.4). The medication is infiltrated after triangulation and identification of the needle in the screen. The idea is to place the PRP around the tendon and within the tendon (very low quantity) to improve high-grade degeneration or partial tearing. A series of three PRP infiltrations are recommended, about 2 weeks apart. In general, as the tendon becomes stronger in its intratendinous area, the infiltration of PRP becomes more difficult. This is a positive sign, as it represents an improvement of the tendon, especially if the first injection inside the tendon offered only mild resistance. It is important to feel at least mild resistance during the first injection. Otherwise, it might be a sign that the tendon has an important tear or that the needle is not in the correct location.

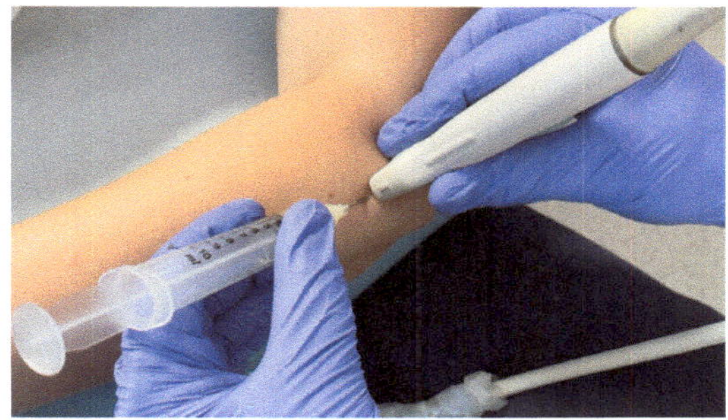

**Fig. 23.4** Detail of lateral epicondylitis PRP infiltration through a lateral approach. The transducer is placed in the long axis of the tendon, and the needle advanced form distal to proximal until it is seen in the screen

### 23.3.3 Aftercare

After the injection, the patient is observed for 15 min for development of any early symptoms. Depending on the injected agent, if no contraindication is present, icing for 15 min may be helpful. Then, the patient is informed about relative rest and low activity in first 24–48 h, especially avoiding wrist and finger extension movements. And patient should be warned about red flags for complications such as redness, swelling, or drainage from the injection zone.

## 23.4 Medial Epicondylitis

### 23.4.1 Anatomy, Diagnosis, Indications, and Type of Injection Agents

Medial epicondylitis is a flexor-pronator tendon degeneration as a result of chronic repetitive movements (overuse). As in any other chronic tendon disorders, injections include corticosteroids, hyaluronic acid, collagen, and PRP. The existing evidence for medial epicondylitis is way inferior than for lateral epicondylitis. However, the effects of corticosteroid injections on the common extensor-supination tendons should also apply to medial epicondylitis because both conditions deal with the same type of tissue (tendons). Therefore, tendon tissue quality impairment with corticosteroid injections prevents its recommendation also for medial epicondylitis.

Accuracy issues with "blinded" injections are also relevant in medial epicondylitis. Ultrasound guidance will also improve the precision of the infiltration, which is even more important considering that if corticosteroid injection is decided, it is paramount that the medication is not placed within the tendon. Therefore, ultrasound direct observation in the screen is very convenient. In addition, the close proximity of the ulnar nerve makes the ultrasound even more important than in other pathologies.

Stahl and Kaufman reported the outcomes of corticosteroid injections in 60 elbows with medial epicondylitis [36]. An improvement of pain during the first 6 weeks was noted, but not maintained at 3 months. Similar outcomes were observed by Lee et al., with initial improvement at 2 weeks followed by a plateau for up to 8 weeks [37]. PRP has an added potential benefit. Because the medication can be administered inside the tendon, the mechanical effect of needle trephination may stimulate bleeding and tendon healing [38]. Suresh et al. observed improved VAS for pain and Nirschl scores at 10 months after a combination of needle stimulation and autologous blood injected into the disrupted tendon [38]. Although the authors did not inject PRP itself, it is likely that the benefits are also observed after PRP. In fact, there are two studies comparing the effectiveness of PRP and surgical treatment (one was the Tenex procedure) for medial epicondylitis [39, 40]. Both studies obtained similar conclusions: improvement in pain scores and functional outcomes (Mayo Clinic Performance Score, Oxford Elbow Score) were comparable between the two groups (PRP and surgery).

Despite the evidence for medial epicondylitis regarding PRP or corticosteroid injections are significantly inferior compared to lateral epicondylitis, its similar etiopathogenesis makes conclusions from the latter potentially applicable to medial epicondylitis. In fact, the outcomes and conclusions from the existing evidence seem to point out toward the same direction, although to a different extend in terms of the amount of studies published.

### 23.4.2 Technique/Tricks/Pitfalls

Injections for medial epicondylitis may be conducted in the supine or lateral decubitus. Lateral decubitus has the advantage to provide a very good access to the medial side of the elbow in a comfortable position for both the patient and physician in cases where limited shoulder external rotation is present. In general, the transducer is positioned in the long axis of the tendon and the needle advanced distal to proximal following the long axis (Fig. 23.5). Once the ulnar nerve is

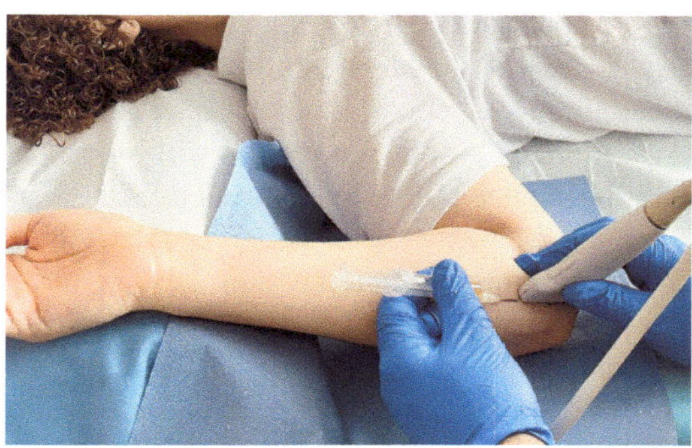

**Fig. 23.5** Infiltration of PRP for medial epicondylitis. The patient is placed lying supine with shoulder abduction and external rotation. The medial aspect of the elbow is prepared and evaluated with the ultrasound. The transducer is placed following the lines of tendon fibers, and the PRP administered in the long axis around and/or inside the tendon. Special care should be taken to locate and avoid the ulnar nerve

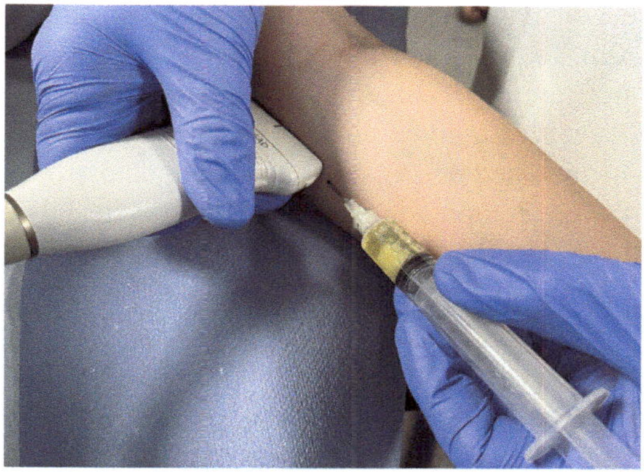

**Fig. 23.7** Detail of medial epicondylitis infiltration using the alternative patients' positioning (lying supine with the arm at 40° of abduction). The ulnar nerve and the tendons are identified, and the transducer is placed in the long axis of the tendons. The needle is then advanced from distal to proximal in the long axis

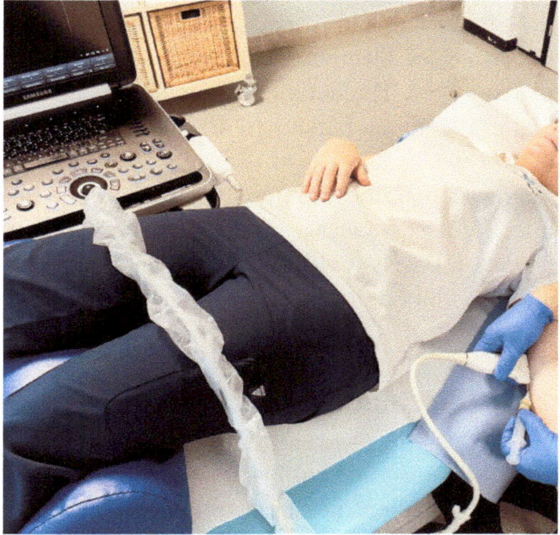

**Fig. 23.6** General view of the alternative patients' positioning, lying supine with the arm at the side with 40° of shoulder abduction

located, and the needle is seen on the screen, the PRP can be infiltrated around the tendon, or inside the tendon for a small amount if high-grade degeneration or partial tearing is present. If the patient has shoulder problems preventing adequate abduction and external rotation, medial epicondylitis can also be injected with the arm at the side as an alternative position (Figs. 23.6 and 23.7). If access to the medial side of the elbow is

difficult, the patient can move the elbow toward the edge of the bed so that the clinician can perform the injection easily.

### 23.4.3 Aftercare

After the injection, the patient is observed for 15 min for development of any early symptoms. Depending on the injected agent, if no contraindication is present, icing for 15 min may be helpful. Then, the patient is informed about relative rest and low activity in first 24–48 h, especially avoiding wrist and finger flexion movements. And patient should be warned about red flags for complications such as redness, swelling, or drainage from the injection zone.

## 23.5  Olecranon Bursitis

### 23.5.1 Anatomy, Diagnosis, Indications, and Type of Injection Agents

Aseptic olecranon bursitis is not an uncommon inflammatory process due to direct trauma, repetitive microtrauma, or an inflammatory disorder. In this condition, fluid effusion is accumulated in the olecranon bursa, causing increase in volume, pain, redness, and limited elbow function. The first line of treatment is noninvasive modalities like ice, rest, immobilization (to avoid friction on the inflamed bursa), oral anti-inflammatory medication, compression dressing, and physical therapy. If this conservative treatment fails, US-guided drainage with or without injections would be indicated.

The outcomes of infiltrations for olecranon bursitis are controversial. To the best of our knowledge, there are no studies evaluating the effects of PRP for olecranon bursitis. The only agent tested in a decent amount of studies is corticosteroid. In a systematic review, Sayegh and Strauch found five studies (of 29 included) using corticosteroid injections for the treatment of aseptic olecranon bursitis [41]. The authors found that corticosteroid injections were associated with increased overall complications and skin atrophy. However, when evaluating the studies in a more individual way good outcomes have also been reported. Weinstein et al. observed that those patients treated with corticosteroid injections (25 patients) had a faster reduction in bursal effusion compared to those treated with fluid aspiration (22 patients) [42]. As stated by Sayegh and Strauch, Weinstein et al. observed long-term (mean 31 months, range 6 to 62 months) adverse effects including infection in three patients, skin atrophy in five patients, and local pain in seven patients. Kim et al. observed that patients treated with aspiration and corticosteroid injection into the bursa was associated with the faster resolution of symptoms, at an average of 2.3 weeks [43]. However, the authors failed to find any other significant difference in the outcomes, maybe explained by a limited sample size. These outcomes were similarly reported by Jaffe and Fetto, who reported a 100% clinical resolution of aseptic olecranon bursitis after aspiration and corticosteroid injection in five patients [44]. In a double-blind prospective study comparing corticosteroid injections (with or without oral anti-inflammatory medications) to oral anti-inflammatory or placebo treatments, Smith et al. found a clear benefit in the injection groups

[45]. Those patients receiving injections had the fastest decrease in bursal swelling as soon as at 1 week, improvements that were kept at 6 weeks. At 6 months, groups not treated with injections had the higher rate of re-aspirations. One confounding factor for the outcomes of this study is the application of a compression dressing in the injection groups, as this treatment has been shown to improve the outcomes of olecranon bursitis [43]. Interestingly, Smith et al. did not find infection or skin atrophy. The authors reasoned that the use of a thin needle entering laterally and placing a sterile dressing after the injection could be related to the absence of septic complications. The careful and meticulous use of sterile injections with bursal access through a normal, healthy-looking skin area is paramount to decrease post-injection infection. A more recent investigation conducted by Wu et al. evaluated the efficacy of ultrasound-guided corticosteroid injections for olecranon bursitis in a group of 45 patients [46]. The authors observed a recurrence rate of olecranon bursitis of 40% after one injection (2 weeks later), and recurrence rate of 13% after a second injection (at 4 weeks from the first injection). There was a significant decrease in the depth of the synovial effusion, synovial thickness, and blood flow signal at 4 weeks.

## 23.5.2 Technique/Tricks/Pitfalls

For an appropriate infiltration of the olecranon bursa, the patient lies either supine or prone. In the supine position, the elbow is flexed with some shoulder abduction, so that the hand lies in the patient's belly. Ideally, the patient should lie close to the edge of the table so that a medial to lateral approach with the needle can be made (Fig. 23.7). The entry point in an olecranon bursitis depends on the most affected area. On occasions, there is redness that has to be avoided. The recommendation is that the entry point of the needle is made through normal-looking skin, to prevent complications with the infiltration. The bursitis is visualized using axial and longitudinal views. The area of the bursa mostly affected is

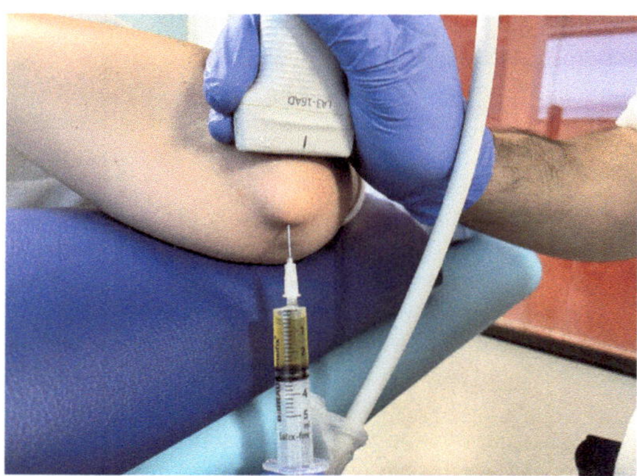

**Fig. 23.8** Infiltration of olecranon bursa under ultrasound guidance medial to lateral in the short axis. The patient is lying supine with the elbow flexed and the olecranon area at the edge of the table so that a medial approach (desired in this case) is possible

identified and drainage is recommended. There are no structures at special risk for this injection. The needle is advanced in the short axis (Fig. 23.8). Very little pressure with the transducer is advised in order to avoid fluid moving away from the view. The infiltration has to be with low volume of medication and typically using a 16/18 G needle.

In conclusion, it seems that ultrasound-guided corticosteroid injection can be considered a very good option for the treatment of aseptic olecranon bursitis if more conservative therapeutic options do not improve the patients' symptoms. Injections need to be conducted in very strict sterile conditions, performed through normal-looking skin, and followed by a compression dressing.

### 23.5.3 Aftercare

After the injection, the patient is observed for 15 min for development of any early symptoms. Depending on the injected agent, if no contraindication is present, icing for 15 min may be helpful. Then, the patient is informed about relative rest and low activity in first 24–48 h. It is especially important to avoid elbow flexion above 60° (the bursa gets tight and tense with elbow flexion) and the olecranon area in contact with any object or surface, particularly hard surfaces. And patient should be warned about red flags for complications such as redness, swelling, or drainage from the injection zone.

## 23.6 Conclusions

Injections can be very helpful as therapeutic option for a number of elbow conditions. It is highly recommended that all injections are conducted under ultrasound guidance to improve accuracy and precision. For tendon-related disorders (lateral epicondylitis, medial epicondylitis, and distal biceps tendinopathy), the gold standard in terms of injection agents is the PRP, which have demonstrated to be a very good option to improve patients' symptoms and function at the short- and long-term. Corticosteroid injections are only recommended for a short-term improvement of tendon disorders in very selected clinical scenarios, but its repetitive use is not recommended because of potentially significant impairment in tendon tissue quality. In contrast, it seems that corticosteroid injections can be a reasonable option when conservative treatment fail to improve aseptic olecranon bursitis. However, very strict sterile conditions and access through healthy-looking skin, followed by a sterile compressive dressing is recommended in order to avoid infection. During ultrasound-guided injections, it is highly recommended to identify the infiltration side from both the axial and longitudinal views right before entering the needle. This will increase the precision of the injection. Further research is needed for aseptic olecranon bursitis in order to evaluate efficacy and safety of PRP injections into the olecranon bursa.

## References

1. Patel RP, McGill K, Motamedi D, Morgan T. Ultrasound-guided interventions of the upper extremity joints. Skelet Radiol. 2023;52:897–909.
2. Dean BJ, Lostis E, Oakley T, Rombach I, Morrey ME, Carr AJ. The risks and benefits of glucocorticoid treatment for tendinopathy: a systematic review of the effects of local glucocorticoid on tendon. Semin Arthritis Rheum. 2014;43:570–6.

3. Wong MW, Lui WT, Fu SC, Lee KM. The effect of glucocorticoids on tendon cell viability in human tendon explants. Acta Orthop. 2009;80:363–7.
4. Wong MW, Tang YN, Fu SC, Lee KM, Chan KM. Triamcinolone suppresses human tenocyte cellular activity and collagen synthesis. Clin Orthop Relat Res. 2004;2004:421.
5. Wong MW, Tang YY, Lee SK, Fu BS, Chan BP, Chan CK. Effect of dexamethasone on cultured human tenocytes and its reversibility by platelet-derived growth factor. J Bone Joint Surg Am. 2003;85:1914–20.
6. De Mos M, Koevoet W, van Schie HT, Kops N, Jahr H, Verhaar JA. In vitro model to study chondrogenic differentiation in tendinopathy. Am J Sports Med. 2009;37:1214–22.
7. De Mos M, van der Windt AE, Jahr H, van Schie HT, Weinans H, Verhaar JA. Can platelet-rich plasma enhance tendon repair? A cell culture study. Am J Sports Med. 2008;36:1171–8.
8. Kwapisz A, Prabhakar S, Compagnoni R, Sibliska A, Randelli P. Platelet-rich plasma for elbow pathologies: a descriptive review of current literature. Curr Rev Musculoskelet Med. 2018;11:598–606.
9. McCarrel T, Fortier L. Temporal growth factor release from platelet-rich plasma, trehalose lyophilized platelets, and bone marrow aspirate and their effect on tendon and ligament gene expression. J Orthop Res. 2009;27:1033–42.
10. Chen X, Jones A, Park C, Vangsness CT. The efficacy of platelet-rich plasma on tendon and ligament healing: a systematic review and meta-analysis with bias assessment. Am J Sports Med. 2018;46:2020–32.
11. Gosens T, Peerbooms JC, van Laar W, den Oudsten BL. Ongoing positive effect of platelet-rich plasma versus corticosteroid injection in lateral epicondylitis: a double-blinded randomized controlled trial with 2-year follow-up. Am J Sports Med. 2011;39:1200–8.
12. Agostini F, de Sire A, Paoloni M, Finamore N, Ammendolia A, Mangone M, Bernetti A. Effects of hyaluronic acid injections on pain and functioning in patients affected by tendinopathies: a narrative review. J Back Musculoskelet Rehabil. 2022;35:949–61.
13. Khan M, Shanmugaraj A, Prada C, Patel A, Babins E, Bhandari M. The role of hyaluronic acid for soft tissue indications: a systematic review and meta-analysis. Sports Health. 2023;15:86–96.
14. Zinger G, Bregman A, Safran O, Beyth S, Peyser A. Hyaluronic acid injections for chronic tennis elbow. BMC Sports Sci Med Rehabil. 2022;14:8.
15. Sellon JL, Wempe MK, Smith J. Sonographically guided distal biceps tendon injections. Technique and validation. J Ultrasound Med. 2014;33:1461–74.
16. van der Vis J, Janssen SJ, Bleys RLA, Eygendaal D, van den Bekerom MPJ. Distal biceps tendon injection. Clin Shoulder Elbow. 2021;24:93–7.
17. Barker SL, Bell SN, Connell D, Coghlan JA. Ultrasound-guided platelet-rich plasma injection for distal biceps tendinopathy. Should Elb. 2015;7:110–4.
18. Sanli I, Morgan B, van Tilborg F, Funk L, Gosens T. Single injection of platelet-rich plasma (PRP) for the treatment of refractory distal biceps tendonitis: long-term results of a prospective multicenter cohort study. Knee Surg Sports Traumatol Arthrosc. 2016;24:2308–12.
19. Chen XT, Fang W, Jones IA, Heckmann ND, Park C, Vangsness CT. The efficacy of platelet-rich plasma for improving pain and function in lateral epicondylitis. A systematic review and meta-analysis with risk-of-bias assessment. Arthroscopy. 2021;37:2937–52.
20. Kraushaar BS, Nirscgl RP. Tendinosis of the elbow (tennis elbow). Clinical features and findings of histological, immunohistochemical, and electron microscopy studies. J Bone Joint Surg Am. 1999;81:259–78.
21. Behera P, Dhillon M, Aggarwal S, Marwaha N, Prakash M. Leukocyte-poor platelet-rich plasma versus bupivacaine for recalcitrant lateral epicondylar tendinopathy. J Orthop Surg. 2015;23:6–10.
22. Gautam VK, Verma S, Batra S, Bhatnagar N, Arora S. Platelet-rich plasma versus corticosteroid injection for recalcitrant lateral epicondylitis: clinical and ultrasonographic evaluation. J Orthop Surg. 2015;23:1–5.
23. Mi B, Liu G, Zhou W, Lv H, Liu Y, Wu Q, Liu J. Platelet rich plasma versus steroid on lateral epicondylitis: meta-analysis of randomized clinical trials. Phys Sports Med. 2017;45:97–104.
24. Peerbooms JC, Sluimer J, Bruijn DJ, Gosens T. Positive effect of autologous platelet concentrate in lateral epicondylitis in a double-blind randomized controlled trial. Am J Sports Med. 2010;38:255–62.
25. Li A, Wang H, Yu Z, Zhang G, Feng S, Liu L, Gao Y. Platelet-rich plasma vs corticosteroids for elbow epicondylitis. A systematic review and meta-analysis. Medicine. 2019;98:e18358.
26. Yadav R, Kothari SY, Borah D. Comparison of local injection of platelet rich plasma and corticosteroids in the treatment of lateral epicondylitis of humerus. J Clin Diagn Res. 2015;9:5–7.
27. Mishra AK, Skrepnik NV, Edwards SG, Jones GL, Sampson S, Vermillion DA, Ramsey ML, Karli DC, Rettig AC. Efficacy of platelet-rich plasma for chronic tennis elbow: a double-blind, prospective, multicenter, randomized controlled trial of 230 patients. Am J Sports Med. 2014;42:463–71.
28. Lebiedziński R, Synder M, Buchcic P, Polguj M, Grzegorzewski A, Sibiński M. A randomized study of autologous conditioned plasma and steroid injections in the treatment of lateral epicondylitis. Int Orthop. 2015;39:2199–203.
29. Niemiec P, Szyluk K, Jarosz A, Iwanicki T, Balcerzyk A. Effectiveness of platelet-rich plasma for lateral epicondylitis. A systematic review and meta-analysis based on achievement of minimal clinically important difference. Orthop J Sports Med. 2022;10:23259671221086920.
30. Li S, Yang G, Zhang H, Li X, Lu Y. A systematic review on the efficacy of different types of platelet-rich plasma in the management of lateral epicondylitis. J Shoulder Elb Surg. 2022;31:1533–44.

31. Kim C-H, Park Y-B, Lee J-S, Jung H-S. Platelet-rich plasma injection vs operative treatment for lateral elbow tendinosis: a systematic review and meta-analysis. J Shoulder Elb Surg. 2022;31:428–36.
32. Singh A, Gangwar DS, Singh S. Bone marrow injection: a novel treatment for tennis elbow. J Nat Sci Biol Med. 2014;5:389–91.
33. Tarpada SP, Morris MT, Lian J, Rashidi S. Current advances in the treatment of medial and lateral epicondylitis. J Orthop. 2018;15:107–10.
34. Yalcin A, Kayaalp ME. Comparison of hyaluronate & steroid injection in the treatment of chronic lateral epicondylitis and evaluation of treatment efficacy with MRI: a single-blind, prospective, randomized controlled clinical study. Cureus. 2022;14:e29011.
35. Arirachakaran A, Sukthuayat A, Sisayanarane T, Laoratanavoraphong S, Kanchanatawan W, Kongtharvonskul J. Platelet-rich plasma versus autologous blood versus steroid injection in lateral epicondylitis: systematic review and network meta-analysis. J Orthop Traumatol. 2016;17:101–12.
36. Stahl S, Kaufman T. The efficacy of an injection of steroids for medial epicondylitis: a prospective study of sixty elbows. J Bone Joint Surg Am. 1997;79:1648–52.
37. Lee SS, Kang S, Park NK. Effectiveness of initial extracorporeal shock wave therapy on the newly diagnosed lateral or medial epicondylitis. Ann Rehabil Med. 2012;36(5):681–7.
38. Suresh SP, Ali KE, Jones H, Connell DA. Medial epicondylitis: is ultrasound guided autologous blood injection an effective treatment? Br J Sports Med. 2006;40:935–9.
39. Boden AL, Scott MT, Dalwadi PP, Mautner K, Mason RA, Gottschalk MB. Platelet-rich plasma versus Tenex in the treatment of medial and lateral epicondylitis. J Shoulder Elb Surg. 2019;28:112–9.
40. Bohlen HL, Schwartz ZE, Wu VJ, Thon SG, Finley ZJ, O'Brien MJ, Savoie FH. Platelet-rich plasma is an equal alternative to surgery in the treatment of type 1 medial epicondylitis. Orthop J Sports Med. 2020;8:2325967120908952.
41. Sayegh ET, Strauch RJ. Treatment of olecranon bursitis: a systematic review. Arch Orthop Trauma Surg. 2014;134:1517–36.
42. Weinstein PS, Canoso JJ, Wohlgethan JR. Long-term follow-up of corticosteroid injection for traumatic olecranon bursitis. Ann Rheum Dis. 1984;43:44–6.
43. Kim JY, Chung SW, Kim JH, Jung JH, Sung GY, Oh K-S, Lee JS. A randomized trial among compression plus nonsteroidal anti-inflammatory drugs, aspiration, and aspiration with steroid injection for nonseptic olecranon bursitis. Clin Orthop Relat Res. 2016;474:776–83.
44. Jaffe L, Fetto J. Olecranon bursitis. Contemp Orthop. 1984;8:51–6.
45. Smith DL, McAfee JH, Lucas LM, Kumar KL, Romney DM. Treatment of nonseptic olecranon bursitis. A controlled, blinded prospective trial. Arch Intern Med. 1989;149:2527–30.
46. Wu Y, Chen Q, Chen K, He F, Quan J, Chen S, Guo X. Clinical efficacy of ultrasound-guided injection in the treatment of olecranon subcutaneous bursitis. J Xray Sci Technol. 2019;27:1145–53.

# Injections of Anatomical Regions and Diseases: Wrist and Hand

Gamlı Alper [ID] and Gereli Arel

## 24.1 Wrist Joint

Arthritis of the radiocarpal and intercarpal joints are typically secondary to inflammatory diseases or carpal instability due to previous fracture and dislocations. If left untreated, lunate avascular necrosis and lunate impingement progress to wrist arthritis. Primary osteoarthritis of the wrist is rare and occurs when a late-stage trapezio-metacarpal arthritis spreads to the carpal joints.

### 24.1.1 Anatomy

Radiocarpal and intercarpal joints, along with the eight carpal bones and multiple extrinsic and intrinsic ligaments, are responsible for complex wrist biomechanics. The tendons pass close to both palmar and dorsal side of the wrist joints. Lister's tubercle is a prominence on the dorsal side of the distal radius and serves as a pulley for the extensor pollicis longus (EPL) tendon.

### 24.1.2 Indication and Diagnosis

Degenerative wrists present as painful, stiff, and weak joints. Numerous operative options are available for permanent solution, but many patients prefer to delay surgery. In such cases, corticosteroid injections are effective in reducing symptoms. Tenosynovitis can cause diffuse swelling and effusion and should be ruled out. A standard X-ray of the wrist provides sufficient information to prove the arthritis.

### 24.1.3 Appliances

Approximately 1 ml of liquid can be injected by a 23-G or thinner syringe tip.

| Needle | Syringe | Corticosteroid | Local anesthetic |
| --- | --- | --- | --- |
| 23 G | 2–3 mL | 40 mg triamcinolone (1 mL) 5 mg betamethasone (1 mL) | 1% lidocaine (0.5 mL) |

### 24.1.4 Agents

Corticosteroids and local anesthetics.

G. Alper (✉)
Acibadem Altunizade Hospital, Department of Orthopedics and Traumatology, Istanbul, Turkey

G. Arel
Acibadem Mehmet Ali Aydınlar University, Faculty of Medicine, Department of Orthopedics and Traumatology, Istanbul, Turkey


B. Kocaoglu et al. (eds.), *Musculoskeletal Injections Manual*,
https://doi.org/10.1007/978-3-031-52603-9_24

### 24.1.5 Technique/Tricks/Pitfalls

The patient is seated next to the table, with the elbow flexed and the forearm pronated. The patient's hand is relaxed, and the palm faces the table. Lister's tubercle at the distal radius is palpated and marked (Fig. 24.1). EPL is easily found when the first finger is lifted off the table. A soft spot should be discovered 1 cm distal to the Lister's tubercle, just along the ulnar border of the EPL. A 23-G needle is then placed perpendicular and slightly angled caudally at the soft point and aspirated (Fig. 24.2). If the joint is penetrated, some synovial joint fluid should be visible in the syringe. The aspirated material can be examined for crystalline deposition, but if infection is suspected, no further injection should be administered

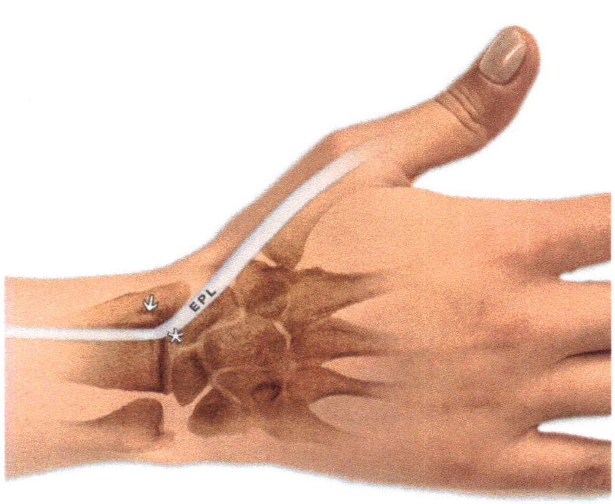


**Fig. 24.1** Wrist joint dorsal soft spot (asterisks) is 1 cm distal to the Lister's tubercle (arrow) and just ulnar border of the extensor pollicis longus

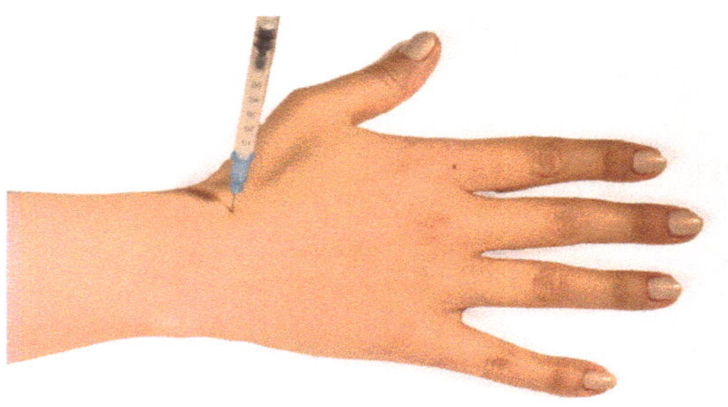

**Fig. 24.2** Wrist joint injection. Needle is perpendicular and slightly angled caudally

until the microbiologic culture results are seen. If the synovial fluid is decided to be sterile, all agents are then administered intraarticularly as a bolus.

### 24.1.6 Aftercare

Ice packs, paracetamol, and nonsteroidal anti-inflammatory drugs (NSAIDs) are needed in the first days until the inflammatory response is resolved. Splinting may help until the steroids take effect. Heavy lifting and repetitive movements should be avoided. Repeated injection may be needed if the osteoarthritis is advanced.

## 24.2 Distal Radioulnar Joint

Distal radioulnar joint (DRUJ) pain with instability and limited motion of the forearm supination and pronation after a distal radius fracture is common. Patients diagnosed with rheumatoid arthritis frequently suffer from DRUJ arthritis. Triangular fibrocartilage complex (TFCC) is an essential support for the DRUJ; their traumatic or degenerative injuries may cause ulnar-sided wrist pain.

### 24.2.1 Anatomy

DRUJ is a synovial joint and responsible from supination and pronation of forearm. Palmar and dorsal radioulnar ligaments accounts for the stability of the joint. The radius has a concavity called sigmoid notch, fits the ulnar head, which should be kept in mind during any injection.

### 24.2.2 Indication and Diagnosis

Snapping, crepitus, and decreased forearm supination-pronation may be seen in advanced arthritis. Grip strength loss may occur. Provocative tests are performed to assess DRUJ tenderness. To rule out the pain of nonunion associated with an ulna styloid fracture, any previous initial trauma should be questioned, where a radiograph can be taken for confirmation.

Repetitive wrist flexion and extension and overuse may result with extensor carpi ulnaris (ECU) tendinopathies, whereas acute trauma may bring out instability and subluxation of the tendon. The ECU synergy test is useful in diagnose [1]. Magnetic resonance imaging (MRI) is useful in identifying pathologies where the diagnosis is uncertain (Fig. 24.3). Ulnar neuropathy should be evaluated in the differential diagnosis. All these situations may benefit more or less from steroid injection; therefore, injection may be considered in the first visit for ulnar-sided wrist pain.

### 24.2.3 Appliances

Approximately 1 ml of liquid can be injected by a 23-G or thinner syringe tip.

| Needle | Syringe | Corticosteroid | Local anesthetic |
| --- | --- | --- | --- |
| 23 G | 2–3 mL | 20 mg triamcinolone (0.5 mL) 2.5 mg betamethasone (0.5 mL) | 1% lidocaine (0.5 mL) |

### 24.2.4 Agents

Corticosteroids and local anesthetics.

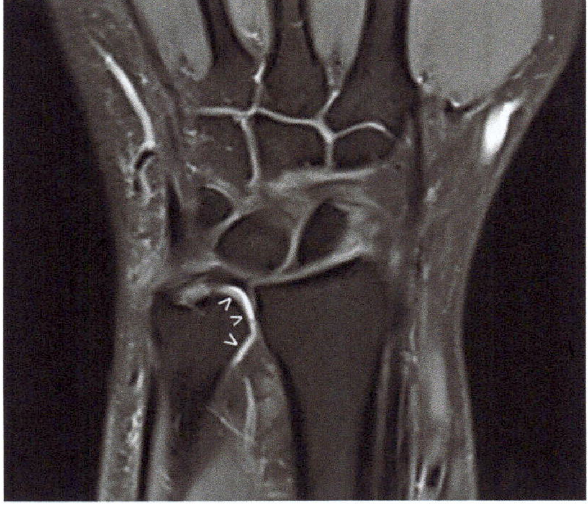

**Fig. 24.3** Magnetic resonance imaging right left wrist with distal radioulnar joint effusion (arrows)

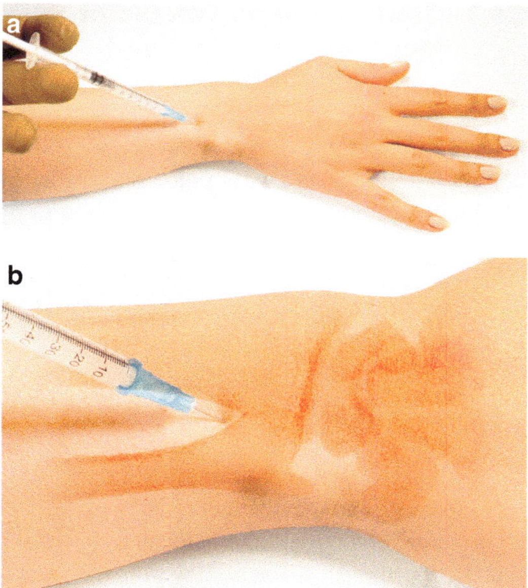

**Fig. 24.4** (a, b) Distal radioulnar joint (DRUJ) injection

### 24.2.5 Technique/Tricks/Pitfalls

With the patient sitting comfortably, the patient's hand is placed flat on the examination table. The forearm is in pronation and the elbow is in flexion. The dorsal medial border of the ulnar head is palpated and marked. In a sterile fashion, a 23-G needle is advanced in medial of the ulnar head (Fig. 24.4a, b). Ultrasonography-guided methods are more accurate, but the outcomes are similar to palpation-guided methods [2]. If an ECU tendinopathy accompany the DRUJ arthropathy, second injection to the ECU sheath may attempt. ECU is palpated dorsal site of the ulnar head and injection is performed with 27-G needle.

### 24.2.6 Aftercare

Since the splint is a reminder to avoid overuse, it is needed in the first days until the inflammatory response is resolved. Ice pack and paracetamol may help the steroids effect. Repeated injection may be needed if the osteoarthritis is advanced.

## 24.3 Trapeziometacarpal Joint (Carpometacarpal Joint of the Thumb)

Osteoarthritis of the first carpometacarpal joint is common and presents with a pain and difficulty by pinching and grasping. A contracture by adduction of the thumb may be seen in severe arthrosis. Scaphotrapeziotrapezoidal (STT) arthrosis accompanies in the late-stage disease.

### 24.3.1 Anatomy

The trapeziometacarpal joint (TMTJ) between the trapezium and the first metacarpal is a "biconcave saddle" type joint and is located just proximal to the thenar eminence, within the "fovea radialis" anatomically. The fovea radialis or the anatomical snuffbox is the area between the tendons extensor pollicis brevis (EPB) and extensor pollicis longus (EPL) tendons. The deep branch of the radial artery and the terminal branches of the radial nerve are lying in the base of this region (Fig. 24.5).

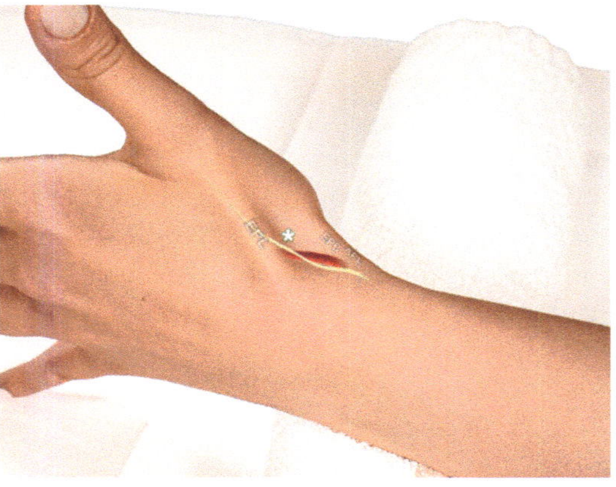

**Fig. 24.5** Left hand fovea radialis (asterisks) injection site, radial artery (red), *n* radialis digitalis dorsalis (yellow). *EPL* extensor pollicis longus tendon, *EPB* extensor pollicis brevis tendon, *APL* abductor pollicis longus tendon

### 24.3.2 Indication and Diagnosis

A progressive degenerative process is generally observed. During the consultation, patients often complain of pain around the joint that increases with activity. Grind test at the TMTJ is painful. An X-ray is sufficient to show the TMTJ arthritis. Injection is indicated at the first visit.

### 24.3.3 Appliances

Approximately 0.5–1 ml of liquid can be injected by a 23-G or thinner syringe tip.

| Needle | Syringe | Corticosteroid | Local anesthetic |
|---|---|---|---|
| 23 G | 1–3 mL | 20 mg triamcinolone (0.5 mL) 2.5 mg betamethasone (0.5 mL) | 1% lidocaine (0.5 mL) |

### 24.3.4 Agents

Corticosteroids, local anesthetics, viscosupplements, and orthobiologic agents.

### 24.3.5 Technique/Tricks/Pitfalls

The patient should be supported in a sitting position and the forearm should be placed on the table. Wrist rests on a towel in semi-supination. Ulnar deviation of the wrist and flexion and adduction of the first finger place the TMTJ more superficial. The injection may be more difficult in these patients, as osteophytes around the degenerative joint narrow the joint space. Flexion and traction may be applied from the thumb to the joint during injection, to make the joint wider and more superficial (Fig. 24.6a, b). The first metacarpal is palpated, and the joint gap is determined and marked after the anatomical "fovea radialis" is tapped. The needle should be aimed to the TMTJ perpendicularly proximal to distal with the 30 degrees of angle (Fig. 24.7a, b).

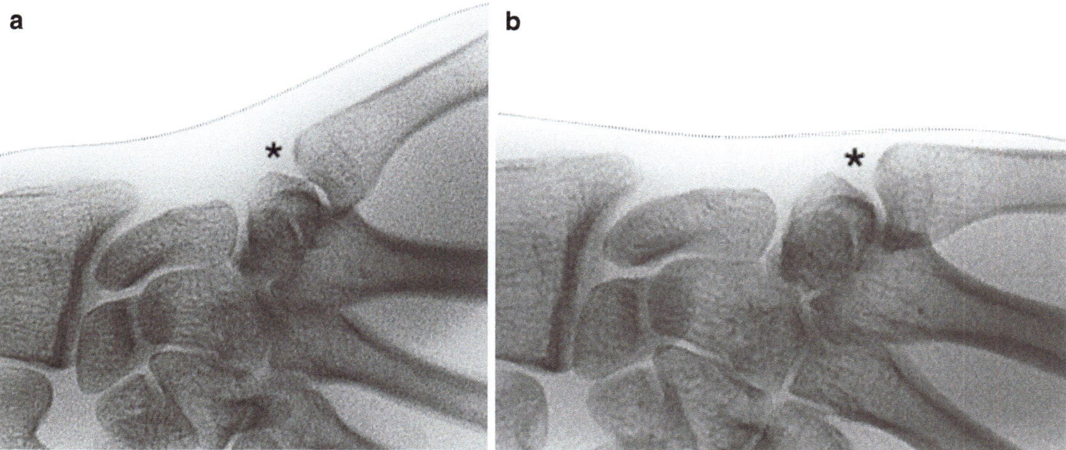

**Fig. 24.6** TMTJ (asterisks) under fluoroscopy, dotted line shows the skin. (**a**) Neutral position without traction. (**b**) Flexion position with traction

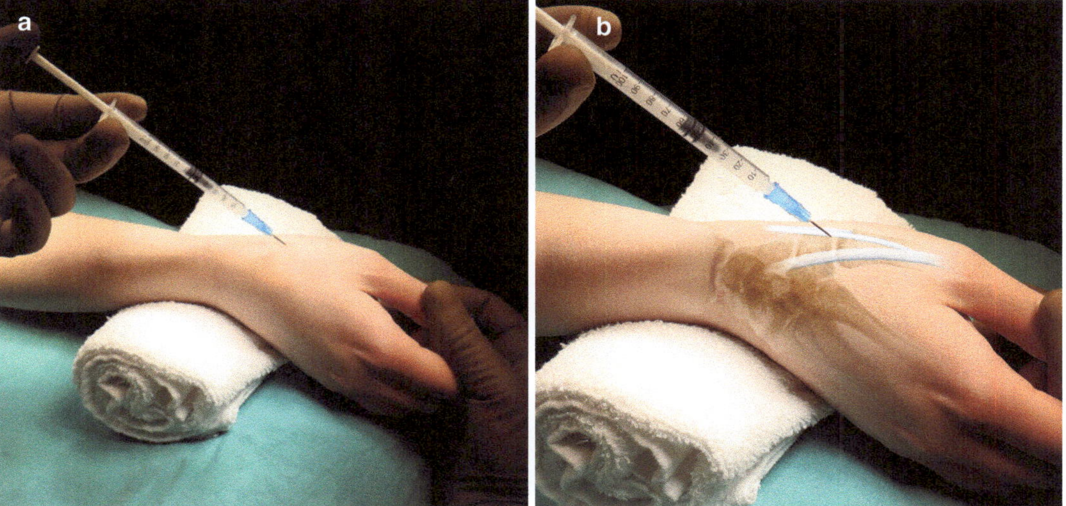

**Fig. 24.7** (**a, b**) Right hand first carpometacarpal joint injection

To protect the radial artery, the needle should be placed on the dorsal ulnar side of the EPB tendon. The superficial branch of the radial nerve should also be kept in mind. First, an injection is made to the proximal of the first metacarpi under the skin and is waited for 20–30 s. As soon as the joint border is identified, the needle is advanced perpendicular to the capsule from the dorsal surface. The content should be given at once after aspiration. Fluoroscopy or ultrasonography (USG) may assist for better accuracy and efficacy of the injection (Figs. 24.8 and 24.9). During the injection "thumbs-up" sign may be seen with proper injection [3].

### 24.3.6 Aftercare

Patient should be motivated for gentle active motion within the pain-free range and is advised against overuse of the thumb. In severe pain, splinting may help in the first days until the inflammatory response resolves. Patient should be informed that late-stage osteoarthritis may

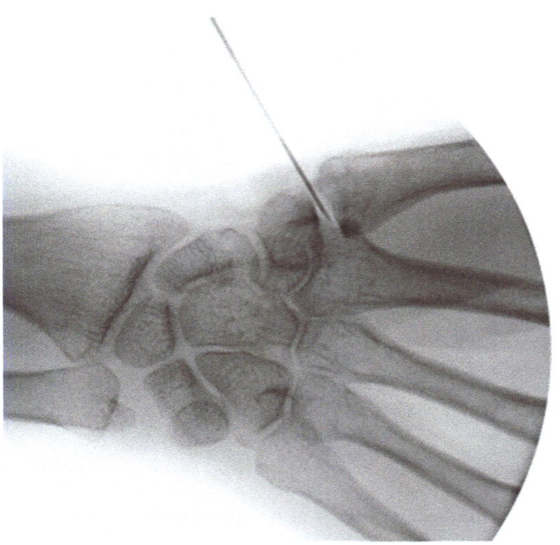

**Fig. 24.8** Fluoroscopy-guided TMTJ injection

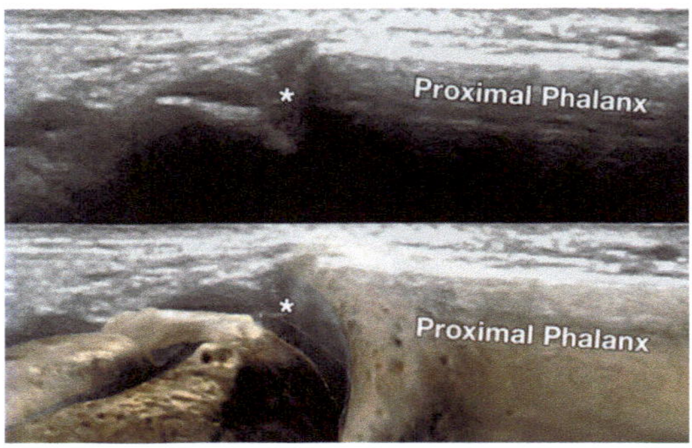


**Fig. 24.9** USG-guided injection (asterisks) TMTJ

have little benefit or only in short period [4]. After 6–12 weeks, the need for further treatment should be discussed.

## 24.4 Metacarpophalangeal and Interphalangeal Joints

The most common causes of pain or limitation of movement in the metacarpophalangeal and interphalangeal joints are degenerative and inflammatory arthritis. Rheumatology patients with pain in their fingers are seeking for pain killers. Corticosteroid injections is one way to suppress the inflammation, but access to an interphalangeal joint is challenging for any injection or aspiration.

### 24.4.1 Anatomy

In each hand, there are five metacarpophalangeal joints and nines interphalangeal joints. There is a single interphalangeal joint in the first finger, and there are two interphalangeal joints, each proximal and distal in the other fingers. Joints are strengthened by collateral ligaments. Flexor and extensor tendons pass through volar and dorsal aspects of the phalangeal bones. Digital artery and nerve packages located in the medial and lateral of the joints should be kept in mind. The volar plate is a thick fibrocartilaginous tissue close to the palmar side of the joints.

### 24.4.2 Indication and Diagnosis

The main purpose of steroid injections to the phalangeal joints is to reduce the synovitis and pain and improve the range of motion. The patient can precisely localize the affected joint. The limitation of motion and tenderness in the joint during physical examination confirms the diagnosis. In some cases, edema is evident in the affected joint. In the interphalangeal joints, flexion and extension are affected. Injection may be indicated in patients who do not benefit from nonsteroidal anti-inflammatory drug (NSAID) therapy.

### 24.4.3 Appliances

Approximately 0.2 ml of liquid can be injected by a 27-G syringe tip. Local anesthetic may cause a numbness distal to the injection area due to the neighborhood of the digital nerves to the capsule. Due to the limited volume of these joints, additional amounts of local anesthetic may be unnecessary.

| Needle | Syringe | Corticosteroid | Local anesthetic |
| --- | --- | --- | --- |
| 27–30 G | 1 mL | 20 mg triamcinolone (0.5 mL) | 1% lidocaine |
|  |  | 2.5 mg betamethasone (0.5 mL) |  |

## 24.4.4 Agents

Corticosteroids and local anesthetics.

## 24.4.5 Technique/Tricks/Pitfalls

The patient is placed or seated with support. Sterile technique is applied. The dorsolateral edge of the joint line is palpated, identified, and marked. Needle is advanced from the radial or ulnar side of the extensor tendon into the joint as the finger is in the full extension position (Fig. 24.10). The space between joint surfaces under the capsule should be identified with the tip of the needle. As osteophytes from degenerative arthritis further narrow the joint space, the possibility of a viable intracapsular injection is further reduced. Unfortunately, the joint interval cannot be expanded by traction as desired to be. Experienced clinicians cannot penetrate into the joint [5]. Periarticular injections may still be

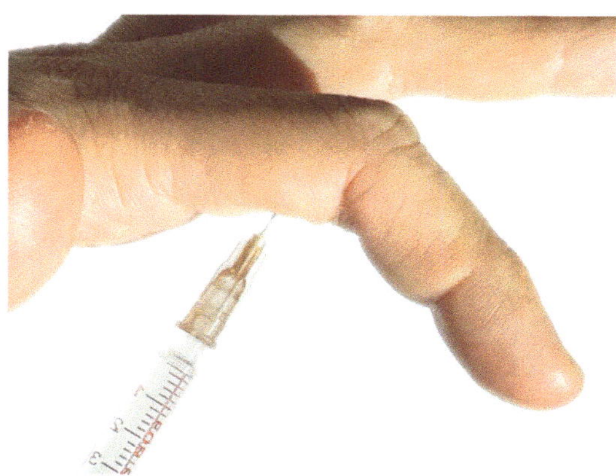

**Fig. 24.11** Volar injection of the proximal interphalangeal joint of the second finger with 27-G needle

effective, but better result has been shown with the intraarticular injection [6]. A palmar approach has been shown for proximal interphalangeal joint (PIPJ) (Fig. 24.11) [7]. With the flexion of the joint, it is possible to relax the volar plate, and empty space is created between the volar cortex of the bone and the capsule (Fig. 24.12). Various angles for injection have been suggested in the literature, but since the measurement of these angles is not practical during any attempt, 40°–60° flexion of the PIPJ and the same angulation of the needle will allow access into the capsular space [8]. Since the position of the needle is affected by the motion of the flexor tendons, the finger should flex before entering the needle (Fig. 24.13). Dorsal side of the joint is well medicated with the approach (Fig. 24.14a–c).

## 24.4.6 Aftercare

Patients should be reminded that there may be an increase in their complaints in the first several hours after injection. High-volume injection may limit full joint movement for a while. It is suggested that the injected finger should rest for 2 weeks. It is not recommended to use ice packs on fingers. Repeated steroid injections that miss the joint may damage the pericapsular soft tissue and tendon.

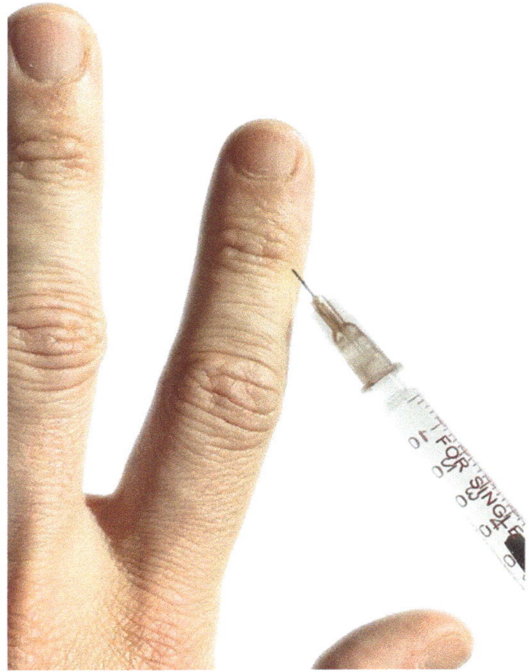

**Fig. 24.10** Dorso-radial injection of the distal interphalangeal joint of the second finger with 27-G needle

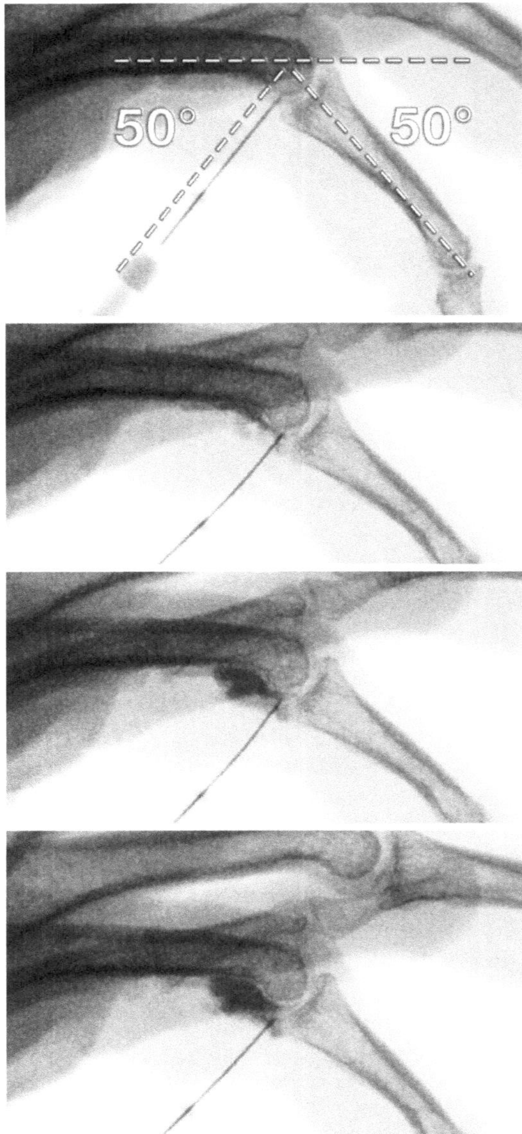

**Fig. 24.12** Volar approach of the proximal interphalangeal joint of the second finger under fluoroscopy with 27-G needle. Proximal interphalangeal joint should be held in 50° flexion, and the volar plate is relaxed to create a capsular space. The needle is 50° angled to the proximal phalanx and the joint fulfilled with radiopaque material to demonstrate joint space

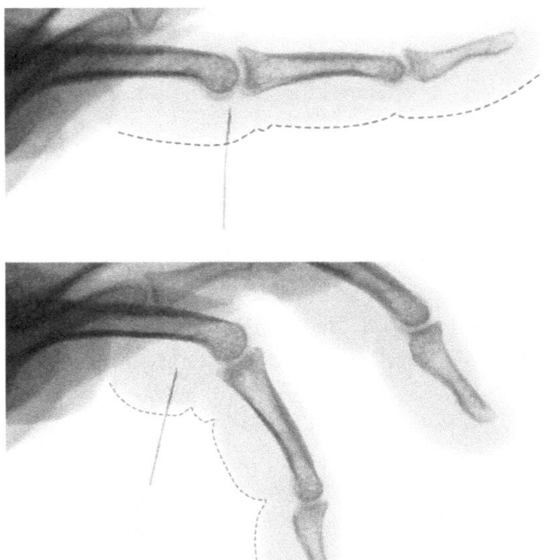

**Fig. 24.13** Volar injection of the proximal interphalangeal joint. The effect of the flexor tendon motion to the needle position and angle with flexion of the finger is demonstrated

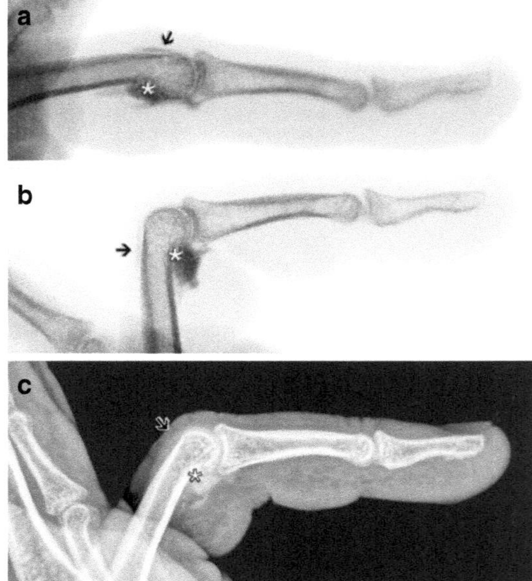

**Fig. 24.14** Fluoroscopy (**a**, **b**) and arthrography (**c**) of the proximal interphalangeal joint after the volar injection. (arrow) The dorsal escape of the volar injection material proves an appropriate joint infiltration. The capsular space (asterisks) created with the relaxation of the volar plate demonstrated with radiopaque material. (**a**) Extension of the finger brings the volar plate close to the proximal phalanx. (**b**) Flexion of the finger relaxes the volar plate and forms a capsular gap. (**c**) Arthrography, 30 min after the injection. Material is well dispersed within the joint

## 24.5 Ganglion Cyst

Ganglion cysts are among the most common tumors in the wrist and hand. These synovial cysts are filled with mucin and can originate from any type of a synovial joint. It is found mostly on the dorsal side of the wrist; however, it can be seen less frequently on the volar side. Ganglion cysts are also associated with the tendon and the tendon sheaths conditions like de Quervain's disease. They are identified as mucous cysts, if attached dorsolateral side of the distal interphalangeal joint.

### 24.5.1 Anatomy

Ganglion cysts are collagen fiber sacs connecting with a duck to the joint or tendon sheath. Dorsal-sided cysts arise from scapholunate ligament. Volar ganglions originate from radiocarpal joint and may wrap the radial artery or its branches.

### 24.5.2 Indication and Diagnosis

Patients refer with a pain on the ganglion with the full flexion or extension of the wrist like when they push the door or do push-ups. Inability to exercise, cosmetic concerns, worries of malignancy, or exacerbation of pain with intense use of the arm are the main complaints of admission. Big cysts may compress the adjacent median or ulnar nerve and cause neuropathies. Transillumination of the cysts, anechoic lesion with well-defined margins in USG without any vessel, and the aspiration of the viscous jelly-like mucin without any blood are efficient for diagnosis (Fig. 24.15). If any doubt arises in differential diagnosis, magnetic resonance imaging (MRI) is needed before any intervention.

### 24.5.3 Appliances

Eighteen-gauge needle is needed to aspirate the highly viscous and sticky content. It is easy to apply force to a smaller syringe due to hydraulics

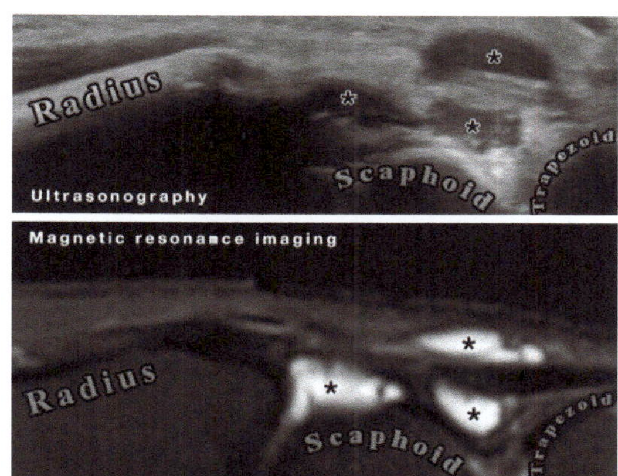


**Fig. 24.15** Dorsal wrist ganglion cyst (asterisk) aspiration and steroid injection

principles. After the aspiration, corticosteroids injection through the same needle reduces the symptoms for a while. Injected steroid will disperse all over the joint through its pedicle.

| Needle | Syringe | Corticosteroid | Local anesthetic |
|---|---|---|---|
| 18 G | 2 mL | 40 mg triamcinolone (1 mL) 5 mg betamethasone (1 mL) | 1% lidocaine |

### 24.5.4 Agents

Corticosteroids.

### 24.5.5 Technique/Tricks/Pitfalls

The patient is placed or seated with support. Sterile technique is applied. Ganglion cysts is palpated. Dorsal wrist ganglions become evident with the wrist flexion (Fig. 24.16). Needle is inserted into the cyst without any hesitation for dorsal large cyst. Small ganglia, especially those

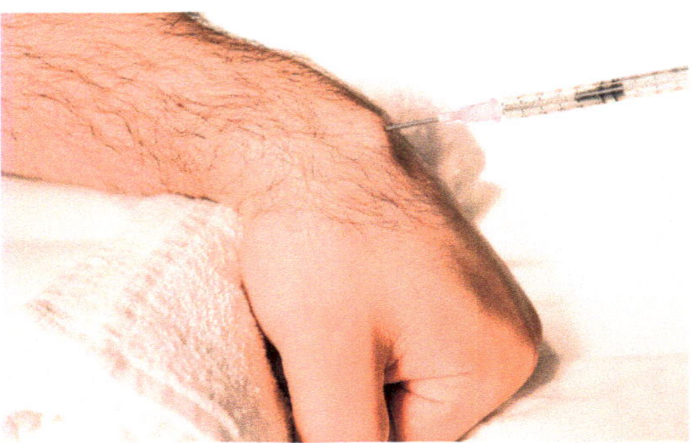

**Fig. 24.16** Dorsal wrist ganglion cyst aspiration and steroid injection

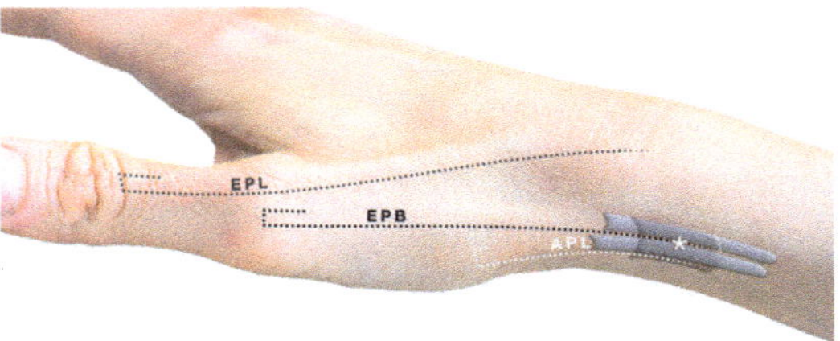


**Fig. 24.17** First extensor compartment with separate tendon sheaths (asterisks). White dotted line: abductor pollicis longus tendon (APL). Black dotted line: Extensor pollicis brevis tendon (EPB). Gray dotted line: Extensor pollicis longus tendon (EPL)

originating from the tendon sheath, may slip off the tip of the needle. The syringe should be kept at negative pressure, and the procedure should be repeated until the aspiration material is visible in the syringe. After the cyst is emptied, the syringe is changed, and steroid injection is applied if desired.

### 24.5.6 Aftercare

Aspiration, perforation of the cyst wall, and injection of steroids can manage pain for weeks. The patient should be told that most of the cyst will relapse. Wrist splint will help control the symptoms.

## 24.6 De Quervain's Tenosynovitis

De Quervain's tenosynovitis is the entrapment of the abductor pollicis longus (APL) and the extensor pollicis brevis (EPB) at the radial side of the wrist. Repeated wrist bending and twisting activities are the main risk factors for tendinopathy.

### 24.6.1 Anatomy

Both APL and EPB tendons run through the first dorsal extensor compartment on the radial styloid. Distal to the bone edge, both tendons build the anterior border of the fovea radialis. This retinacular compartment may have a septation for each tendon (Fig. 24.17). This septum may inhibit the injected steroid from spreading into both sheaths.

### 24.6.2 Indication and Diagnosis

A sharp pain is felt on the radial side of the wrist and aggravated by the ulnar deviation of the wrist while holding the thumb fist. Physical examination is sufficient for diagnosis and further imaging studies are not needed. Corticosteroid injection attempt at the first visit is logical since it is more efficacious than the splinting alone [9].

### 24.6.3 Appliances

Approximately 1 ml of liquid can be injected by a 27-G syringe tip. Local anesthetic may cause a numbness in the innervation area of the superficial branch of the radial for several hours.

| Needle | Syringe | Corticosteroid | Local anesthetic |
| --- | --- | --- | --- |
| 25–27 G | 1–3 mL | 20 mg triamcinolone (0.5 mL) | 1% lidocaine (0.5 mL) |
| | | 5 mg betamethasone (1 mL) | |

### 24.6.4 Agents

Corticosteroids.

## 24.6.5 Technique/Tricks/Pitfalls

The patient is seated; the forearm is held in the neutral position. The wrist is supported with a towel in ulnar deviation. The two tendons are palpated on the styloid process of radius. The needle inserted along the line of the tendon to the distal edge of the extensor compartment and advanced retrogradely toward into the tendon sheaths (Fig. 24.18). With the injection of the liquid, a bump will appear above the proximal edge of the first compartment. The needle should be pulled slowly before draining all the solution, and it should hit the neighbor tendon sheath by passing the septum. The patient may benefit from the allocation of the steroid to both tendon sheaths.

## 24.6.6 Aftercare

Superficial injection of the corticosteroid injection has the risks of adipose tissue atrophy, skin hypopigmentation, and thinning. The patient must be informed about these risks and repeated multiple injection should be avoided. Because the local anesthetics will affect superficial branch of the radial nerve over tendon, numbness of the dorsal skin of the thumb will be felt for an hour.

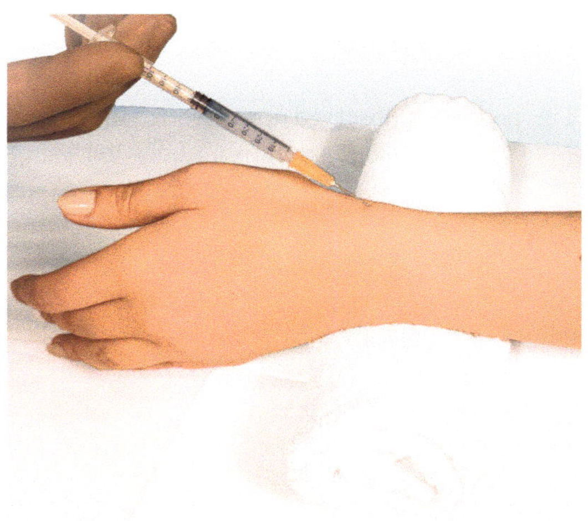

**Fig. 24.18** Injection position for de Quervain's tenosynovitis

Patients should be warned that the effect of the steroid begins after a few days. The patient should rest and avoid painful movements. Additional splinting may reduce the symptoms till the steroid take effect [10]. A splint should limit the wrist and thumb motion for proper care.

## 24.7 Carpal Tunnel Syndrome

Carpal tunnel syndrome is the most diagnosed site of nerve compression in the upper extremity. Nocturnal paresthesia of the median nerve distribution area is almost pathognomonic with the weakness of the thenar muscles.

### 24.7.1 Anatomy

Carpal tunnel is bordered dorsally by the carpal bones and the volar by the transverse carpal ligament (flexor retinaculum). Flexor retinaculum extend from the hamate and triquetrum to the scaphoid and trapezium and beneath the roof pass the median nerve and the flexor tendons of the hand. Palmaris longus tendon is just above and slightly on the ulnar side of the median nerve.

### 24.7.2 Indication and Diagnosis

The Phalen test is a provocation test for the evaluation of carpal tunnel syndrome. Flexion of the wrist for a minute increases the pressure on the median nerve, revealing the symptoms of pain and numbness. Direct digital pressure or percussion on the median nerve at the wrist by the examiner may give similar findings. The patient may complain of a tingling sensation that radiates into the sensory distribution of the median nerve. Electrodiagnostic studies are used to confirm physical findings and determine the extent of disease. Carpal tunnel injection can be applied in patients with median nerve compression, when in physiotherapy, resting splint and when NSAIDs are unresponsive.

### 24.7.3 Appliances

Approximately 0.5 ml of liquid can be injected by a 23-G syringe tip. Local anesthetic may cause an uncomfortable numbness in the fingers and palm for several hours.

| Needle | Syringe | Corticosteroid | Local anesthetic |
|---|---|---|---|
| 23–25 G | 1–3 mL | 20 mg triamcinolone (0.5 mL) | Nil |

### 24.7.4 Agents

Corticosteroids.

### 24.7.5 Technique/Tricks/Pitfalls

The patient is seated with some support, the forearm is kept in supination, and the wrist is extended. Before the injection, the proximal wrist is folded, and the palmaris longus tendon are palpated. The palmaris longus tendon is best noticed when all fingertips are clenched in the neutral position of the wrist (Fig. 24.19). Digital flexor tendons are palpated with the fingers in flexion and extension. The median nerve is at the radial of the palmaris longus and between this tendon and the flexor carpi radialis.

Sterile technique is applied; structures are marked. The injection is applied from the proximal wrist fold right from the ulnar edge of the palmaris longus tendon (Fig. 24.20). The angle of the needle should be 30° to the flexor tendons in the carpal tunnel.

The needle is advanced gently and the injection is continued. Swelling in the skin means that

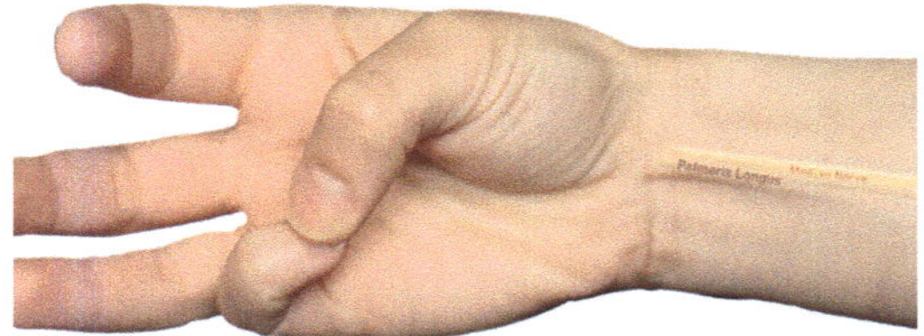

**Fig. 24.19** Opposition of the thumb and fifth finger distinct the palmaris longus tendon

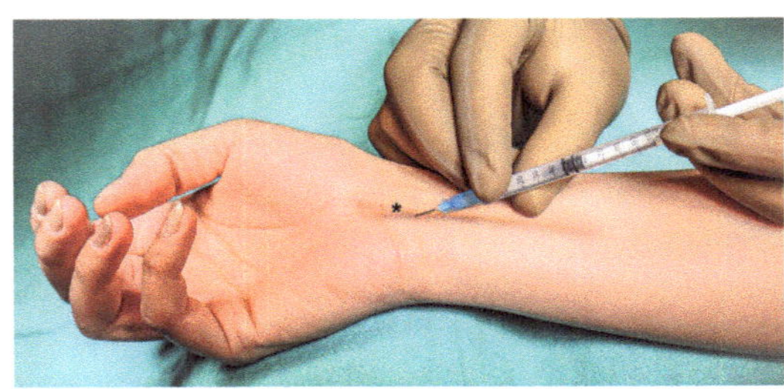

**Fig. 24.20** Injection site for median nerve, ulnar of the (asterisks) palmaris longus tendon (demonstrated)

the needle is in the superficial of the deep fascia. The sense of the flexor tendons motion toward the tip of the needle indicates that the needle is at the proper depth. If the needle moves with finger movements, the needle is inside the tendon and must be removed. Corticosteroid should be injected slowly, without any resistance. If the needle encounters resistance or patients experience paresthesia, the needle should be retracted and redirected a little bit more toward the ulna. The injection is administered slowly but with continued pressure. In patients without palmaris longus tendon, the midline formed between the thenar and hypothenar regions during the thumb and fifth finger opposition helps the location of the medial nerve within the carpal tunnel.

### 24.7.6 Aftercare

The patient is observed for a while. In the first 24–48 h, information is given about the symptoms that may worsen due to corticosteroids. Similarly, NSAIDs are recommended. Normal activities can be resumed after resting for a few days after the injection. Night splinting helps reduce the symptoms.

## 24.8 Trigger Finger (Stenosing Tenosynovitis)

The snapping or locking of the flexor tendons of the thumb or other fingers as a result of stenosing tenosynovitis is called "trigger finger," named by the analogy with the movement of the finger on the trigger of a gun. Mostly the ring finger is affected, and the thumb comes second.

### 24.8.1 Anatomy

The tendon pathology is observed at the A1 pulley level. A1 pulley is a rigid, fibrous annular band located volar to the metacarpophalangeal joint. A1 pulley projection to the skin is about 1 cm distal to the distal palmar crease.

The digital nerves are located near both sides of the tendon. The radial digital nerve may intersect the A1 pulley, specifically when the thumb is held in pronation. To protect the nerve from an injury of the needle tip, first wrist should be flexed to face the volar side to the examiner correctly (Fig. 24.21a, b).

### 24.8.2 Indication and Diagnosis

Patients typically blame their proximal interphalangeal joint for catching and pain. A1 pulley should be palpated to identify the source of complaints. Diagnosis is usually made only by physical examination. Differential diagnosis with Dupuytren's contracture and tendon sheath tumors should be made before any injection.

### 24.8.3 Appliances

Approximately 0.5 ml of liquid can be injected by a 27-G syringe tip. Patients may experience an exacerbation of symptoms with the high volume of injections.

| Needle | Syringe | Corticosteroid | Local anesthetic |
| --- | --- | --- | --- |
| 25–27 G | 1–3 mL | 10 mg triamcinolone (0.25 mL) 2,5 mg betamethasone (0.5 mL) | 1% lidocaine (0.25 mL) |

### 24.8.4 Agents

Corticosteroids and local anesthetics.

### 24.8.5 Technique/Tricks/Pitfalls

Unlike the aforementioned diseases, corticosteroid injection can be applied for treatment in the early phase of this stenosing tenosynovitis in the first visit. The patient is supported in a sitting position; the forearm is comfortably held in the supine position. The flexor sheath is palpated on

**Fig. 24.21** Digital nerves are adjacent to the A1 pulley (asterisks) of the thumb. (**a**) In relaxed position of the hand, the radial digital nerve overlaps the A1 tendon from the examiners view. (**b**) Entry site is clear when all of the volar thumb area faces the examiner

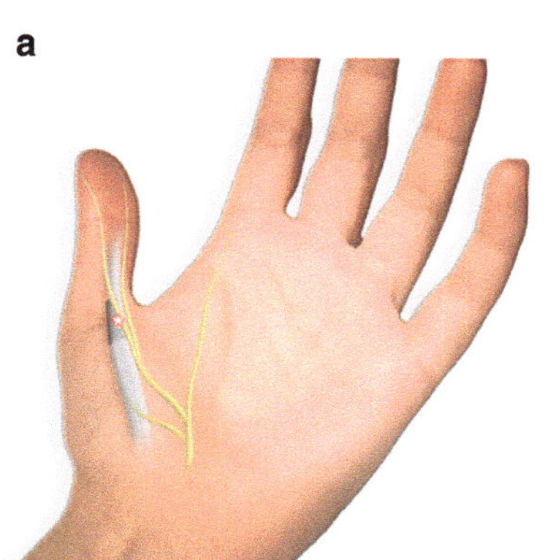
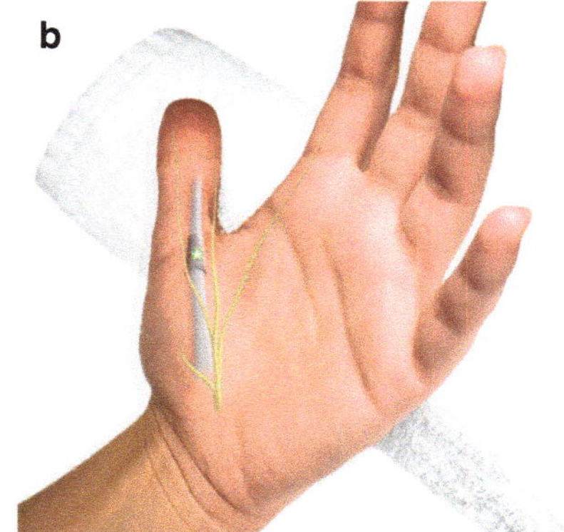

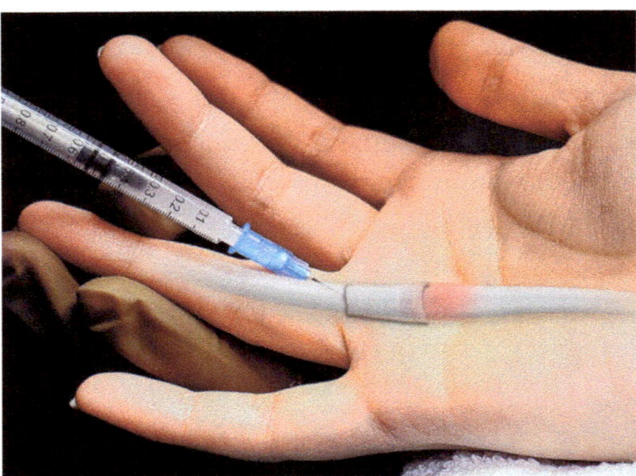

**Fig. 24.22** Corticosteroid injection to the triggering tendon of the fourth finger

the affected finger, and its line is determined with the tip of the finger. The nodule due to tenosynovitis can often be palpated at the level of the metacarpal bone head where the affected tendon passes. If not found, active flexion and extension of the fingers may uncover the area of triggering. The needle is inserted with the bevel down, distal to a1 pulley and at an angle of about 30° to the palmar face and advanced retrogradely toward the nodule (Fig. 24.22). If the needle is advanced too deep, the flexor tendon will be penetrated. In this situation, the needle will move along with the finger movements. Some authors suggest this test to be sure that the needle bevel is under the pulley in the tendon sheath (Fig. 24.23).

After the penetration of the tendon, move the needle bevel away from the tendon into the tendon sheath with a pressure of the syringe, where it can be observed that the fluid passes through when the resistance is gone. After the penetration of the tendon, move the needle bevel away from the tendon into the tendon sheath with a slight pressure to the syringe, where it can be observed that the fluid passes through the space between the tendon and its sheath, when the resistance gone. It is still effective, if the sheath is passed and the steroid has been applied over the A1 pulley under the skin [11].

Percutaneous release can be tried after the effect of local anesthetics is seen. The needle should be replaced with 21 G or lower (thicker) (Fig. 24.24) [12].

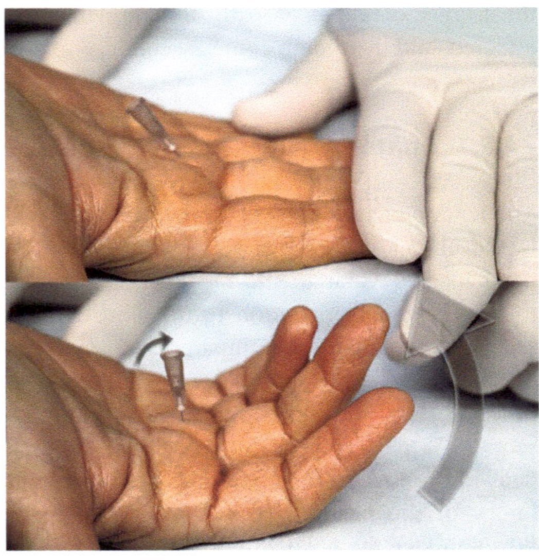

**Fig. 24.23** The needle passes the A1 pulley penetrates the tendon. If the needle is deep enough, it accompanies the finger movement

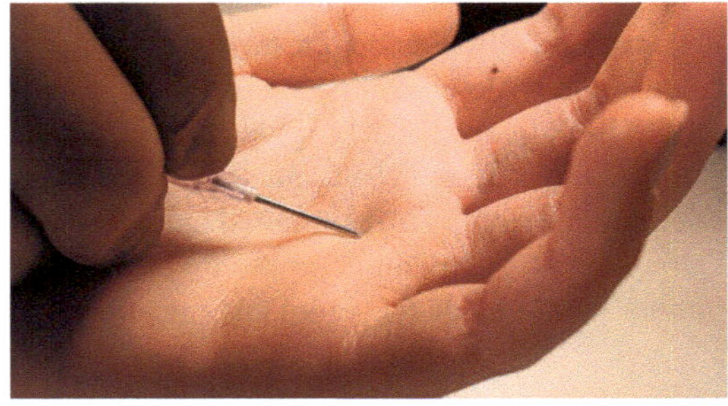

**Fig. 24.24** Percutaneous release of the A1 pulley

## 24.8.6 Aftercare

Patients may experience an exacerbation of symptoms that resolve within the first 48 h. The local anesthetics will affect the digital nerve beside the tendon; because of that numbness of the finger will be felt for an hour. Patients should be warned that the effect of the steroid begins after a few days. Pain and triggering resolve completely by 6 weeks. No restriction is mandatory. Surgical A1 pulley release is a definite solution if the symptoms persist.

# References


1. Ruland RT, Hogan CJ. The ECU synergy test: an aid to diagnose ECU tendonitis. J Hand Surg Am. 2008;33:1777–82. https://doi.org/10.1016/j.jhsa.2008.08.018.
2. Nam SH, Kim J, Lee JH, Ahn J, Kim YJ, Park Y. Palpation versus ultrasound-guided corticosteroid injections and short-term effect in the distal radioulnar joint disorder: a randomized, prospective single-blinded study. Clin Rheumatol. 2014;33:1807–14. https://doi.org/10.1007/s10067-013-2355-7.
3. Erpelding JM, Shnayderman D, Mickschl D, Daley RA, Grindel SI. The "thumbs-up" sign and trapeziometacarpal joint injection: a useful clinical indicator. Hand. 2015;10:362–5. https://doi.org/10.1007/s11552-014-9683-1.
4. Fowler A, Swindells MG, Burke FD. Intra-articular corticosteroid injections to manage trapeziometacarpal osteoarthritis—a systematic review. Hand. 2015;10:583–92. https://doi.org/10.1007/s11552-015-9778-3.
5. Pichler W, Grechenig W, Grechenig S, Anderhuber F, Clement H, Weinberg AM. Frequency of successful intra-articular puncture of finger joints: influence of puncture position and physician experience. Rheumatology. 2008;47:1503–5. https://doi.org/10.1093/rheumatology/ken295.
6. Sibbitt WL, Peisajovich A, Michael AA, Park KS, Sibbitt RR, Band PA, Bankhurst AD. Does sonographic needle guidance affect the clinical outcome of intraarticular injections? J Rheumatol. 2009;36:1892–902. https://doi.org/10.3899/jrheum.090013.
7. McClelland WB, McClinton MA. Proximal interphalangeal joint injection through a volar approach: anatomic feasibility and cadaveric assessment of success. J Hand Surg Am. 2013;38:733–9. https://doi.org/10.1016/j.jhsa.2013.01.014.
8. Saito S, Suzuki S, Ishikawa K. Letter regarding "proximal interphalangeal joint injection through a volar approach: anatomic feasibility and cadaveric assessment of success". J Hand Surg Am. 2013;38:1261–2. https://doi.org/10.1016/j.jhsa.2013.03.062.
9. Ashraf MO, Devadoss VG. Systematic review and meta-analysis on steroid injection therapy for de Quervain's tenosynovitis in adults. Eur J Orthop Surg Traumatol. 2014;24:149–57. https://doi.org/10.1007/s00590-012-1164-z.
10. Mardani-Kivi M, Karimi Mobarakeh M, Bahrami F, Hashemi-Motlagh K, Saheb-Ekhtiari K, Akhoondzadeh N. Corticosteroid injection with or without thumb spica cast for de quervain tenosynovitis. J Hand Surg Am. 2014;39:37–41. https://doi.org/10.1016/j.jhsa.2013.10.013.
11. Kazuki K, Egi T, Okada M, Takaoka K. Clinical outcome of extrasynovial steroid injection for trigger finger. Hand Surg. 2006;11:1–4. https://doi.org/10.1142/S0218810406003115.
12. Eastwood DM, Gupta KJ, Johnson DP. Percutaneous release of the trigger finger: an office procedure. J Hand Surg Am. 1992;17:114–7. https://doi.org/10.1016/0363-5023(92)90125-9.

Bruno Capurro, Francesco Vecchi,
Beatriz Álvarez de Sierra, Alex Ortega,
Laura Gimeno-Torres, and Eva Llopis


**Technical Note**

As a common denominator for all the techniques described in the next chapter, an informed consent must be read and signed by the patient prior to the infiltrations. A sterile technique will be required, where the patient's skin area to be treated is sterilized with a chlorhexidine swab (avoid Betadine as it can stain the ultrasound transducer); sterile ultrasound gel is used just above the target injection site. In addition, it is advisable to capture the image before and after the procedure where the correct path of the needle is observed. For anesthetic purposes, ethyl chloride can be applied to the skin, but it should not come into direct contact with the transducer as it can damage it. It should always be aspirated prior to the infiltration to ensure that it is not in a blood vessel, and afterward the infiltration can be proceeded. Once the injection is completed, the needle should be withdrawn maintaining visualization and a small bandage applied so that the patient can mobilize.


B. Capurro (✉)
Department of Orthopaedics and Sports Traumatology, IMSKE Hospital - European Musculoskeletal Institute, Valencia, Spain

European Hip Preservation Associates, ESSKA–EHPA, Eich, Luxembourg

Iberian Group of Hip Preservation Surgery (GIPCA), Portugal, Spain

Muscle and Tendon Study Group (GELMUT) Asociación Española de Artroscopia – AEA, Madrid, Spain

F. Vecchi
Department of Orthopaedics and Sports Traumatology, IMSKE Hospital - European Musculoskeletal Institute, Valencia, Spain

B. Álvarez de Sierra
Department of Radiology, Clínica Universidad de Navarra, Madrid, Spain

A. Ortega
Department of Orthopaedics and Sports Traumatology, IMSKE Hospital - European Musculoskeletal Institute, Valencia, Spain

Hospital Clínico Universitario de Valencia, Valencia, Spain

L. Gimeno-Torres
Department of Orthopaedics and Traumatology, Centro Hospitalario Durango, Ciudad de México, México, Mexico

E. Llopis
Department of Orthopaedics and Sports Traumatology, IMSKE Hospital - European Musculoskeletal Institute, Valencia, Spain

Department of Radiology, IMSKE Hospital - European Musculoskeletal Institute, Valencia, Spain



© The Author(s), under exclusive license to Springer Nature Switzerland AG 2024
B. Kocaoglu et al. (eds.), *Musculoskeletal Injections Manual*,
https://doi.org/10.1007/978-3-031-52603-9_25

## 25.1 Hip Joint

The intra-articular ultrasound-guided (US) injection technique in the hip performed in-office has shown superiority compared to injections guided by fluoroscopy [1]. It presents high reliability, an excellent degree of satisfaction and convenience for patients, in addition to requiring a minimal learning curve for its execution [1].

This technique can be used to infiltrate interarticular corticosteroids for the conservative treatment of hip osteoarthritis and in non-arthritis diseases such as femoroacetabular impingement, where its use has also increased, mainly for diagnostic purposes [2]. It can also be used for intra-articular viscosupplementation and in the expanding role of orthobiologics [3]. The advantages and disadvantages of this infiltration technique are indicated in Table 25.1.

### 25.1.1 Procedure

Aim: To enter the joint capsule on the anterolateral surface of the femur neck at the femoral head-neck junction to avoid chondral damage.

Position: Supine position, neutral rotation of the leg, or slight internal rotation 15° (Fig. 25.1).

US preview: A linear probe can be used with a frequency between 8 and 14 MHz and preferably a convex probe with a frequency between 3 and 6 MHz. Inspection of the hip anatomy is performed with the long axis allowing visualization of the anterior edge of the acetabulum, femoral head, femoral neck, and joint capsule (if there is effusion, its aspiration can be planned in advance). In addition, routine Doppler visualization of the femoral neurovascular bundle (located medially) and the lateral circumflex femoral artery is recommended to avoid iatrogenesis (Fig. 25.2a) [5].

Injection technique: Using a sterile and aseptic method, the needle should enter the joint capsule as lateral as possible while maintaining long-axis visualization of the femoral neck and joint capsule to avoid contact of the needle with the femoral vessels (Fig. 25.3) [5]. Under in-plane visualization with the transducer, a 3.5-inch 22-G sterile spinal needle is inserted, bevel up, approximately 1 cm from the distal portion of the ultrasound transducer until it pierces the joint capsule at the femoral head neck junction. When the needle is in the correct position within the capsule, the drug can be injected and visualized entering the joint capsule (Fig. 25.2b).

Technical note: If the patient is obese or if it is difficult to immediately find the femoral neck in a long-axis view, the greater trochanter can be visualized in the short axis and followed to the femoral neck, and then rotate the axis 35–45° proximal and then allow the femoral head neck junction to be seen on the long axis.

**Table 25.1** Ultrasound-guided intra-articular injection: advantages and disadvantages (adapted from Bardowsky et al. [4])

| Advantages | Disadvantages |
| --- | --- |
| 1. Fast learning curve | 1. User dependent |
| 2. Cost effective and convenient for the patient | 2. The provider uses more in-office time to perform |
| 3. More accurate than landmark injections | 3. Upfront cost of ultrasound equipment and supplies |
| 4. Less painful than fluoroscopic guided and no radiation exposure | 4. Limited by the characteristics of the patient's body |
| 5. Can visualize joint effusion and aspirate if needed | |
| 6. Allows for immediate postinjection reassessment/ real-time information | |

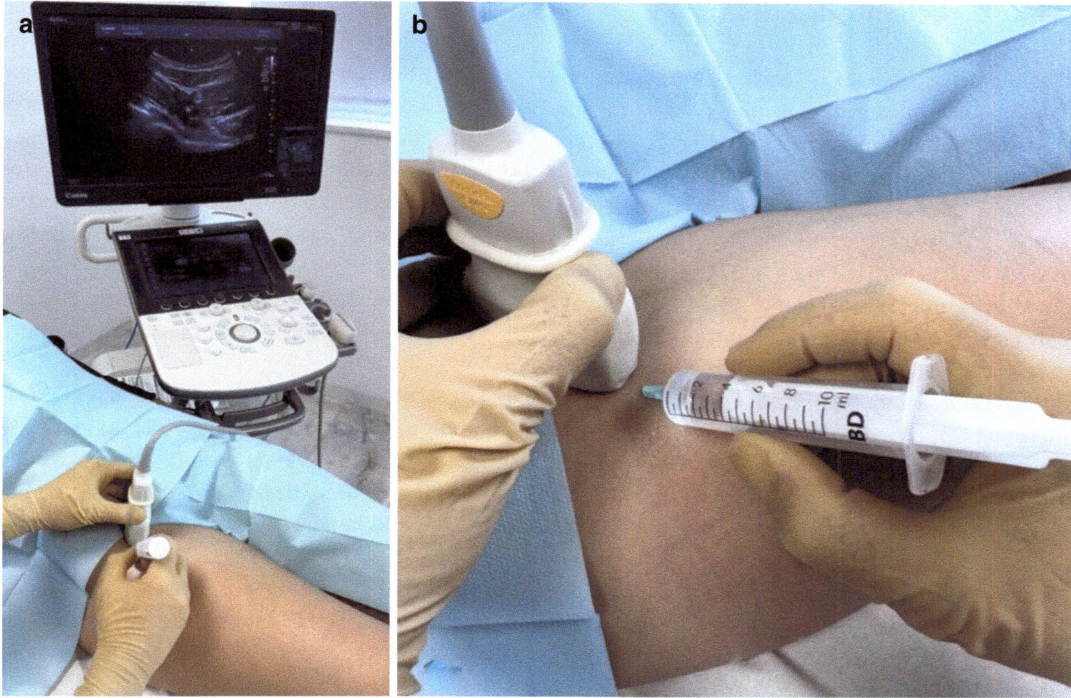

**Fig. 25.1** Position for optimal US hip joint visualization. (**a**) Infiltration position with US preview. (**b**) Long axis and position for infiltration of a normal hip

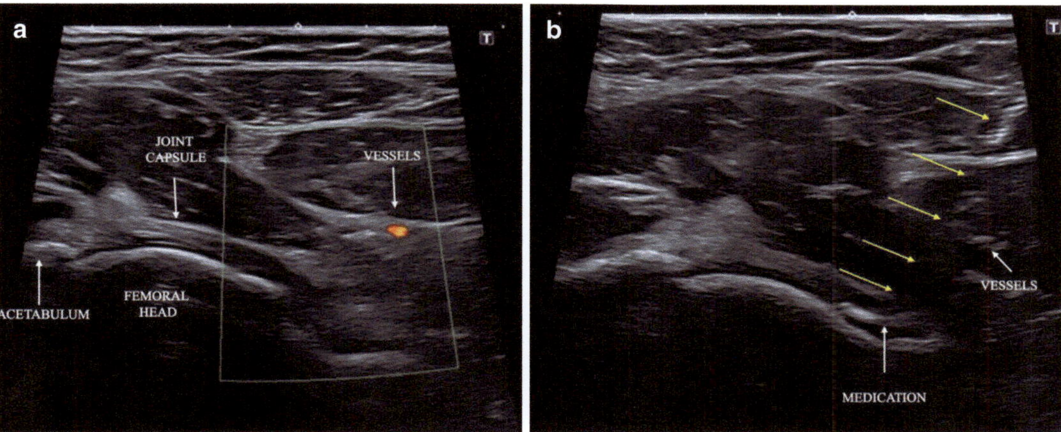



**Fig. 25.2** Intra-articular US-guided hip injection. (**a**) Ultrasound-guided approach in the anterior joint recess of the hip (anechoic), it is recommended to look for the circumflex vessels (red Doppler area). (**b**) Intra-articular hip ultrasound-guided injection technique. A flash effect is observed as the intra-articular fluid has penetrated the anterior recess, elevating the capsule (white arrow). Needle direction (yellow arrow) avoiding damaging the circumflex vessels

## 25.2   Trochanteric Bursitis

The greater trochanter is located on the proximal femur and is a large bulge that arises from the junction of the femoral neck and shaft [6]. The anatomy of the greater trochanter is made up of four facets: anterior, lateral, posterior, and superoposterior [7] and seven muscles attach to the greater trochanter, but their anatomical relationships are complex to understand and not always constant due to anatomical variants of the tendons. Schematically the gluteus medius is inserted on the lateral and superoposterior facets, the gluteus minimus is inserted on the anterior facet, the piriformis is located superomedially without presenting a specific facet insertion, the external obturator is more medial, and the internal obturator is adjacent together with the superior and inferior Geminus insert into the trochanteric fossa [6, 8, 9].

There are three trochanteric bursae described, and it is important to understand that they are not visualized in all cases unless they are filled with synovial fluid (Fig. 25.3) [7]:

(a) The subgluteus maximus (greater trochanteric) bursa is the largest and is located over the posterior and lateral facets, being separated from the trochanter by the insertions of the gluteus medius.
(b) The subgluteus medius bursa is located between the lateral facet and gluteus medius tendon and can extend posteriorly in the trochanter.
(c) The subgluteus minimus bursa is located between the anterior facet and gluteus minimus tendon.

Greater trochanteric pain syndrome (GTPS) is a common pathology that can affect up to 17.6% of the population [10]. It mainly affects women in their fourth to sixth decade, presenting lateral hip pain and tenderness, which worsens with walking, lying on the affected side and/or stair climbing [11]. Multiple diagnoses are included within the GTPS; US and MRI can be used for diagnosis and most often reveal pathologic conditions involving the gluteus medius and minimus tendons, including tendinosis, calcifications, and tears. Although, the gluteal tendinopathy with or without bursitis has been identified as the primary source of pain and dysfunction, other causes such as trochanteric bursitis, external snapping hip, proximal iliotibial band syndrome should also be considered [12].

Initial conservative treatment for GTPS includes activity modification, weight loss, ice, and physical therapy to address strength and flexibility deficits [6, 13]. Corticosteroid injections are frequently used concomitantly with conservative treatment, resulting in pain reduction in the first 4–8 weeks in mild and moderate tendinopathy and improvement in function and pain at rest and during activity. However these effects do not last more than 3–6 months, and their effectiveness decreases as time goes by [14, 15]. Corticosteroid US-guided infiltration targeting the greater trochanteric bursae carries a low risk of side effects and is shown to be more effective than blind or fluoroscopy infiltrations; however, prolonged steroid use can result in possible tendon degeneration and rupture, which is why it is not recommended to repeat it more than once every 3–4 months [16]. Additionally, studies have

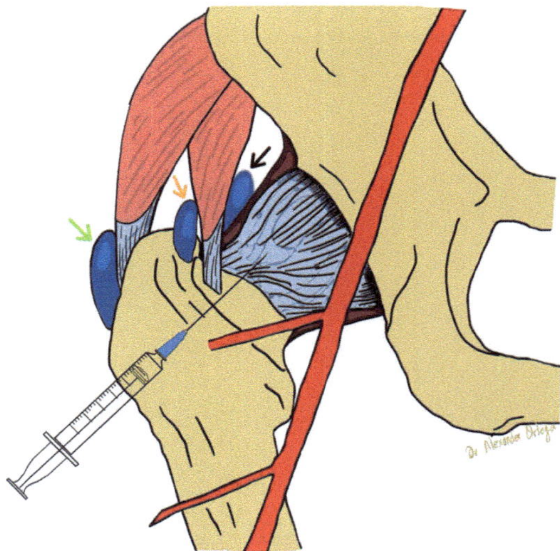

**Fig. 25.3** Hip joint and trochanteric bursae illustration. Illustration of the intra-articular infiltration of the hip and the direction where the needle must go in relation to the anatomy to avoid contact of the needle with the femoral vessels and the three trochanteric bursae: subgluteus maximus bursa (green arrow); subgluteus medius bursa (orange arrow); and subgluteus minimus bursa (black arrow)

demonstrated inhibitory effects of tendon repair and delayed tendon sheath healing [12, 13]. Platelet-rich plasma (PRP) infiltrations have increased their use compared to corticosteroids in gluteal tendinopathy, as they promote tissue healing through a high concentration of platelet-derived growth factors that help activate the healing cascade and reverse the degenerative process [17]. Systematic reviews have been reported showing improved outcomes for PRP at 2 years compared to a CSI as measured by the Harris Hip Score [18] and reported a sustained benefit through 2 years following PRP in randomized controlled trials [12, 19]. Other conservative treatment modalities, such as ultrasound-guided percutaneous needle tenotomy, prolotherapy, and shock waves, have been described and may be considered [8, 20]. Although all of these injections and therapeutic options can improve symptoms, physical therapy should be used after corticosteroids and PRP injections as it improves modifiable factors, including hip abductor weakness, band tension iliotibial pain, and loss of pelvic control in the frontal plane, which are present in the pathophysiology of GTPS [12]. Also when conservative treatment fails, surgical management should be considered [11, 12]. However, more randomized controlled trials are needed to compare the efficacy of these treatments to determine the best treatment for GTPS.

## 25.2.1 Procedure

Aim: To infiltrate the greater trochanteric bursa and the others when they are inflamed; in the case of corticosteroids associated with local anesthetic outside the tendon or with PRP that allows intra-tendinous infiltration if the tendon is affected and associated with ultrasound needling to seek resolution of enthesopathy and dissolution of enthesophytes.

Position: Lateral decubitus on the unaffected side, hip and knee semi-flexed for patient comfort and visualization in both axes with the US (Fig. 25.4).

US preview: A linear probe can be used with a frequency between 8 and 14 MHz and preferably in obese patients a convex probe with a frequency between 3 and 6 MHz. It is recommended to visualize systematically in the long axis and in the short axis on the trochanter maximus and to

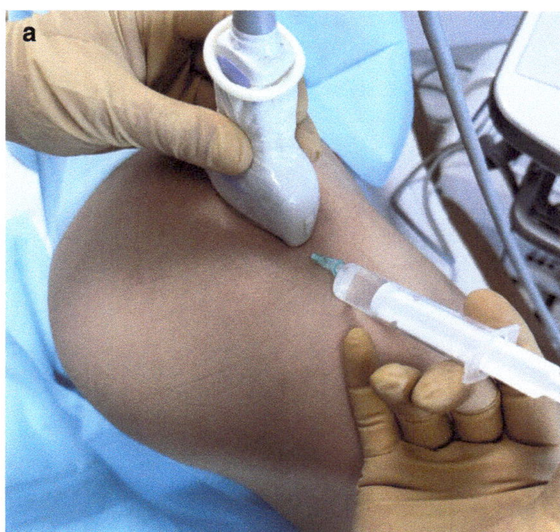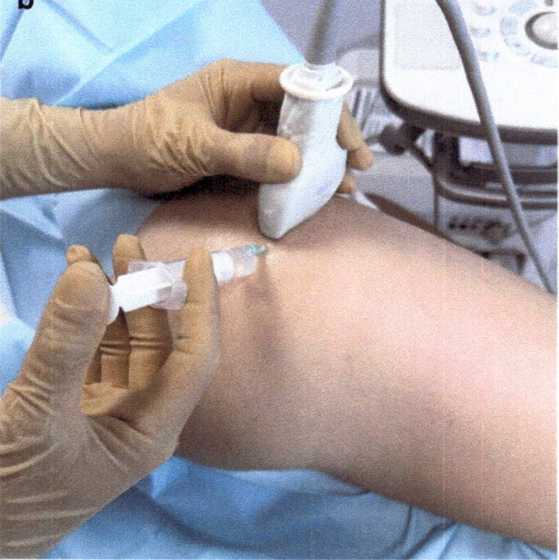

**Fig. 25.4** Ultrasound positions for greater trochanteric infiltration techniques. (**a**) Visualization of the longitudinal axis of the greater trochanter. (**b**) Position in the transverse axis of the greater trochanter, which is the recommended position to carry out the infiltrations due to less sensitivity in the posterior skin area

search on the facet anterior to the gluteus minimus and on the facet posterior to the gluteus medius. It is also recommended to visualize the posterior area given the extension of the trochanteric bursa so that the bursitis can be previously drained if necessary (Fig. 25.5).

Injection technique: US is placed in the transverse axis of the trochanter, with an oblique path toward the posterosuperior facet and the needle is placed under the plane until it reaches the gluteal major bursa. Following a needle (22 G, 64–89 mm) is advanced in plane with the transducer using a posterior to anterior approach under direct US guidance into the tissue plane between the superficial gluteus maximus-iliotibial band and the deep gluteus medius tendon, where the injection is delivered (Fig. 25.5). It is not recommended to infiltrate more than 2–3 ml.

Technical note: For optimal ultrasound visualization (in the short axis) of the gluteal tendons in the greater trochanter, it is recommended to start with the leg in a neutral position from proximal to distal to find the insertion of the gluteus medius in the posterosuperior facet. Then, to see the insertion of the gluteus minimus on the anterior facet of the greater trochanter, it is recommended to perform external rotation of 15° and slide the transducer slightly anteriorly in the short axis (Fig. 25.6).

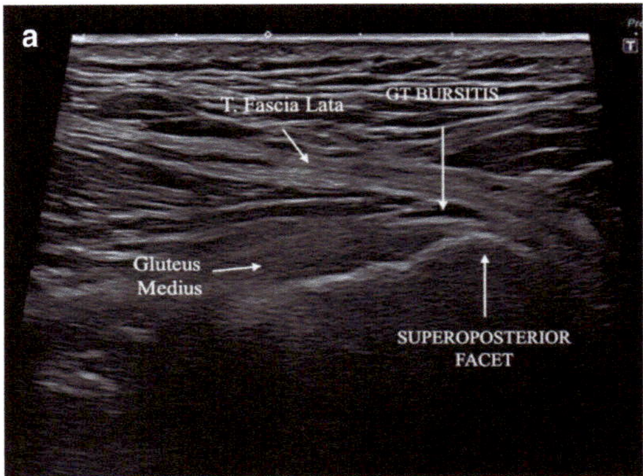


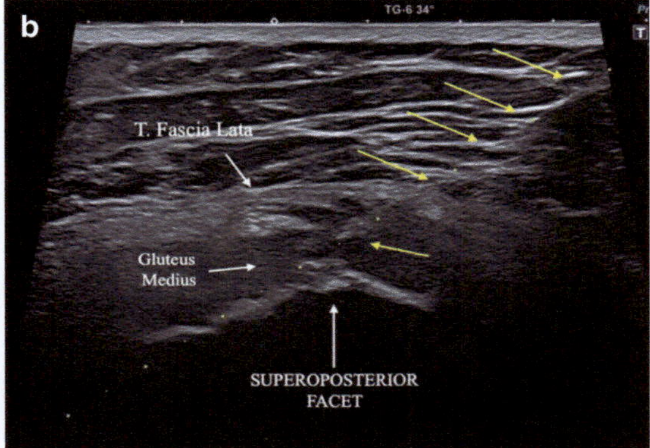


**Fig. 25.5** Transverse axis view of the greater trochanter. (a) An insertional tendinopathy of the gluteus medius and an inflamed greater subgluteus maximus bursa (greater trochanteric bursae) where the peritendon fluid is visualized (anechoic). (b) Infiltration of PRP in the gluteus medius; previously, the aspiration of the bursal fluid has been carried out, followed by the intratendinous infiltration together with the needling technique

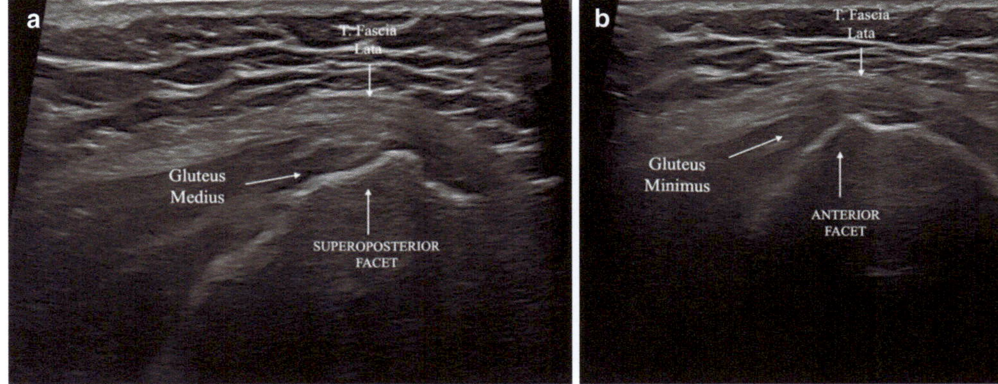


**Fig. 25.6** Gluteus medius and minimus trochanteric insertions. (a) Gluteus medius: US short-axis view. (b) Gluteus minimus: US short-axis view

## 25.3 Osteitis Pubis

Osteitis pubis (OP) is an idiopathic inflammatory condition that affects the pubic symphysis and surrounding soft tissues and is caused by overuse or trauma [21]. The etiology of OP is not fully understood [22], and the imbalance between the adductors of the abdomen and hip (most often the adductor muscles) is currently considered the most important pathogenic factor in the development of osteitis pubis (Fig. 25.7). A chronic imbalance between these muscles causes abnormal forces across the pubic symphysis, affecting the biomechanics of the joint and resulting in damage to bone and cartilage degeneration. For these reasons, OP is more common in high-level athletes, such as football, rugby, distance running, and ice hockey players, with a prevalence ranging from 0.5% to 8% [23].

OP is a self-limiting disease characterized by pain in the pubic symphysis and medial groin area, which radiates to the adductors, suprapubic, and lumbar regions and worsens with physical activity. Tenderness upon palpation in the symphysis area is common; however, clinical examination is not standardized and includes various tests such as lateral compression and the pubic symphysis gap test with isometric adductor contraction. Limited hip range of motion, positive

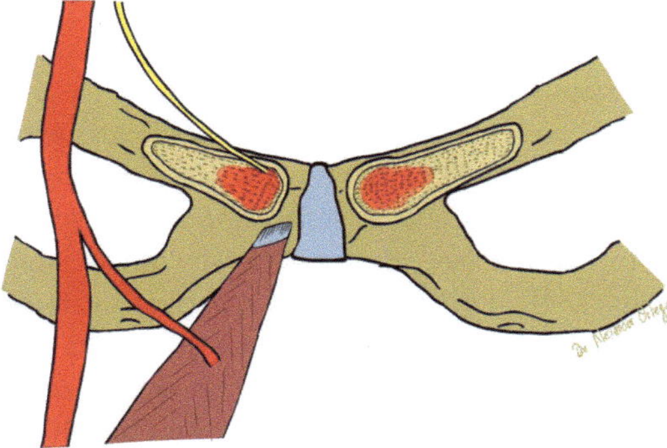

**Fig. 25.7** Illustration of osteitis pubis. On the left side, it is shown the anatomical relationship of the insertion of the adductor longus and the structures at risk, which are the genitofemoral and ilioinguinal nerves together with the spermatic cord in men that emerge from the inguinal canal and the external pudendal artery. On the right side, there is an illustration in red of osteitis pubis

FABER test results, sacroiliac joint dysfunction, and abductor or adductor weakness may be associated with clinical findings [24].

Diagnosis is challenging because of the anatomical complexity of the groin area, biomechanics of the pubic symphysis area, and large number of potential sources of groin pain. Differential diagnoses include intra-articular pathologies such as femoroacetabular impingement syndrome (FAI), acetabular labral tears, and chondral lesions, as well as extra-articular pathologies such as insertional adductor and rectus abdominis tendinopathy, adductor injuries, sports hernia, inguinal hernia, urinary tract infection, and prostatitis.

Imaging is not pathognomonic, but radiography, three-phase scintigraphy, and MRI can aid physical examination and confirm the diagnosis and/or exclude other pathological conditions and possible sources of groin pain [24]. As the gold standard, MRI provides a more detailed view of the pubic symphysis and surrounding soft tissues as well as the bony pelvis and hip. The most common finding in athletic osteitis pubis lasting less than 6 months is the presence of a hyperintense signal on T2-weighted images in the symphysis and adjacent parasymphyseal region, while subchondral sclerosis, subchondral resorption with bony irregularities, and osteophytosis or pubic beaks are characteristic of the chronic phase [25]. Some studies report similar bone marrow edema in asymptomatic athletes as well, so correlation between MRI and clinical examination is mandatory.

Osteitis pubis is described as a self-limiting condition that improves with activity modification and individualized conservative treatment, while surgical treatment is required in approximately 5–10% of patients. However, not all athletes are eligible for conservative treatment due to difficulties in pain management and the long or unpredictable time frame of conservative treatment [23]. Conservative treatment includes rest, limited activity, ice, and anti-inflammatory medications, followed by an individualized progressive multimodal rehabilitation program [26]. Noninvasive treatment aims to correct muscle imbalances around the pubic symphysis and usually consists of a progressive exercise program involving stretching and strengthening of the pelvic floor muscles.

If the symptoms do not improve with conservative measures, local injections can be administered. Corticosteroid injections into the symphysis and surrounding tissues have been used in various studies; however, there is little evidence to support this. Choi et al. described pain relief at short-term follow-up with corticosteroid injections; however, a high proportion of patients did not respond [27]. Despite a successful return to sports, a large percentage of these patients continue to report pain and/or require multiple injections [27].

Surgical intervention is required in 5–10% of patient's refractory to conservative approaches and may be indicated after a minimum of 3 months of a well-conducted rehabilitation protocol if conservative treatment fails [28].

### 25.3.1 Procedure

Aim: To perform an intra-articular infiltration of symphysis pubis.

Patient position: Patient in supine or lateral position, with the thigh abducted and externally rotated so that the adductors are relaxed and the adductor longus tendon is accessible if needed (Fig. 25.8).

US preview: A linear probe with a frequency between 8 and 14 MHz is used; the sensor is placed longitudinally at the level of the pubic bone. The examination begins with a study of the pubic symphysis, which appears as a darker hypoechoic line between the two pubic bones. The articular surfaces are oval in shape with a mean length of 30–35 mm and a mean width of 10–12 mm [29]. The examination of the adductors is then carried out in both the superficial and deep plane of the inner thigh with special focus on the adductor longus, which is the most common site of tendinopathy at this level. Ultrasound

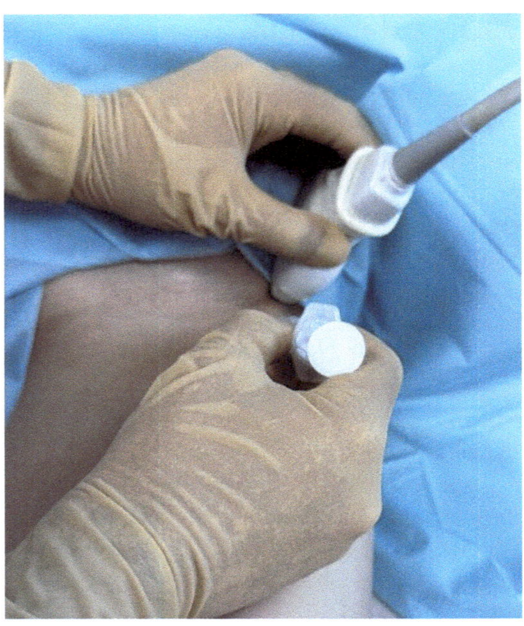

**Fig. 25.8** Position to visualize the proximal adductor longus insertion on the long axis and how the infiltration of the adductor longus would be performed

imaging of long adductor tendinopathy is characterized by the presence of hypoechoic images associated with small intratendinous tears as well as thickening of the tendon and the presence of small calcifications and may also show increased vascularity in the area of tendon degeneration on a Doppler US scan. In other patients, there is a rupture of the myotendinous junction, which presents as a large anechoic defect at the site of interruption.

Injection technique: Using a sterile and aseptic method, a 21-G needle is advanced using an out-of-plane technique superior to the transducer until it reaches the symphysis pubis joint (Fig. 25.9). Infiltration can be performed under direct ultrasound guidance when the needle is in the correct position inside the capsule. After the needle is removed, the pubic bone should be scanned to ensure there is no bleeding after the procedure.

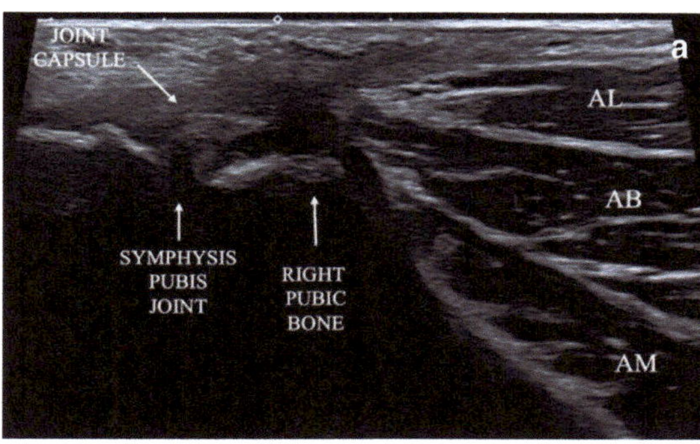


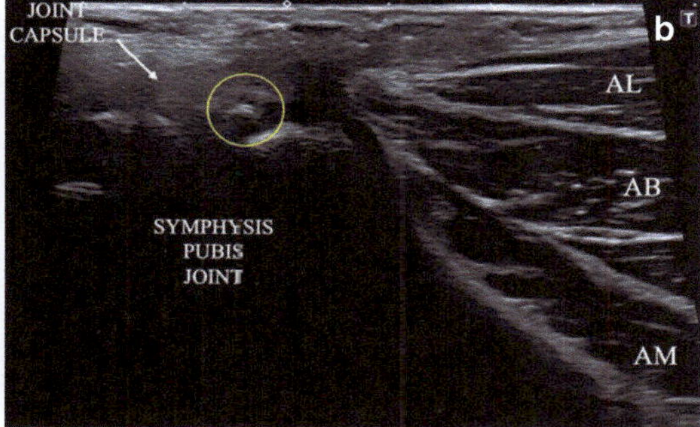


**Fig. 25.9** Symphysis pubis joint infiltration. (**a**) US symphysis pubis anatomy. *AL* adductor longus, *BA* adductor brevis, *AM* adductor magnus. (**b**) Out-of-plane technique for intraarticular symphysis pubis joint. Yellow circle shows the needle out of plane

## 25.4  Iliopsoas Bursitis

The psoas major is a long muscle that originates from the transverse processes, vertebral bodies, and intervertebral discs T5 to L5. The iliacus is a shorter muscle originating from the upper two-thirds of the iliac fossa, ventral lip of the iliac crest, and sacral ala. The medial and lateral bundles are connected to the iliac [30]. The psoas major and iliacus converge to form the iliopsoas muscle at the L5 to S2 levels and are inserted into the lesser trochanter of the femur as the iliopsoas tendon. The psoas major tendon is located medial to the lateral iliac tendon [31]. The deep part of the iliopsoas muscle is anterior and lateral to the labrum of the hip joint. The iliopectineal or iliopsoas bursa is the largest bursa in the human body, located between the iliopsoas muscle, bony surface of the pelvis, and proximal femur.

The iliopsoas is the primary hip flexor that can help tilt the pelvis forward. It also functions as an external hip rotator and is considered a core muscle due to its attachment to the spine. As the iliopsoas muscle connects the spine to the lower limbs, it plays an important role in many activities of daily life, including sports. For example, the psoas major muscle helps in sitting and maintaining an upright spine position. The iliopsoas contributes significantly to running, particularly to the initiation of the swing phase, and plays a crucial role during the kicking and deceleration of the thigh [32, 33].

The iliopsoas musculotendinous unit and iliopsoas bursa are exposed to continuous mechanical stress owing to their proximity to the acetabular rim and hip joint. This can lead to iliopsoas tendinosis or tear. Pathological conditions of the iliopsoas tendon can be accompanied by abnormal tendon movement, and the source of internal snapping hip is described as painful audible and/or palpable snapping of the iliopsoas over the iliopectineal eminence [33]. The tendon may cause anterior hip pain after total hip arthroplasty because of friction between the tendon and the protruding acetabular component or impinging on the collar of the femoral prosthesis (anterior impingement), with a prevalence of 0.4–18% [34, 35]. In this case, nonoperative management of iliopsoas impingement, including physical therapy, nonsteroidal anti-inflammatory drugs (NSAIDs), and ultrasound-guided iliopsoas tendon sheath corticosteroid injections, led to groin pain resolution in 50% of patients [36]. With its location between the deep surface of the iliopsoas tendon and the acetabular rim and hip joint, iliopsoas bursopathy may accompany iliopsoas tendon pain. Because of the connection between the hip joint and iliopsoas bursa in some individuals, iliopsoas distention is often associated with pathological conditions of the hip joint, including rheumatoid arthritis, osteoarthritis, villonodular synovitis, synovial chondromatosis, and septic arthritis [37]. Iliopsoas bursitis can also indicate an underlying pathological condition typically associated with a previous injury or overuse syndrome [37]. In this sense, it is important to remember that injury to the iliopsoas muscle is the second most frequent among professional soccer players [38].

Image-guided injections can help in the diagnosis and treatment of iliopsoas disorders. Accurate diagnostic iliopsoas injections should be administered to identify the source of the pain (suppression test). US-guided iliopsoas bursa injection provides pain relief and predicts good outcomes after iliopsoas tendon release surgery in patients with anterior iliac crest pain and suspected iliopsoas tendon rupture. In addition, ultrasound-guided iliopsoas peritendon injections have been described in patients with previous hip pain following total hip arthroplasty.

## 25.4.1 Procedure

Aim: It is crucial to identify the interval between the deep layer of the iliopsoas tendon and upper margin of the iliacus muscle, which is the most common site of bursitis. It is essential to remain extracapsular and avoid the neurovascular bundle by using a lateral-to-medial in-plane technique (Fig. 25.10).

Position: The patient is placed in a supine position with the hip in neutral rotation. Infiltration can be done in the longitudinal axis or out of plane (short or transverse axis) (Fig. 25.11).

US preview: US evaluation of the iliopsoas region is usually performed using a convex probe with a frequency of 3–6 MHz. The sensor is initially placed in the transverse plane, above the femoral head. The transducer is then translated superiorly and at an angle parallel to the inguinal ligament in the oblique axial plane. On the proximal side of the femoral head, the bony contours of the femoral head and acetabulum can be visualized, together with the iliopsoas muscle and tendon. Transducer switching may be necessary to optimize the visualization of the iliopsoas tendon secondary to anisotropy. Moving the transducer superiorly allows visualization of the hip joint at the level of the iliopectineal eminence and allows continued imaging of the iliopsoas muscle and tendon. If a snapping iliopsoas is suspected, a US scan of the iliopsoas tendon can be performed, while the patient is performing provocative maneuvers. If the patient cannot reproduce snapping, US can be performed as the hip moves from flexion, external rotation, and abduction to full extension, adduction, and internal rotation. Preprocedural scanning includes evaluation of anechoic or hypoechoic distention of the iliopsoas bursa and, if present, assessment of communication between the bursa and hip joint.

Injection technique: The transducer is placed transverse to the iliopsoas tendon in an oblique axial plane, parallel to the inguinal ligament and above the femoral head. The skin at the lateral edge of the traducer is marked with a marking pen, and the area is prepared in a sterile manner. After application of local anesthesia, a 22-G 89-mm needle is advanced in plane with the transducer using a lateral-to-medial approach to

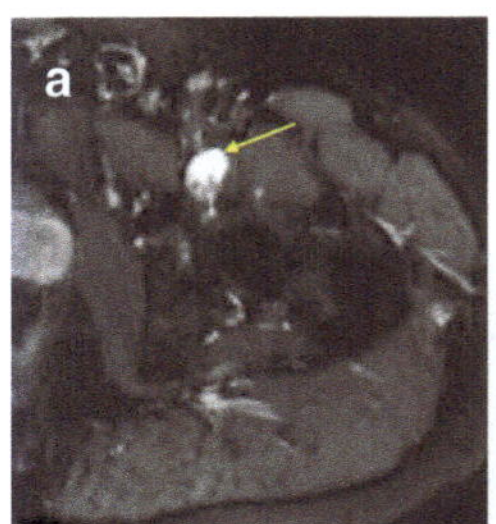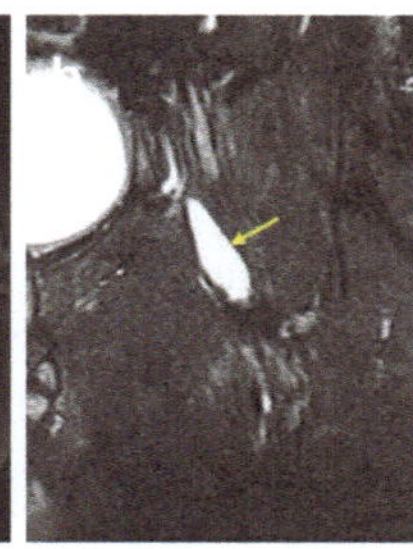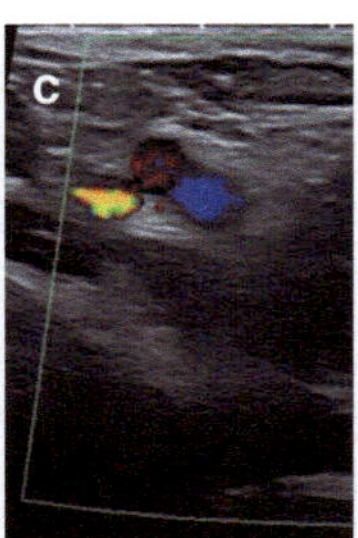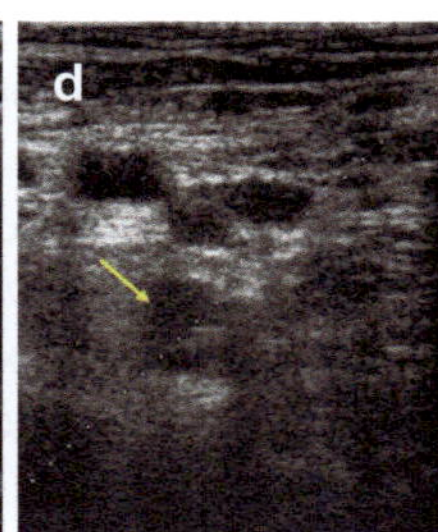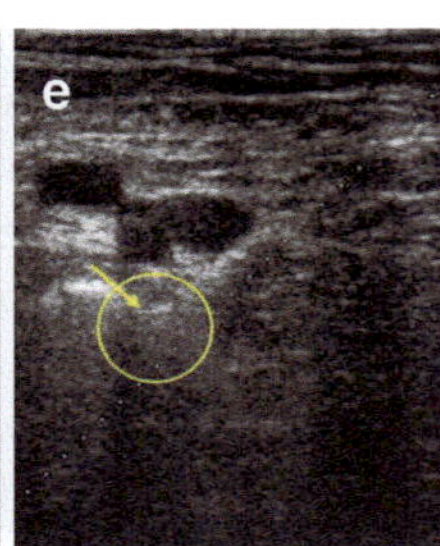

**Fig. 25.10** Clinical case of iliopsoas bursitis (yellow arrow). (**a**) MRI axial section. (**b**) MRI coronal section. (**c**) Visualization of the femoral neurovascular bundle with Doppler. (**d**) Visualization of bursitis. (**e**) Ultrasound-guided bursitis aspiration avoiding the neurovascular bundle and where the posterior infiltration is performed (yellow circle)

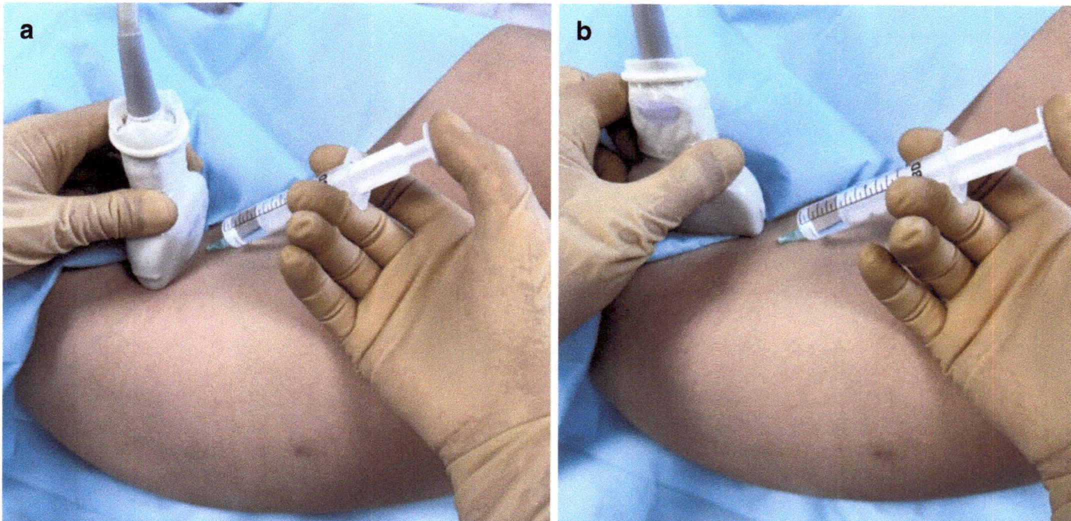

**Fig. 25.11** Position for the US-guided techniques of infiltration of psoas bursitis. (**a**) Out of plane infiltration (transverse axis). (**b**) Infiltration in plane (longitudinal axis)

avoid vessels. The needle is then advanced under direct US guidance into the deep lateral part of the iliopsoas tendon, where it is directed between the deep surface of the iliopsoas tendon and the superficial surface of the ilium at the level of the iliopectineal eminence, or alternatively, between the iliopsoas tendon and the rim of the acetabulum. When the injection is administered, fluid is seen between the iliopsoas tendon and the ilium, as well as on the medial side of the iliopsoas tendon (Fig. 25.11).

Technical note: Hydrodissection may be useful for identifying a plane deep to the iliopsoas tendon but superficial to the hip capsule to avoid inadvertent penetration of the capsule (Fig. 25.12; Table 25.2).

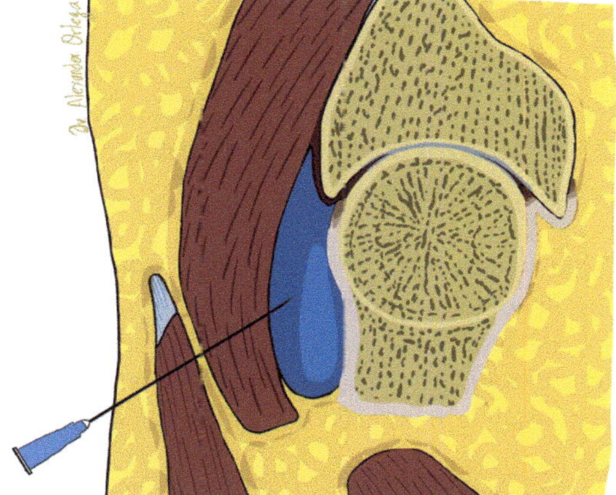

**Fig. 25.12** Iliopsoas bursitis in plane hydrodissection/infiltration illustration

**Table 25.2** Ultrasound-guided iliopsoas bursa: advantages and disadvantages [39, 40]

| Advantages | Disadvantages |
| --- | --- |
| 1. Safe and low-impact procedure as it does not involve the use of radiation | 1. User dependent |
| 2. It can be both diagnostic and therapeutic at the same time | 2. The provider uses more in-office time to perform |
| 3. It provides greater anatomical detail compared to fluoroscopy-guided infiltration | 3. Upfront cost of ultrasound equipment and supplies |
| 4. It is possible to visualize the target area and neurovascular bundle in real-time | 4. Limited by the characteristics of the patient's body |
| 5. Visualization of joint effusion is possible and aspiration can be done if needed | |
| 6. It allows for immediate post injection reassessment/real-time information | |

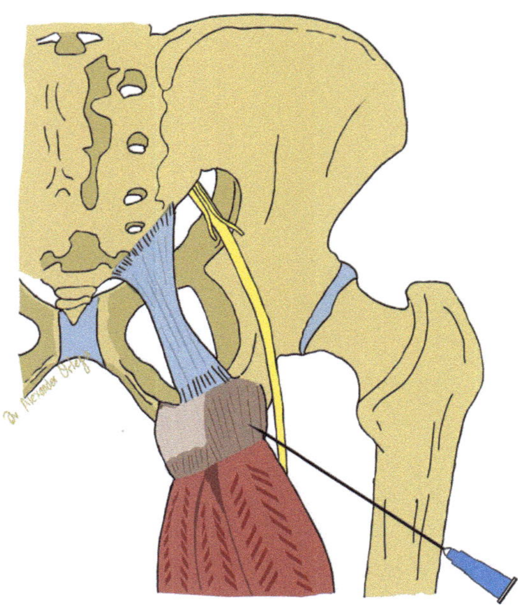

**Fig. 25.13** Hamstring infiltration illustration. The close relationship and precaution that must be taken with the sciatic nerve is observed laterally

## 25.5 Hamstring Origin and Ischial Bursitis

Ultrasound-guided (US) injection solutions for the treatment of proximal hamstring tendinopathy and hamstring bursitis is a good alternative to treat chronic hamstring tendinopathy where other conservative treatments have failed (Fig. 25.13) [41].

Surgical intervention has been shown to have good results in the treatment of pain and function of hamstring tendinopathy, but has potential risks such as sciatic nerve injury, infection, and re-rupture, as well as a longer time to return to normal physical activity than conservative treatments [42].

### 25.5.1 Procedure

Aim: To perform intratendinous injections at the origin of the hamstrings and ischial bursitis.

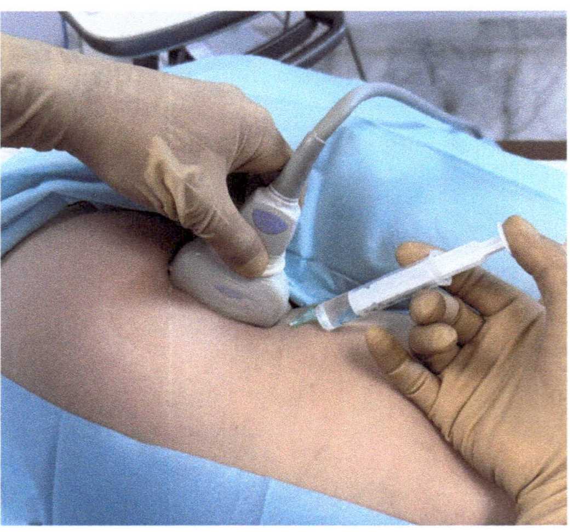

**Fig. 25.14** Ultrasound-guided position for approach of the proximal hamstring insertion infiltration

Position: Patients in prone position and neutral rotation of the leg (Fig. 25.14).

US preview: A linear probe can be used with a frequency between 8 and 14 MHz and preferably a convex probe with a frequency between 3 and

6 MHz. The examination focuses on the ischium, specifically the origin of the common BF/ST (biceps femoris/semitendinosus) tendon, which is located lateral to the ischial tuberosity, and close to the section of the sciatic nerve (Fig. 25.15). Additionally, it is necessary to conduct routine Doppler visualization of the sciatic vascular bundle. Images were obtained on the long and short axes (Fig. 25.16).

Injection technique: Using a sterile and aseptic method, the needle should enter the origin of the hamstrings. Under in-plane visualization with the transducer, a sterile 3.5-inch, 22-G spinal needle is inserted at a 45° angle to the skin in the longitudinal plane, due to the lateral-medial approach, until the spinal trocar reaches the lateral aspect of the ischium, avoiding the sciatic nerve pathway. At this level, at the origin of the proximal tendon, 3 ml or 5 ml of intratendinous solution was injected, while real-time imaging was performed.

Technical note: If it is difficult to detect the common hamstrings, we can see the fibers of the ST muscle there, because it is the only hamstring with muscle fibers that go directly to the ischial tuberosity.

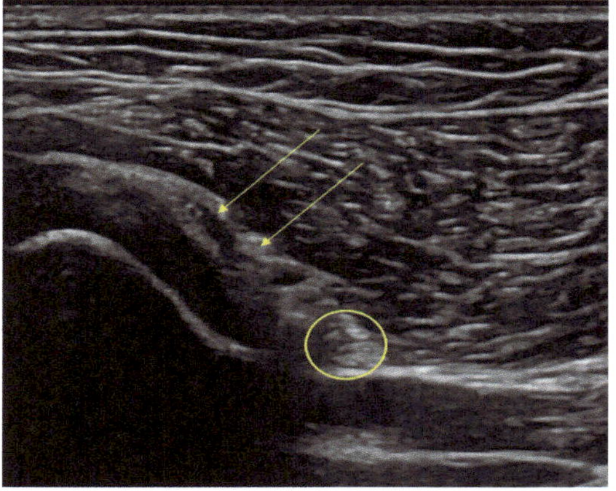

**Fig. 25.15** Ultrasound-guided approach of the hamstrings (yellow arrows) and the ischium. The sciatic nerve (yellow circle) is also seen close to the semimembranosus tendon

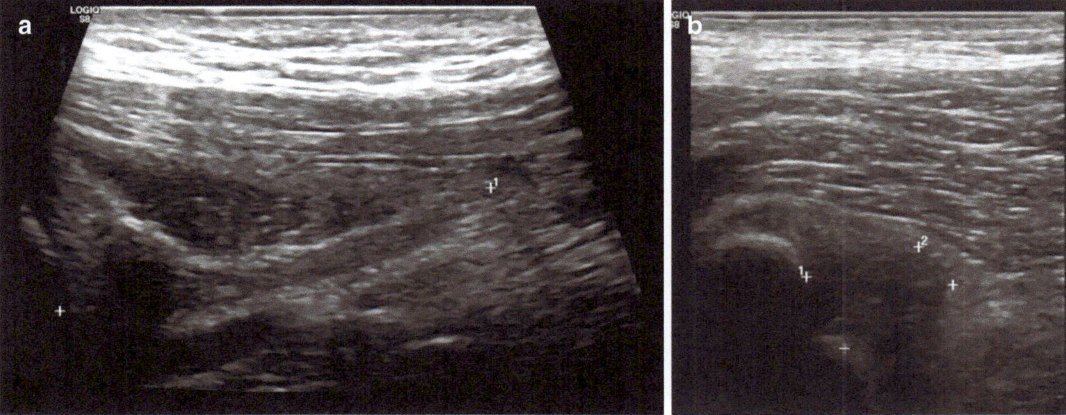

**Fig. 25.16** Proximal hamstring tendinopathy. (**a**) Longitudinal and (**b**) transversal ultrasound section of the proximal semimembranosus tendon on the ischial tuberosity

## 25.6　Deep Gluteal Space

Deep gluteal syndrome (DGS) is a posterior hip pain and/or radicular pain due to non-discogenic entrapment of the sciatic nerve in the subgluteal space [43].

The ultrasound-guided (US) injection in the subgluteal space makes it possible to detect the nerve along its entire length and to perform a dynamic study, which is particularly interesting in the case of entrapment in the ischiofemoral space. This technique can be used to achieve a greater therapeutic effect in the piriformis muscle or the perisciatic region [44].

### 25.6.1　Procedure

Aim: To enter into the deep gluteal space between the greater trochanter (GT) and the ischial tuberosity (IT) and identify the sciatic nerve in order to avoid its damage.

Position: Patient in prone position and neutral rotation (Fig. 25.17).

US preview: A linear probe can be used with a frequency between 8 and 15 MHz. During US examination, the patient lies in prone position. In a morphometric cadaveric study, it is described for a safer approach to identifying the SN in the deep gluteal space, the use as a constant landmarks relative to the SN the reference between the GT-SN and IT-SN, showing that the distance between tip of the GT to SN = 7.23 cm (±0.83) and the center of the IT to SN = 5.28 cm (±0.73) [45]. The sciatic nerve appears as an oval-shaped hypoechoic (Fig. 25.18).

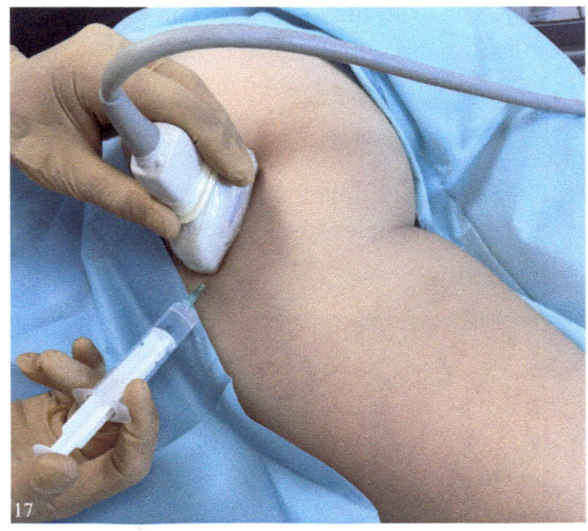

**Fig. 25.17** Illustrates the positioning utilized for a DGS Ultrasound-guided injection employing a lateral to medial approach with a spinal needle. The patient is positioned prone, maintaining a neutral rotation

The transducer is moved caudally along the sciatic nerve with the IT on its medial aspect, the origin of the hamstrings on its lateral aspect, the gluteus maximus on its superior aspect, and the pelvitrochanteric muscles (superior gemellus, obturator internus, inferior gemellus, and quadratus femoris) on its inferior aspect (Fig. 25.17).

Injection technique: Using a sterile and aseptic method, the needle should enter the deep gluteal space between the GT and the IT (Fig. 25.18). Under in-plane visualization with the transducer, a sterile 3.5-inch 22-G spinal needle is inserted at a 45° angle to the skin in the longitudinal plane, due to the lateral-medial approach, until the spinal trocar reaches the medial aspect of the sciatic nerve (Figs. 25.19 and 25.20).

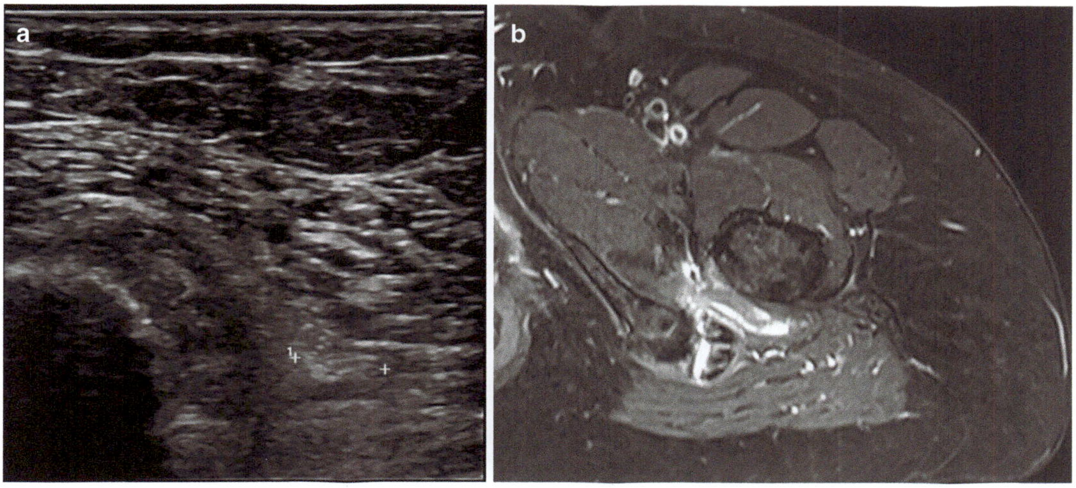

**Fig. 25.18** Clinical case: DGS secondary to hamstrings proximal enthesopathy. Acute ischiofemoral impingement in a 40-year-old female shows a narrowing ischiofemoral space causing impingement of the quadratus femoris mus-cle. (**a**) Ultrasound transverse plane with enlargement of hamstrings common tendon. (**b**) MR axial PD FS with secondary muscle edema. Sciatic neuritis is also shown

Technical note: If the patient is obese or if it is difficult to immediately find the deep gluteal space, the sciatic nerve was located lateral to the semimembranosus origin in the ischial tuberosity (Table 25.3).

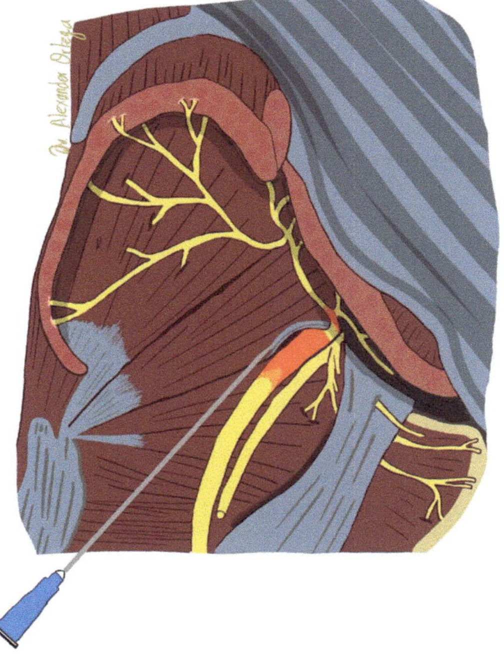

**Fig. 25.19** Illustration of the DGS of the hip and the direction where the needle must go in relation to the anatomy

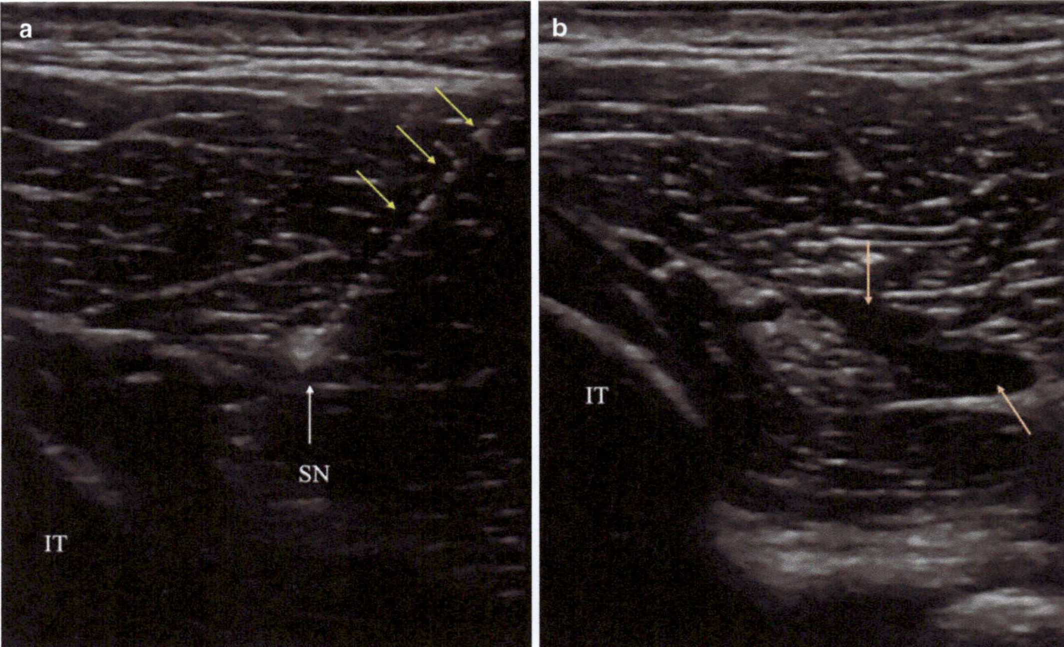



**Fig. 25.20** DGS ultrasound-guided injection technique. (**a**) Ultrasound-guided injection with the spinal needle lateral to medial approach (yellow arrows). (**b**) An anechoic halo is observed as hydrodissection of the sciatic nerve (orange arrows). *IT* ischial tuberosity; *SN* sciatic nerve

**Table 25.3** Ultrasound-guided deep gluteal space injection: advantages and disadvantages [39, 40]

| Advantages | Disadvantages |
| --- | --- |
| 1. Less expensive than the endoscopic technique | 1. Operator dependent (long learning curve) |
| 2. Fast and safe compared to surgery | 2. Variable progression compared to surgery |
| 3. More accurate than landmark injections | 3. The volume of injection should be limited because affect multiple structures |
| 4. Less painful than fluoroscopic-guided and no radiation exposure | 4. Limited by the characteristics of the patient's body |
| 5. Can visualize the perisciatic infiltration in real-time | |
| 6. Allows for immediate postinjection reassessment/real-time information | |

# References


1. Byrd JWT, Potts EA, Allison RK, Jones KS. Ultrasound-guided hip injections: a comparative study with fluoroscopy-guided injections. Arthrosc J Arthrosc Relat Surg. 2014;30(1):42–6.
2. Chandrasekaran S, Lodhia P, Suarez-Ahedo C, Vemula SP, Martin TJ, Domb BG. Symposium: evidence for the use of intra-articular cortisone or hyaluronic acid injection in the hip. J Hip Preserv Surg. 2016;3(1):5–15.
3. Dallari D, Stagni C, Rani N, Sabbioni G, Pelotti P, Torricelli P, et al. Ultrasound-guided injection of platelet-rich plasma and hyaluronic acid, separately and in combination, for hip osteoarthritis: a randomized controlled study. Am J Sports Med. 2016;44(3):664–71.
4. Bardowski EA, Byrd JWT. Ultrasound-guided intra-articular injection of the hip: the Nashville sound. Arthrosc Tech. 2019;8(4):383–8.
5. Zhang M, Pessina MA, Higgs JB, Kissin EY. A vascular obstacle in ultrasound-guided hip joint injection. J Med Ultrasound. 2018;26(2):77–80.
6. Williams BS, Cohen SP. Greater trochanteric pain syndrome: a review of anatomy, diagnosis and treatment. Anesth Analg. 2009;108(5):1662–70.

7. Pfirrmann CW, Chung CB, Theumann NH, Trudell DJ, Resnick D. Greater trochanter of the hip: attachment of the abductor mechanism and a complex of three bursae–MR imaging and MR bursography in cadavers and MR imaging in asymptomatic volunteers. Radiology. 2001;221(2):469–77.

8. Payne JM. Ultrasound-guided hip procedures. Phys Med Rehabil Clin N Am. 2016;27(3):607–29.

9. Lespasio MJ. Lateral hip pain: relation to greater trochanteric pain syndrome. Perm J. 2022;26(2):83–8.

10. Segal NA, Felson DT, Torner JC, Zhu Y, Curtis JR, Niu J, et al. Greater trochanteric pain syndrome: epidemiology and associated factors. Arch Phys Med Rehabil. 2007;88(8):988–92.

11. Ladurner A, Fitzpatrick J, O'Donnell JM. Treatment of gluteal tendinopathy: a systematic review and stage-adjusted treatment recommendation. Orthop J Sports Med. 2021;9(7):23259671211016850.

12. Disantis A, Andrade AJ, Baillou A, Bonin N, Byrd T, Campbell A, et al. The 2022 International Society for Hip Preservation (ISHA) physiotherapy agreement on assessment and treatment of greater trochanteric pain syndrome (GTPS): an international consensus statement. J Hip Preserv Surg. 2023;10(1):48–56.

13. Reid D. The management of greater trochanteric pain syndrome: a systematic literature review. J Orthop. 2016;13(1):15–28.

14. Brinks A, van Rijn RM, Willemsen SP, Bohnen AM, Verhaar JAN, Koes BW, et al. Corticosteroid injections for greater trochanteric pain syndrome: a randomized controlled trial in primary care. Ann Fam Med. 2011;9(3):226–34.

15. Bolton WS, Kidanu D, Dube B, Grainger AJ, Rowbotham E, Robinson P. Do ultrasound guided trochanteric bursa injections of corticosteroid for greater trochanteric pain syndrome provide sustained benefit and are imaging features associated with treatment response? Clin Radiol. 2018;73(5):505.e9–505.e15.

16. Stout A, Friedly J, Standaert CJ. Systemic absorption and side effects of locally injected glucocorticoids. PM R. 2019;11(4):409–19.

17. Lee JJ, Harrison JR, Boachie-Adjei K, Vargas E, Moley PJ. Platelet-rich plasma injections with needle tenotomy for gluteus medius tendinopathy: a registry study with prospective follow-up. Orthop J Sports Med. 2016;4(11):2325967116671692.

18. Migliorini F, Kader N, Eschweiler J, Tingart M, Maffulli N. Platelet-rich plasma versus steroids injections for greater trochanter pain syndrome: a systematic review and meta-analysis. Br Med Bull. 2021;139(1):86–99.

19. Fitzpatrick J, Bulsara MK, O'Donnell J, Zheng MH. Leucocyte-rich platelet-rich plasma treatment of gluteus medius and minimus tendinopathy: a double-blind randomized controlled trial with 2-year follow-up. Am J Sports Med. 2019;47(5):1130–7.

20. Jacobson JA, Rubin J, Yablon CM, Kim SM, Kalume-Brigido M, Parameswaran A. Ultrasound-guided fenestration of tendons about the hip and pelvis: clinical outcomes. J Ultrasound Med. 2015;34(11):2029–35.

21. Lynch TS, Bedi A, Larson CM. Athletic hip injuries. J Am Acad Orthop Surg. 2017;25(4):269–79.

22. Fricker PA, Taunton JE, Ammann W. Osteitis pubis in athletes. Infection, inflammation or injury? Sports Med. 1991;12(4):266–79.

23. Via AG, Frizziero A, Finotti P, Oliva F, Randelli F, Maffulli N. Management of osteitis pubis in athletes: rehabilitation and return to training - a review of the most recent literature. Open Access J Sports Med. 2019;10:1–10.

24. Maffulli N, Giai Via A, Oliva F. Groin pain. In: Volpi P, editor. Football traumatology: new trends. Cham: Springer; 2015. p. 303–15.

25. Omar IM, Zoga AC, Kavanagh EC, Koulouris G, Bergin D, Gopez AG, et al. Athletic pubalgia and 'sports hernia': optimal MR imaging technique and findings. Radiogr Rev Publ Radiol Soc N Am. 2008;28(5):1415–38.

26. Jardí J, Rodas G, Pedret C, Til L, Cusí M, Malliaropoulos N, et al. Osteitis pubis: can early return to elite competition be contemplated? Transl Med. 2014;10:52–8.

27. Choi H, McCartney M, Best TM. Treatment of osteitis pubis and osteomyelitis of the pubic symphysis in athletes: a systematic review. Br J Sports Med. 2011;45(1):57–64.

28. Amer ML, Omar K, Malde S, Nair R, Thurairaja R, Khan MS. The challenges in diagnosis and management of osteitis pubis: an algorithm based on current evidence. BJUI Compass. 2022;3(4):267–76.

29. Becker I, Woodley SJ, Stringer MD. The adult human pubic symphysis: a systematic review. J Anat. 2010;217(5):475–87.

30. Tatu L, Parratte B, Vuillier F, Diop M, Monnier G. Descriptive anatomy of the femoral portion of the iliopsoas muscle. Anatomical basis of anterior snapping of the hip. Surg Radiol Anat. 2001;23(6):371–4.

31. Philippon MJ, Devitt BM, Campbell KJ, Michalski MP, Espinoza C, Wijdicks CA, et al. Anatomic variance of the iliopsoas tendon. Am J Sports Med. 2014;42(4):807–11.

32. Fitzgerald P. The action of the iliopsoas muscle. Ir J Med Sci. 1969;8(1):31–3.

33. Lifshitz L, Bar Sela S, Gal N, Martin R, Fleitman KM. Iliopsoas the hidden muscle: anatomy, diagnosis, and treatment. Curr Sports Med Rep. 2020;19(6):235–43.

34. Ala Eddine T, Remy F, Chantelot C, Giraud F, Migaud H, Duquennoy A. Anterior iliopsoas impingement after total hip arthroplasty: diagnosis and conservative treatment in 9 cases. Rev Chir Orthop Reparatrice Appar Mot. 2001;87(8):815–9.

35. Bartelt RB, Yuan BJ, Trousdale RT, Sierra RJ. The prevalence of groin pain after metal-on-metal total hip arthroplasty and total hip resurfacing. Clin Orthop. 2010;468(9):2346–56.

36. Chalmers BP, Sculco PK, Sierra RJ, Trousdale RT, Berry DJ. Iliopsoas impingement after primary total hip arthroplasty: operative and nonoperative treatment outcomes. J Bone Joint Surg Am. 2017;99(7):557–64.

37. Tormenta S, Sconfienza LM, Iannessi F, Bizzi E, Massafra U, Orlandi D, et al. Prevalence study of iliopsoas bursitis in a cohort of 860 patients affected by symptomatic hip osteoarthritis. Ultrasound Med Biol. 2012;38(8):1352–6.

38. Werner J, Hägglund M, Waldén M, Ekstrand J. UEFA injury study: a prospective study of hip and groin injuries in professional football over seven consecutive seasons. Br J Sports Med. 2009;43(13):1036–40.

39. Lento PH, Primack S. Advances and utility of diagnostic ultrasound in musculoskeletal medicine. Curr Rev Musculoskelet Med. 2008;1(1):24–31. https://doi.org/10.1007/s12178-007-9002-3. PMID: 19468895; PMCID: PMC2684149.

40. Finnoff JT. The Evolution of Diagnostic and Interventional Ultrasound in Sports Medicine. PM R. 2016;8(3 Suppl):S133–8. https://doi.org/10.1016/j.pmrj.2015.09.022. PMID: 26972262.

41. Davenport KL, Campos JS, Nguyen J, Saboeiro G, Adler RS, Moley PJ. Ultrasound-guided intratendinous injections with platelet-rich plasma or autologous whole blood for treatment of proximal hamstring tendinopathy: a double-blind randomized controlled trial. J Ultrasound Med. 2015;34(8):1455–63.

42. Benazzo F, Marullo M, Zanon G, Indino C, Pelillo F. Surgical management of chronic proximal hamstring tendinopathy in athletes: a 2 to 11 years of follow-up. J Orthop Traumatol. 2013;14(2):83–9.

43. Hernando MF, Cerezal L, Pérez-Carro L, Abascal F, Canga A. Deep gluteal syndrome: anatomy, imaging, and management of sciatic nerve entrapments in the subgluteal space. Skelet Radiol. 2015;44(7):919–34.

44. Rosales J, García N, Rafols C, Pérez M, Verdugo MA. Perisciatic ultrasound-guided infiltration for treatment of deep gluteal syndrome: description of technique and preliminary results. J Ultrasound Med. 2015;34(11):2093–7.

45. Capurro B, Tey M, Monllau JC, Carrera A, Marques F, Reina F. Anatomic landmarks for a safe arthroscopic approach to the deep gluteal space: a cadaveric study. Int J Morphol. 2021;39(2):359–65.

# Injections of Anatomical Regions and Diseases: Knee

Sarper Gursu, Ahmet Sukru Mercan, Anıl Erbas, Serda Duman, and Ozgur Ismail Turk

## 26.1 Knee Joint

Knee is the body's largest joint and has a very complex structure composed of bone, cartilage, meniscus, and the ligaments.

### 26.1.1 Anatomy

The knee joint is a complex diarthrodial joint. Medial and lateral femoral condyles, plateau tibia, and the patella make up the joint, and the joint surfaces are covered with hyaline cartilage. The knee joint has three compartments as follows: medial tibiofemoral compartment, lateral tibiofemoral compartment, and patellofemoral compartment. Between the femur and the tibia, there are two crescent-shaped structures known as meniscus which absorb the shock and reduce

the friction between the bones. The knee joint is supported by a couple of ligaments and tendons which are located both inside and outside of the joint space. The knee joint capsule surrounds the joint and consists of two layers: the outer layer being made up of fibrous tissue and the inner layer also known as synovial membrane. The synovial membrane produces the knee joint fluid and helps reduce the friction between the joint surfaces and nourishes the cartilage tissue, covering the articular surfaces [1, 2].

### 26.1.2 Indication and Diagnosis

Knee joint injections are mostly performed for patients with osteoarthritis. Other common indications are rheumatoid arthritis, crystal arthropathies, psoriatic arthritis, and other inflammatory diseases affecting the knee joint. Degeneration of the cartilage, along with changes in subchondral tissues, is usually encountered in these patients, leading to many different symptoms. Pain is the most prominent symptom for most diseases in the knee joint, and relieving pain is the main target of most agents used for injections. Relieving pain usually improves functional status and increases range of motion. Pathologies of the knee may be diagnosed on the basis of clinical presentation and findings in various imaging studies. Simple AP (anteroposterior) and lateral X-rays reveal the damage occurring in the joint

S. Gursu (✉)
Baltalimani Bone and Joint Diseases Hospital, Istanbul, Turkey

Health Sciences University, Athlete's Health and Sports Sciences Institute, Istanbul, Turkey

A. S. Mercan
Nisantasi University, Istanbul, Turkey

A. Erbas · S. Duman
Baltalimani Bone and Joint Diseases Hospital, Istanbul, Turkey

O. I. Turk
Cevre Hastanesi, Istanbul, Turkey

very clearly in most cases; however, true differential diagnose may require specific blood tests or further imaging studies such as MRI (magnetic resonance imaging) [3–5].

Inserting a needle to the knee joint may also be required for aspirating joint fluid or blood in cases with a suspicion of septic arthritis or hematoma. Samples for further evaluation may also be obtained by aspiration [6].

### 26.1.3 Appliances

Approximately 5 ml of liquid can be injected by a 21-gauge (G) or 22-G, 2-inch-long syringe tip and a 5-ml syringe. For aspiration an 18-G or 20-G syringe would be more appropriate.

### 26.1.4 Agents

Corticosteroids, local anesthetics, viscosupplements, and orthobiologic agents (PRP, PRGF, stem cell solutions, etc.)

### 26.1.5 Technique/Tricks/Pitfalls

There are various approaches for infiltrating or aspirating the knee joint. The best approach is the one with least obstruction and easier access to the synovial cavity.

If the superomedial approach is preferred, the patient either sits while the knee is extended or lays supine with a thin pad under the knee for support. The medial edge of the patella is palpated and marked. Marking the edge is strongly recommended especially in overweight individuals. The clinician may push the patella's lateral edge gently, widening the medial entrance of the suprapatellar pouch and also stabilizing the patella. Sterile technique is applied, and the needle is inserted 1 cm medial to the patella at the intersection point of upper one third of the patella and the middle part. Then it is further inserted laterally at an angle of 45° directing to the center of the knee joint and into the suprapatellar pouch,

beneath the insertion site of the quadriceps tendon. The injection is done gently and not against resistance. If any resistance is encountered, this could mean that the tip of the needle is within soft tissues and not in the pouch. In this case, the needle should be retracted and re-inserted at an angle directing more toward the center of the joint [6–8].

Another very well-known and highly preferred injection site is the superolateral aspect of the knee joint. Similar to the superomedial approach, the patient holds the knee fully extended with a pad underneath for relaxation. After stabilizing the patella from the medial side with the clinician's thumb, the needle is inserted underneath the superolateral part of the patella, directing toward to the center of the joint space posterior and inferomedial. Both the superolateral and the superomedial approaches are known as rather safe for performing intraarticular injections to the knee joint [6–8].

Intraarticular knee injections may also be performed through the anteromedial or anterolateral approaches (Figs. 26.1 and 26.2). These two approaches are also known as the arthroscopic approaches. For these two approaches, the patient should either be sitting or lying with the knees flexed to 90°. The needle is inserted to the lateral side of the patellar tendon (for the anterolateral approach) or medial side of the patellar tendon (for the anteromedial approach) 1 cm above the tibial plateau. The needle should be directed to the space between the medial and lateral femoral condyles at an angle of 30–45°. If the medial approach is preferred for injection, the track of the saphenous nerve should be considered in order to avoid saphenous neuropathy [6–8].

Lateral and medial mid-patellar approaches may also be preferred for either infusions or aspirations of the knee. The patient is seated with the knees fully extended. The needle is directed to the center of the knee, while the patella is pushed medially or laterally [6–8]. Ultrasonography may be used to increase the accuracy of injections and helps giving the medication properly to the synovial joint [9]. The size of the needle to be used during the injection should be decided according

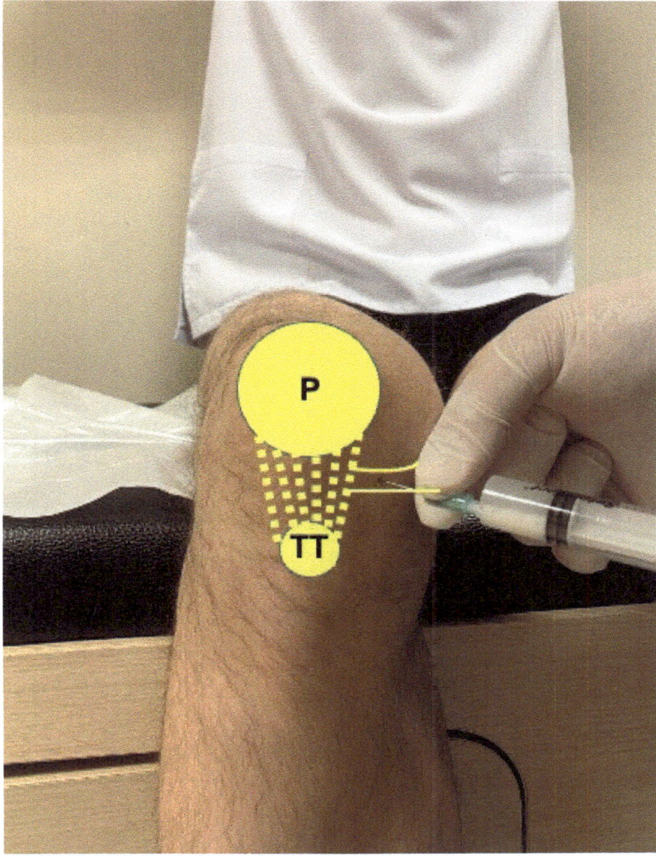


**Fig. 26.1** The use of the anterolateral approach for intraarticular knee injection. Patient should either be sitting or lying with the knees flexed to 90°. The needle is inserted to the lateral side of the patellar tendon 1 cm above the tibial plateau. The needle should be directed to the space between the medial and lateral femoral condyles at an angle of 30–45° (*P* patella, *TT* tuberositas tibia)

**Fig. 26.2** The use of the anteromedial approach for intraarticular knee injection. The needle is inserted to the medial side of the patellar tendon 1 cm above the tibial plateau. The needle should be directed to the space between the medial and lateral femoral condyles at an angle of 30–45 degrees (*P* patella, *TT* tuberositas tibia)

to the viscosity of the fluid to be injected. For the purpose of arthrocentesis, smaller sizes (18–20 G) should be preferred for easier flow of the joint fluid.

### 26.1.6 Aftercare

The patient is observed for a while and let for walking immediately, although undue weight-bearing should be avoided for at least 5–7 days. NSAID use is recommended. Normal activities can be resumed afterward along with strengthening exercises to be performed at home. Further repeat injections can be performed to the same knee after 3 months, at earliest.

## 26.2    Pes Anserine Bursa

Pes anserine bursitis occurs usually as an overuse injury and may be significantly painful in some cases. It is an inflammatory condition. Frequent and forced flexion of the knee increases the friction, leading to irritation of the bursa. Trauma, in the form of direct blow to the pes anserine, may also lead to bursitis [10].

### 26.2.1 Anatomy

The pes anserine, also known as goose foot, is the name given to the insertion site of the conjoined medial knee tendons. These tendons are sartorius, gracilis, and semitendinosus, and they are supplied by three different lower extremity nerves: femoral, obturator, and tibial nerves. They are located superficial to the medial collateral ligament of the knee. The tendons making up the pes anserine are primarily flexors of the knee, but they also contribute to the rotatory stability of the knee. The insertion site of the pes anserine is just below the knee joint line, on the medial aspect of proximal tibia. The bursa, known with the same name, lays between the conjoined tendon and the medial collateral ligament on the medial tibia surface [10–12].

### 26.2.2 Indication and Diagnosis

Pes anserine bursitis is usually known as a self-limiting injury. Conservative treatment modalities lead to very satisfactory results and complete healing in most cases. Injection of pes anserine bursa should be regarded as a second-line treatment and reserved for patients not responding to NSAIDs, rest, use of ice, and physical rehabilitation. Pain located in the medial aspect of the proximal tibia is the most common symptom of the disease. Pain which is increasing while arising from seated position is a typical finding. Tenderness over the bursa and local swelling may also be encountered. History of recent athletic activity is present in some cases. Runners, swimmers, and dancers are prone to pes anserine bursitis. During a proper physical examination, it is possible to palpate the painful bursa, and the symptoms may be reproduced with resisted flexion of the knee. Pes anserine bursitis is usually accompanied by degenerative diseases of the knee making further imaging studies necessary. If there is suspicion about the diagnosis, a diagnostic injection with a local anesthetic may be performed, with alleviated symptoms being in favor of present pes anserine bursitis [13].

### 26.2.3 Appliances

Approximately 2–3 ml of liquid can be injected by a 22-G or 23-G, 1.5-inch-long syringe tip and a 5-ml syringe.

### 26.2.4 Agents

Corticosteroids, local anesthetics, and orthobiologic agents (PRP, PRGF, stem cell solutions, etc.)

### 26.2.5 Technique/Tricks/Pitfalls

The patient sits with the knees extended or slightly flexed and supported. The pes anserine is palpated, while the patient is asked to flex the

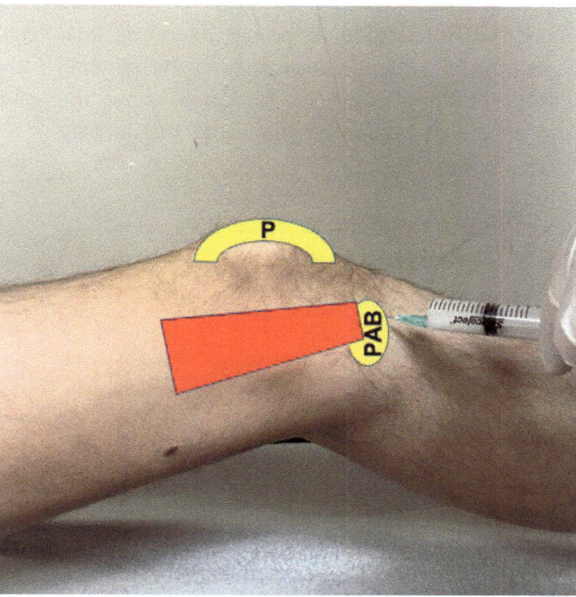



**Fig. 26.3** Injection technique for the pes anserine bursitis. The patient sits with the knees extended or slightly flexed and supported. The bursa is located immediately before the insertion point of the tendons and this point is marked. Making the injection 0.5–1 cm higher than the hamstring tendons is important as in this location it is easier to involve the MCL within the injection. The needle is inserted with an angle of 45° to the point of maximal tenderness until it touches bone. The needle is slightly retracted and injection is done (*P* patella, *PAB* pes anserine bursa)

knee against resistance. The bursa is located immediately before the insertion point of the tendons, and this point is marked. Making the injection 0.5–1 cm higher than the tendons is important as in this location it is easier to involve the MCL (medial collateral ligament) within the injection site as MCL may be contributing to the current pain. Sterile technique should be used. The needle is inserted with an angle of 45° to the point of maximal tenderness until it touches bone. The needle is slightly retracted and injection is done (Fig. 26.3). Little resistance during injection is possible; however, if there is significant resistance, the tip of the needle may be within the tendons, and in this case, the angle of the needle should be rearranged for better access to the bursa. Injection to the tendons may not only worsen the pain but also affects them. Repeated injections may be performed for resistant pain, but a number of injections should not exceed

three in 1 year. Furthermore, it should be kept in mind that if the initial injection does not lead to any improvements in the symptoms, the possibility of getting better results with further injections is very low [7, 8, 14].

Injection for pes anserine bursitis is generally regarded as a safe method of treatment; however, care should be given not to injure the saphenous nerve which lies on the medial side of the knee, descending vertically between the tendons of sartorius and the gracilis muscles [15].

### 26.2.6 Aftercare

After the injection an elastic bandage is applied and use of local cold is initiated. NSAID could be given if needed. The patient is observed for a while and let for walking immediately. Overuse activities are avoided 7–10 days. Graded hamstring stretching and strengthening exercises are initiated afterward. Patients must be educated about the importance of proper exercises, and especially in athletic settings, both the patients and the trainers must be informed about the importance of gradually increasing the activity levels.

## 26.3 Prepatellar Bursa

Prepatellar bursa lies just below the skin and in front of the patella. Similar to all other bursae, prepatellar bursa also decreases the friction arising during the flexion-extension movements of the knee. It is important to make a discrimination between non-septic bursitis and septic bursitis as the treatment differs accordingly. Prepatellar bursitis, also known as housemaid's knee or carpenter's knee, may occur due to acute trauma; however, repeating microtrauma is a more common cause. Septic bursitis usually arises from skin lesions but may also be due to spread of infection from other sources within the body. Other rare causes include some inflammatory conditions like gout, pseudogout, rheumatoid arthritis, and tuberculosis [16, 17].

### 26.3.1 Anatomy

The prepatellar bursa is located between the skin and patella with a thin synovial lining, and it is rather superficial. The bursa has no contact with the joint space. Typically, there is a minimum amount of fluid within the bursa, but in cases of prepatellar bursitis, it significantly increases. Usually there is only one sac, but it is possible to encounter lobulated structures as well [18, 19].

### 26.3.2 Indication and Diagnosis

Prepatellar bursitis is the second most common bursitis after olecranon bursitis and usually presents with excessive swelling and erythema in front of the patella. In cases with infected bursitis, pain is accompanied with other findings. A detailed history and physical examination are enough for a true diagnosis. However, laboratory tests may be required for diagnosing infection, and imaging studies may be required for patients with possible patella fractures.

Prepatellar bursitis usually heals by conservative means and does not require any injections or aspirations in most cases. Aspiration may be indicated in cases with a suspicion about the diagnosis for further evaluation of the fluid within the bursa. Aspiration for septic bursitis may help faster recovery [19]. Corticosteroid injection is reserved for cases with recurrent bursitis and should be performed only when the possibility of infection is eliminated and the symptoms of the bursitis does not respond to ice and NSAIDs.

### 26.3.3 Appliances

Approximately 2 ml of liquid can be injected by a 22-G or 23-G, 1.5-inch-long syringe tip and a 5-ml syringe. For aspiration 18-G or 20-G syringes should be preferred for better flow of the fluid.

### 26.3.4 Agents

Corticosteroids and local anesthetics.

### 26.3.5 Technique/Tricks/Pitfalls

The injection or the aspiration may be performed, while the patient is either sitting with 90° of flexed knees or lying supine with the knees slightly flexed and supported. The sterile technique is applied, and the needle is inserted to the center of the fluctuated bursa, perpendicular to the skin. If an injection is also indicated after aspiration, the needle is left in the bursa, and the corticosteroid is given with a new syringe. The injection should be made without any resistance, and if any resistance is encountered, the needle should be pulled slightly (Fig. 26.4) [8].

It should be always kept in mind that injection to the prepatellar bursa is rarely indicated and the

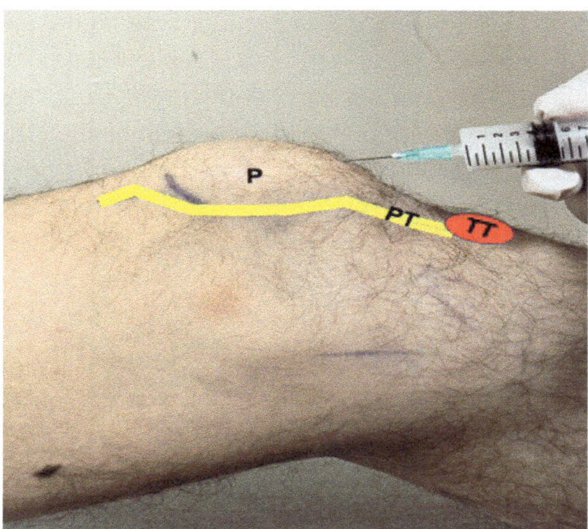


**Fig. 26.4** Injection to the prepatellar bursa. The injection may be performed while the patient is either sitting with 90° of flexed knees or lying supine with the knees slightly flexed and supported. The needle is inserted to the center of the fluctuated bursa, perpendicular to the skin. If an injection is also indicated after aspiration, the needle is left in the bursa and the corticosteroid is given with a new syringe (*P* patella, *PT* patellar tendon, *TT* tuberositas tibia)

bursitis observed in this localization is usually healed without any invasive procedures. Corticosteroid injection to the prepatellar bursa has significantly higher risks of complications when compared to injections to most other bursa within the body as the subcutaneous tissue at this site is very thin, making the bursa prone to infection or skin atrophy [20–22].

When doing aspiration, there is the possibility of aspirating less fluid than expected due to the multiloculate nature of the bursa. In such cases ultrasonography may be helpful to assist aspiration for better approach to various sacs. Aspiration may be more successful if the bursa is milked toward the needle by the tips of the clinician's other hand [22].

The injection or the aspiration may also be performed from the bottom of the fluctuated bursa instead of the center, with the needle again targeting the center of the bursa (Fig. 26.4).

### 26.3.6 Aftercare

After the injection an elastic compressive bandage is applied, and the use of local cold is initiated. Compressive bandage helps preventing re-accumulation of fluid in traumatic bursitis. NSAID could be given if it is needed. The patient is observed for a while and let for walking immediately. Kneeling or putting pressure on the patella should be avoided for at least 48–72 h, and occupational modifications should be made especially for patients with specific jobs like carpenters, housemaids, and gardeners. Special caution is recommended to the patient for possible complications.

## 26.4 Iliotibial Band

Iliotibial band is one of the major stabilizers of the lateral part of the knee joint. It also helps in the stabilization of the hip. In some individuals, excessive numbers of knee extension and flexion like in runners lead to iliotibial band syndrome, which is due to friction between the iliotibial band and the lateral femoral epicondyle. Iliotibial band syndrome is an overuse injury. The result is a painful chronic inflammation [23, 24].

### 26.4.1 Anatomy

It starts from the iliac crest and the capsule of the hip and includes parts of tensor fascia lata and the gluteus maximus. It extends through the lateral aspect of the thigh and passes two major joints: the hip and the knee. The insertion point is the Gerdy tubercle, located on the anterolateral aspect of the proximal tibia [25, 26]. Along with the stabilization of the hip and the knee, the iliotibial band has a postural function and helps maintain the erected position. The bursa, known as the iliotibial bursa, lies between the band and the lateral femoral condyle and helps to decrease the friction between the band and the bony structures during motion.

### 26.4.2 Indication and Diagnosis

The iliotibial band syndrome is typically more common in young active adults. Most important symptom is pain arising during motion. The pain is worst at 30 degrees of knee flexion as the friction, and the tension on the band is highest at this level of action [27]. Pain usually starts with 5–10 min of exercise and improves with rest. Clinical findings are enough for a diagnosis; however, further imaging studies are helpful in confirming the diagnosis and making a differential diagnosis especially in repetitive cases. An MRI will reveal local inflammatory findings and presence of a small amount of fluid at the friction site.

Nonsurgical treatment is the mainstay of the treatment for iliotibial band syndrome. The symptoms may be treated successfully by NSAIDs, rest, activity modulation, and physiotherapy focusing on strengthening of gluteal muscles. If conservative treatment fails, injection therapy may also be preferred. Corticosteroid injection helps improve swelling and ease rehabilitation [27–30]. Aspiration may also be performed through the same approach for cases with significant swelling.

### 26.4.3 Appliances

Approximately 2 ml of liquid can be injected by a 23-G, 1.5-inch-long syringe tip and a 5-ml syringe.

### 26.4.4 Agents

Corticosteroids, local anesthetics, and orthobiologic agents (PRP, PRGF, stem cell solutions, etc.).

### 26.4.5 Technique/Tricks/Pitfalls

The patient sits with the knees extended or slightly flexed and supported. The injection is performed using the sterile technique. Localization of the iliotibial band bursa may be detected by palpation as a tender area on the lateral aspect of the distal femur. Another method of finding the iliotibial band is tracing it backward from its insertion point at the Gerdy tubercle. The needle is inserted at this point to touch the bone and retracted slightly. Both the anterolateral and posterolateral approaches may be used for injection. The solution to be injected is deposited. If the insertion point and the distal end of the band are also tender, this part may also be infiltrated at the same time without injecting the corticosteroid to the tendon (Fig. 26.5) [7, 8]. The injection is done gently and not against resistance. If any resistance is encountered, this could mean that the tip of the needle is within the band and not in the bursa. In this case, the needle should be retracted and reinserted at an angle directing more toward the bursal space. The tract of common fibular nerve should be considered during the injection (Fig. 26.6).

Ultrasonography may be used to increase the accuracy of injections and helps in giving the medication properly to the iliotibial band bursa. Making the injection under ultrasonography guidance is proved to have superior outcomes and better pain relief when compared to injections without ultrasonography guidance [23, 31].

### 26.4.6 Aftercare

Resting after injection is strongly recommended for about a week along with avoiding strenuous activity for the first 48 h. NSAID could be given

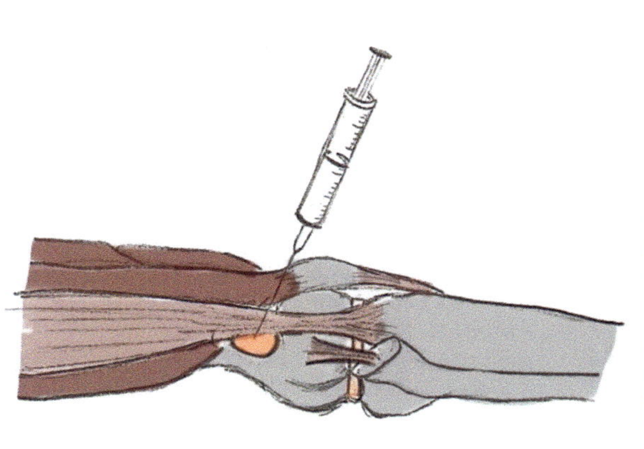

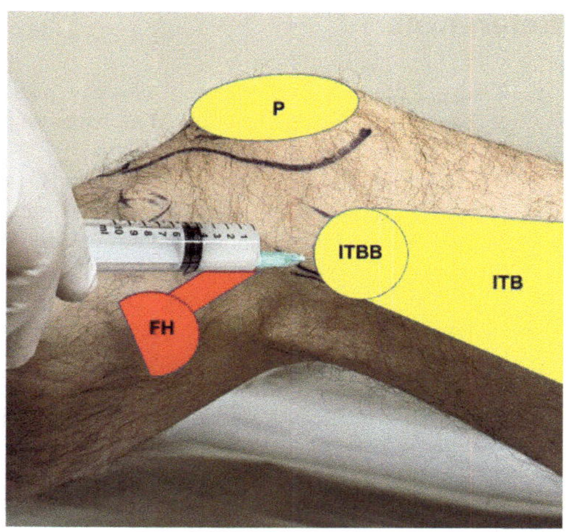



**Fig. 26.5** Iliotibial bursa injection. The patient sits with the knees extended or slightly flexed and supported. Localization of the iliotibial band bursa may be detected by palpation as a tender area on the lateral aspect of the distal femur. Another method of finding the iliotibial band is tracing it backward from its insertion point at the Gerdy tubercle. The needle is inserted at this point to touch the bone and retracted slightly. The solution to be injected is deposited (*FH* fibular head, *P* patella, *ITBB* iliotibial band bursa, *ITB* iliotibial band)

**Fig. 26.6** Ultrasonography may be used to increase the accuracy of iliotibial band bursa injection (P: patella, ITBB: iliotibial band bursa)

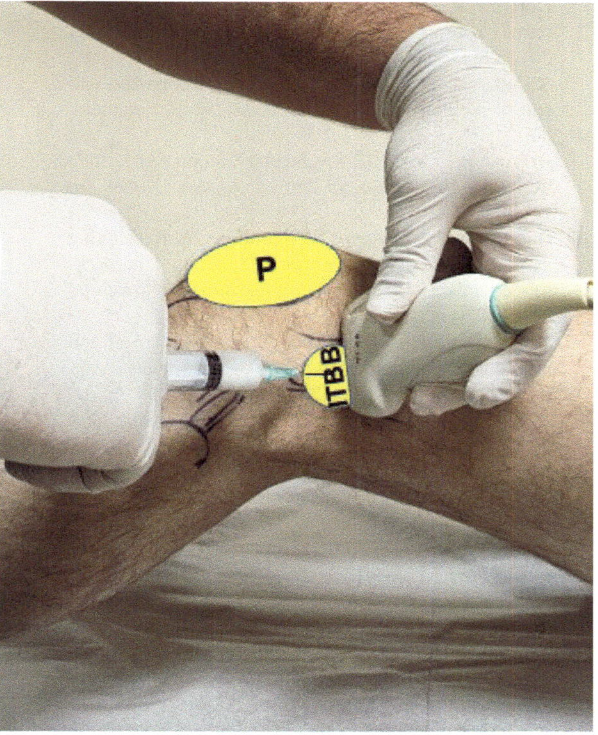



if it is needed. Once the symptoms are relieved, a few sessions of rehabilitation should be initiated in order to achieve stretching and strengthening in most patients. In order to prevent any recurrences, activity modulation and gradual return to fully functional status should be recommended to the patient.

# References

1. Flandry F, Hommel G. Normal anatomy and biomechanics of the knee. Sports Med Arthrosc Rev. 2011;19(2):82–92.
2. Saavedra MA, Navarro-Zarza JE, Villasenor-Ovies P, Canoso JJ, Vargas A, Chiapas-Gasca K, Hernandez-Diaz C, Kalish RA. Clinical anatomy of the knee. Rheumatol Clin. 2012;8(Suppl 2):39–45.
3. Konai MS, Vilar Furtado RNV, Dos Santos MF, Natour J. Monoarticular corticosteroid injection versus systemic administration in the treatment of rheumatoid arthritis patients: a randomized double-blind controlled study. Clin Exp Rheumatol. 2009;27(2):214–21.
4. Micheal JW-P, Schüter-Brust KU, Eysel P. The epidemiology, etiology, diagnosis, and treatment of osteoarthritis of the knee. Dtsch Arztebl Int. 2010;107(9):152–62.
5. Cooper C, Rannou F, Richette P, Bruyere O, Al-Daghri N, Altman RD, Brandi ML, Basset SC, Herrero-Beaumont G, Migliore A, Pavelka K, Uebelhart D, Reginster JY. Use of intraarticular hyaluronic acid in the management of knee osteoarthritis in clinical practice. Arthritis Care Res. 2017;69(9):1287–96.
6. Douglas RJ. Aspiration and injection of the knee joint: approach portal. Knee Surg Relat Res. 2014;26(1):1–6.
7. Saunders S. Injection techniques in orthopaedic and sports medicine. 2nd ed. London: W B Saunders; 2002.
8. Borman P, Erdem HR. Kas Iskelet Sistemi Ağrılarında Terapötik Enjeksiyonlar. 2nd ed. Ankara: Nobel Tıp Kitabevleri; 2015.
9. Lundstrom ZT, Sytsma TT, Greenlund LS. Rethinking viscosupplementation: ultrasound-versus landmark-guided injection for knee osteoarthritis. Ultrasound Med. 2020;3(1):113–7.
10. Joshi AD, Gupta S, Shukla A, Gaur AK. Clinical and ultrasonographic correlations of pes anserinus bursitis. J Musculoskelet Res. 2021;24(3):2150004.
11. Rennie WJ, Saifuddin A. Pes anserine bursitis: incidence in symptomatic knees and clinical presentation. Skelet Radiol. 2005;34(7):395–8.
12. Lee JH, Kim KJ, Jeong YG, Lee NS, Han SY, Lee CG, Kim KY, Han SH. Pes anserinus and anserine bursa: anatomical study. Anat Cell Biol. 2014;47(2):127–31.
13. Sarikafalioglu B, Afsar SI, Yalbuzdag SA, Ustaomer K, Bayramoglu M. Comparison of the efficacy of physical therapy and corticosteroid injection in the treatment of pes anserine tendino-bursitis. J Phys Ther Sci. 2016;28(7):1993–7.
14. Cardone DA, Tallia AF. Diagnostic and therapeutic injection of the hip and knee. Am Fam Physician. 2003;67(10):2147–52.
15. Hemler DE, Ward WK, Karstetter KW, Bryant PM. Saphenous nerve entrapment caused by pes anserine bursitis mimicking stress fracture of the tibia. Arch Phys Med Rehabil. 1991;72(5):336–7.
16. Wilson-MacDonald J. Management and outcome of infective prepatellar bursitis. Postgrad Med J. 1987;63(744):851–3.
17. Sato M, Watari T. Housemaid's knee (prepatellar septic bursitis). Cureus. 2020;12(9):e10398.
18. McAfee JH, Smith DL. Olecranon and prepatellar bursitis. Diagnosis and treatment. West J Med. 1988;149(5):607–10.
19. Parker CH, Leggit JC. Novel treatment of prepatellar bursitis. Mil Med. 2018;183(11-12):e768–70.
20. Khodaee M. Common superficial bursitis. Am Fam Physician. 2017;95(4):224–31.
21. Baumbach SF, Lobo CM, Badyine I, Mutschler W, Kanz KG. Prepatellar and olecranon bursitis: literature review and development of a treatment algorithm. Arch Ortop Trauma Surg. 2014;134:359–70.
22. Huang YC, Wu WT, Chang KV. Ultrasound imaging for a male with anterior knee pain: prepatellar bursitis. J Med Ultrasound. 2021;29(4):300–1.
23. Diaz FJ, Gitto S, Sconfienza LM, Draghi F. Ultrasound of iliotibial band syndrome. J Ultrasound. 2020;23(3):379–85.
24. Khaund R, Flynn SH. Iliotibial band syndrome: a common source of knee pain. Am Fam Physician. 2005;71(8):1545–50.
25. Hyland S, Graefe SB, Varacallo M. Anatomy, bony pelvis and lower limb, iliotibial band (tract). In: StatPearls. Treasure Island: StatPearls Publishing; 2023. Available https://www.ncbi.nlm.nih.gov/books/NBK537097/.
26. Hadeed A, Tapscott DC. Iliotibial band friction syndrome. In: StatPearls. Treasure Island: StatPearls Publishing; 2023. Available https://www.ncbi.nlm.nih.gov/books/NBK542185/.
27. Noble CA. The treatment of iliotibial band friction syndrome. Br J Sports Med. 1979;13(2):51–4.
28. Bolia IK, Gammons P, Scholten DJ, Weber AE, Waterman BR. Operative versus nonoperative management of distal iliotibial band syndrome-where do we stand? A systematic review. Arthtosc Sports Med Rehabil. 2020;2(4):e399–415.
29. Gunter P, Schwellnus MP. Local corticosteroid injection in iliotibial band friction syndrome in runners: a randomised controlled trial. Br J Sports Med. 2004;38(3):269–72.
30. Beers A, Ryan M, Kasubuchi Z, Fraser S, Taunton JE. Effects of multi-modal physiotherapy, including hip abductor strengthening, in patients with iliotibial band friction syndrome. Physiother Can. 2008;60(2):180–8.
31. Hong JH, Kim JS. Diagnosis of iliotibial band friction syndrome and ultrasound guided steroid injection. Korean J Pain. 2013;26(4):387–91.

# Injections of Anatomical Regions and Diseases: Ankle and Foot


Tekin Kerem Ulku and Berhan Bayram


## 27.1 Ankle Joint

### 27.1.1 Anatomy and Biomechanics

Ankle joint is a hinged synovial joint formed by the articulation of the tibia fibula and talus. Medial border is formed by articular facet of medial malleolus; lateral border is formed by articular facet of fibula-lateral malleolus. The upper border of the joint forms the most distal articular surface of tibia. Ankle joint is surrounded by a thin broad and fibrous capsule anteriorly. Capsule thickened on lateral side and transversed on the posterior side. Joint is surrounded by strong ligamentous attachments both on medial and lateral side, which increase the inherent stability of the ankle. Two important bundle of neurovascular structures cross the joint from anterior and posteromedially which innervate most of the anatomic structures of the foot and ankle.

Static and dynamic stabilizers of ankle joint play an important role in biomechanics of the joint. Static stabilizers consist of bony structures and ligamentous structures, while dynamic stabilizers include tendons around the joint.

### 27.1.2 Pathologies and Indications for Injections

Pathologies in the ankle joint mainly consist of impingement syndromes (anteromedial, anterolateral, and posterior), intraarticular cartilage pathologies, rheumatoid conditions such as rheumatoid arthritis and crystalline arthropathies and osteoarthritis. Ankle impingement syndrome is common in football players, track and field athletes, and in ballets [1]. Injections can both be used to diagnose and treat the pathology. Evidence shows that fluoroscopic-guided injections can improve the symptoms of 84% of athletes with ankle impingement syndromes [2, 3]. Soft tissue impingement syndromes have a better response to intraarticular injections. However, surgery may be recommended in nonresponsive population. Surgical options may include open or arthroscopic procedures.

### 27.1.3 Appliances

Since ankle joint is a relatively superficial joint and easy to penetrate, usually a 25- to 27-G needle with 25–38 mm length can be used. Depending


T. K. Ulku (✉)
Acibadem Mehmet Ali Aydınlar University, Faculty of Medicine, Department of Orthopedics and Traumatology, Istanbul, Turkey

Acibadem Altunizade Hospital, Department of Orthopedics and Traumatology, Istanbul, Turkey

B. Bayram
Acibadem Mehmet Ali Aydınlar University, Faculty of Medicine, Department of Orthopedics and Traumatology, Istanbul, Turkey



© The Author(s), under exclusive license to Springer Nature Switzerland AG 2024
B. Kocaoglu et al. (eds.), *Musculoskeletal Injections Manual*,
https://doi.org/10.1007/978-3-031-52603-9_27

on the preference of the treating physician, thinner needles should be preferred for elite athletes to avoid injection site any discomfort or pain. Usually, a 3–5 ml of injectate is enough to avid overdistension of the joint.

### 27.1.4 Agents

Injectable agents include local anesthetics, steroid of different potencies, or combination of these two, viscosupplementations, and orthobiologic agents.

### 27.1.5 Injection Technique

Injections can be performed with freehand direct palpation technique, under fluoroscopy or ultrasound guidance. For anterior technique, the patient is positioned supine, and a small pillow can be placed under the hip joint to prevent external rotation of the lower extremity to get a better orientation. Injection site is prepared for sterile technique and to prevent contamination of the joint. Bony and tendinous landmarks can be marked with a sterile marker for freehand technique. Gentle dorsiflexion can help to prevent tightening of the anterior capsule and avoid extraarticular injection of the fluid (Fig. 27.1). For lateral-sided injections, extensor digitorum tendon is palpated, and the soft spot just lateral to the tendon is used about 2 cm above the level of lateral malleolus. For medial-sided injections tibialis anterior tendon is palpated, and the soft spot just medial to tibialis anterior tendon and 1.5 cm above the medial malleolus is used to penetrate the skin (Fig. 27.2). When performing combined local anesthetics and steroids for treatment purposes, one needle and two separate syringes, one containing local anesthetic agent and the other one containing steroids, can be used. After penetrating the joint capsule, first,

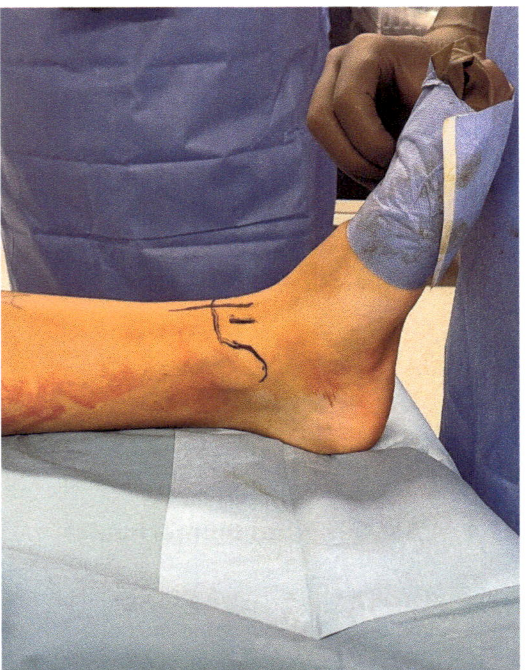

**Fig. 27.1** Patient is placed supine with a pillow under extremity to correct internal rotation and gentle dorsiflexion is applied to increase anterior joint space

syringe containing local anesthetic is infused, and low resistance of joint to fluid is confirmed without removing the needle. Second, syringe containing corticosteroid is infused after confirming the position of the needle in the joint.

### 27.1.6 Aftercare

After the injection the patient is observed for 15 min for development of any early allergic symptoms. Depending on the injected agent, if no contraindication is present, icing for 15 min may be helpful. After, the patient is informed about relative rest, low activity for the first 24–48 h, and warned about red flags for complications such as redness, swelling, or drainage from the joint.

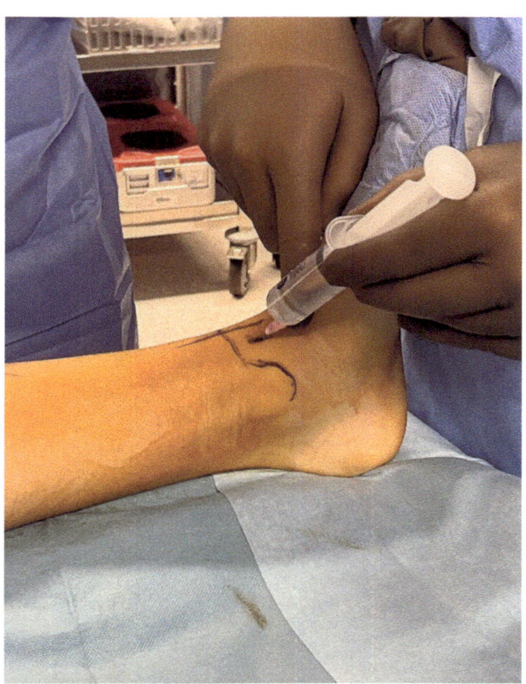

**Fig. 27.2** Tibialis anterior is palpated, and the soft spot which is just medial to the tendon and 1.5 cm above medial malleolus is marked for the entry point

## 27.2 Achilles Tendon Pathologies

### 27.2.1 Anatomy and Biomechanics

Achilles tendon is the thickest tendon in the human body, and it is the most common tendon formed by gastrocnemius, soleus, and plantaris muscle tendons in the posterior side of the ankle joint. After fusion of the tendon fibers, tendon wings about 90° causing soleus-oriented fibers to attach medially and gastrocnemius fibers laterally. Plantaris tendon attaches to most medial side of the tendon just proximal to attachment site. Tendon attaches to posterior surface of calcaneus and is supported by a deep retrocalcaneal bursa. Superficially tendon is separated from the skin by superficial calcaneal bursa [4].

Tendon serves as plantarflexor approximately 93% of all plantarflexor force that is generated. It is very important in ambulation and propulsion [5, 6]. Tendon receives its blood supply from the posterior tibial artery and peroneal artery. Its innervation is from the sural nerve and tibial nerve [7].

### 27.2.2 Pathologies and Indications for Injections

Achilles tendon is susceptible to injury with repetitive overuse. Mostly sports like running soccer and field sports can cause tendon pathologies [8]. Different parts of the tendon can be affected, and different pathologies can cause tendinopathies. Overuse of Achilles tendon injuries can be either insertional or non-insertional. Insertional tendinopathies are classified as retrocalcaneal bursitis, insertional tendinitis, or superficial calcaneal bursitis. Non-insertional tendinopathies are classified as midportion Achilles tendinopathy and Achilles paratendinitis [9]. Injections for the Achilles tendon can be used for both diagnostic and treatment purposes.

### 27.2.3 Appliances

Since the Achilles tendon is a relatively superficial structure and easy to penetrate, usually a 25- to 27-G needle with 25–38 mm length can be used. Depending on the preference of the treating physician, thinner needles should be preferred for elite athletes to avoid injection site any discomfort or pain. Usually, a 3–5 ml of injectate is enough to avid overdistension of the area. For some midportion tendinopathies, high-volume injections of about 10 ml can be used to create hydrodissection effect around the tendon and paratenon.

## 27.2.4 Agents

Injectable agents include local anesthetics, steroid of different potencies, or combination of these two, viscosupplementations, and orthobiologic agents. High-volume injections can be used for midportion Achilles tendinitis.

## 27.2.5 Injection Technique

Injections around the Achilles tendon can both be performed by freehand technique and under ultrasound guidance. Patient is placed prone and a supportive pillow is placed just anterior to ankle joint. Injection area is prepared sterile, and anatomical landmarks can be marked with a sterile marking pen. For midportion injections injection site is identified, and just ventral to the tendon, the needle is inserted oblique to midline and the injection is performed (Fig. 27.3).

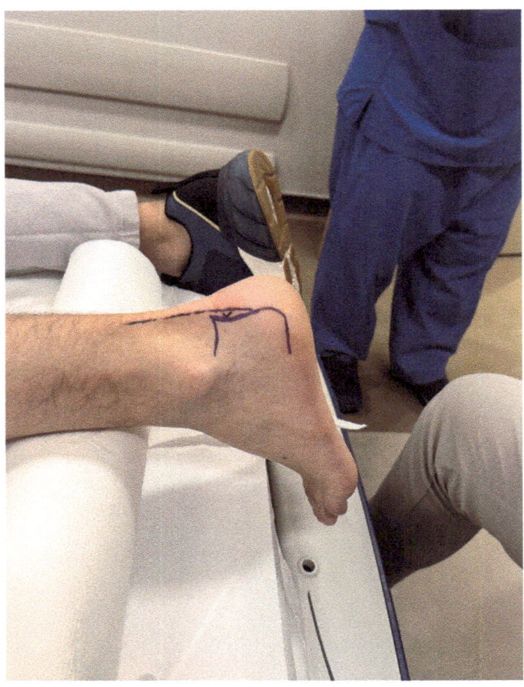

**Fig. 27.4** Skin marking of calcaneus and Achilles attachment site

For retrocalcaneal injections after skin marking, the needle is inserted just ventral to the tendon and dorsal to calcaneus in transverse direction. A lateral entry point is usually preferred (Figs. 27.4 and 27.5).

For superficial calcaneal bursa injections, the most painful zone in posterior bursal region is located, palpated, and skin marked. Then the needle is inserted perpendicular to the skin long enough to penetrate subcutaneous bursa without injecting the tendon itself, and injection is performed at the bursal tissue (Fig. 27.6).

## 27.2.6 Aftercare

After the injection patient is observed for 15 min for development of any early symptoms. Depending on the injected agent if no contraindication is present, icing for 15 min may be helpful. After, the patient is informed about relative rest, low activity for the first 24–48 h, and warned about red flags for complications such as redness, swelling, or drainage from the injection zone.

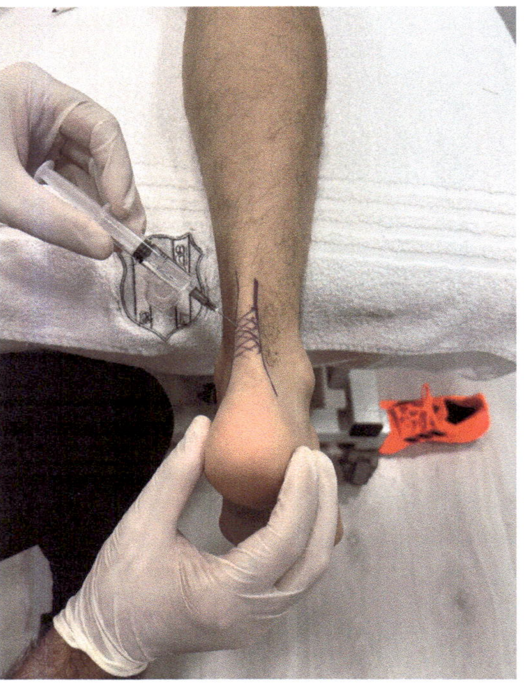

**Fig. 27.3** For midportion tendinopathies after detecting and marking the pathologic zone using the medial and ventral entry point, the needle is progressed

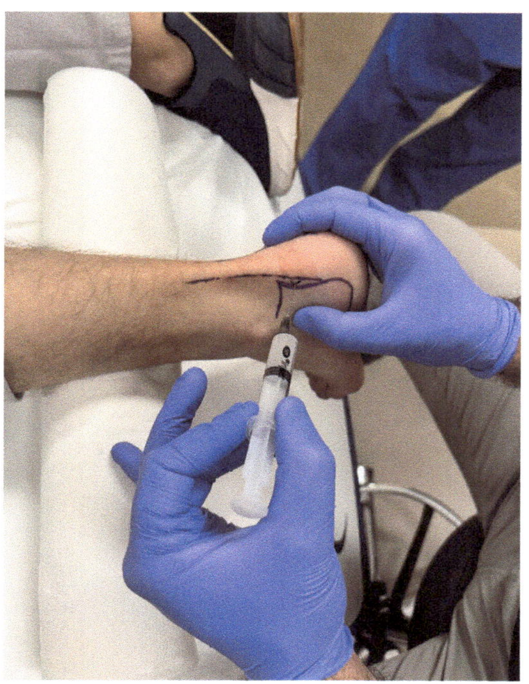

**Fig. 27.5** Penetration of the needle in a transverse fashion under the Achilles insertion

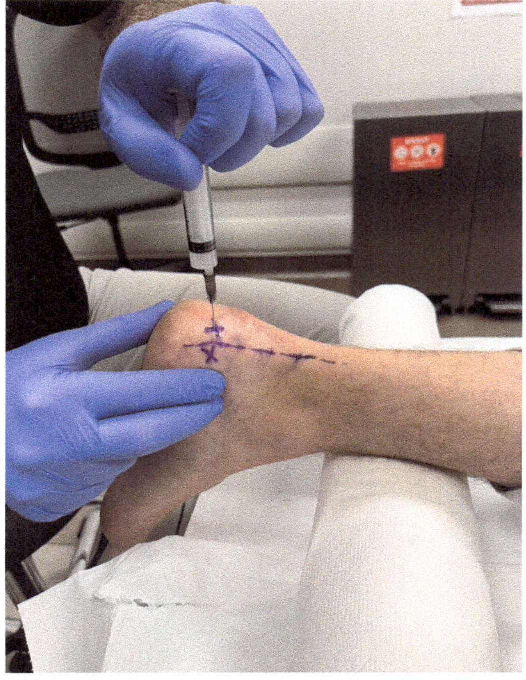

**Fig. 27.6** Superficial bursal injection a perpendicular penetration over the painful bursal site

## 27.3 Plantar Fasciopathy

### 27.3.1 Anatomy and Biomechanics

Plantar aponeurosis (facia) is the modification of the deep fascia starting from calcaneal tubercles to metatarsal head region [10]. It is a thick connective tissue that supports the arch and protects the underlying vital structures of the foot. It can be divided to three main distinct parts: medial part, central part, and lateral part. The central part is the thickest and largest part of the fascia.

Plantar fascia is an important structure biomechanically. It supports the plantar arches, acts as a shock absorber, distributes the plantar contact pressures during static and dynamic loading, acts as a site for muscular attachments, and prevents excessive dorsiflexion [11, 12].

### 27.3.2 Pathologies and Indications for Injections

Plantar fasciopathy is the most common cause of medial heel pain. It constitutes about 1% of all orthopedics outpatient clinics visits. The pathology is generally caused by repetitive microtrauma to plantar fascia and is closely related to obesity, age, gastrocnemius tightness, and planovalgus feet and is common among runners [13].

Pain generally worsens after weight bearing following long periods of resting. The term plantar fasciitis which refers to an inflammatory pathology has increasingly changed with "fasciopathy" or "Fasciosis." Studies showed that there is an absence of inflammatory cells, and the pathology is actually a chronic degenerative process [14].

### 27.3.3 Appliances

Usually, a 25- to 27-G needle with 25–38 mm length can be used. Usually, a 5 ml of injectate is enough to avid overdistension of the area.

### 27.3.4 Agents

Injectable agents include local anesthetics, steroid of different potencies, or a combination of these two and orthobiologic agents such as PRP.

### 27.3.5 Injection Technique

Injections around the plantar fascia can be performed by freehand technique or under ultrasound guidance. Two different techniques can be used.

#### 27.3.5.1 Supine Technique

The patient is placed supine; a supportive pillow beneath the unaffected hip is placed to help the medial side of the affected heel be exposed. Injection area is prepared sterile, and anatomical landmarks can be marked with a sterile marking pen. The needle is inserted from the medial side of the heel at junction of the lines from the posterior side of the medial malleolus and about 1 cm above the plantar surface until it reaches the medial side of the plantar facia. The injection is performed all through the medial side of plantar facia (Fig. 27.7).

#### 27.3.5.2 Prone Technique

The patient is placed prone, the knee is flexed, and the ankle is in neutral position. Injection area is prepared sterile, and anatomical landmarks can be marked with a sterile marking pen. The needle is inserted from the medial side of the heel on the plantar surface to the area that the patient feels most pain with direct palpation. When the resistance of facial tissue is felt with the needle, the injectate is given to the area (Fig. 27.8).

### 27.3.6 Aftercare

After the injection patient is observed for 15 min for development of any early symptoms. Depending on the injected agent, if no contraindication is present, icing for 15 min may be helpful. After, the patient is informed about relative rest, low activity for the first 24–48 h, and warned about red flags for complications such as redness, swelling, or drainage from the injection zone.

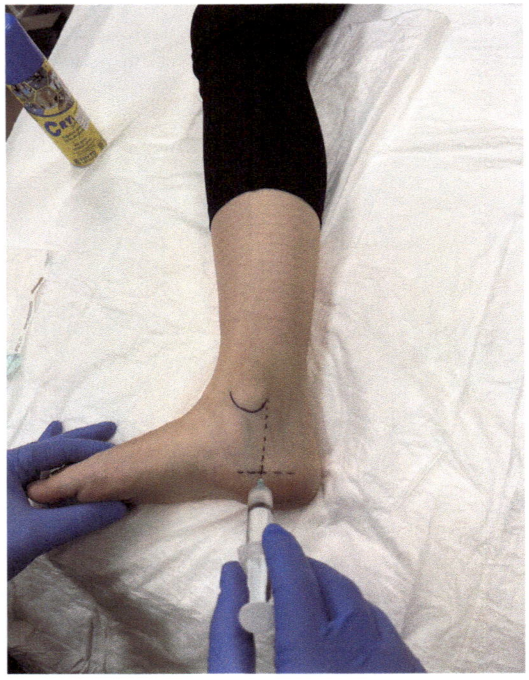

**Fig. 27.7** The needle is inserted from the medial side of the heel at the junction of the lines from posterior side of medial malleolus and about 1 cm above the plantar surface until it reaches the medial side of the plantar facia

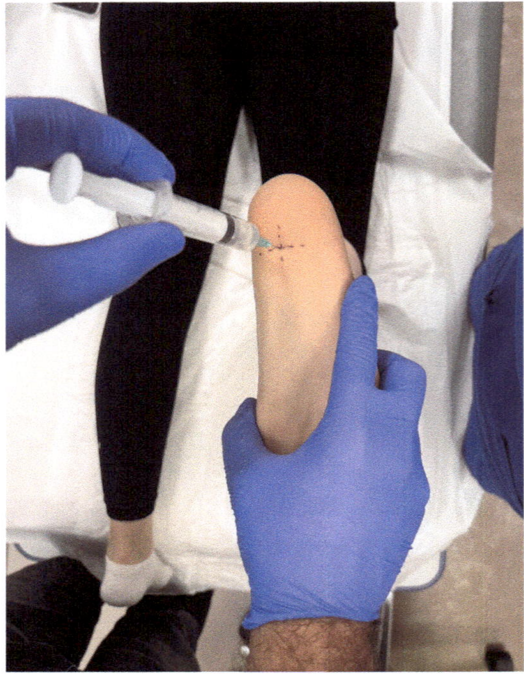

**Fig. 27.8** The patient is placed prone, the knee is flexed, and the ankle is in neutral position

## 27.4  Morton's Neuroma

### 27.4.1  Anatomy and Biomechanics

The tibial nerve forms the medial and lateral plantar nerves. Common plantar digital nerves are branches of medial and lateral plantar nerves. The first, second, and third common plantar nerve arises from the medial plantar nerve, while the fourth and fifth from the lateral plantar nerve. Each common plantar digital nerve then bifurcates into two proper plantar digital nerves supplying opposing sides of the adjacent toes. Morton's neuromas occur at the terminal bifurcation of the common plantar digital nerve. There are four intermetatarsal spaces between metatarsal bones. Each of these spaces is separated to two levels by the deep transverse ligament. Neurovascular bundle is plantar to this transverse ligament layer. Distally the nerve makes a sharp turn dorsally along the anterior free edge of transverse metatarsal ligament. The bursal tissue is present dorsal to transverse ligament. In the first and fourth space, the bursal tissue does not extend distal to transverse ligament. However, in the second and third space, the bursal tissue extends distally [15].

Exact pathomechanism is not clearly established for Morton's neuroma. Entrapment theory, chronic microtrauma theory, intermetatarsal bursitis theory, and ischemic theory are among the most commonly accepted ones.

### 27.4.2  Pathologies and Indications for Injections

Morton's neuroma is very common among the middle-aged female population increased due to high-heeled and/or narrow toe box shoes. Most of the neuromas occur in the third intermetatarsal space followed by second. Patients present with localized pain on the plantar surface of the foot at the level of metatarsal heads usually worsened after shoewear. Pain can radiate to the toes with paresthesia and tingling [16].

On physical examination, characteristic findings of a metatarsal space tenderness on compression, percussion, and a palpable click (Mulder's click) may be felt. Initial treatment modalities include shoe-wear modification, use of metatarsal lifting insoles, and nonsteroidal anti-inflammatories. However, in resistant cases, injection to relevant intermetatarsal space can be considered both for therapeutical and diagnostic purposes as last resort before surgical intervention.

### 27.4.3  Appliances

Usually, a 25- to 27-G needle with 25–38 mm length can be used. Usually, a 5 ml of injectate is enough to avid overdistension of the area.

### 27.4.4  Agents

Injectable agents include local anesthetics, steroid of different potencies, or a combination of these two and orthobiologic agents such as PRP.

### 27.4.5  Injection Technique

The patient is placed supine with a supportive pillow underneath the hip joint to avoid external rotation of the lower extremity and better orientation. Injection area is prepared and made sterile, and anatomical landmarks are marked with a sterile marking pen. Metatarsal heads are palpated from dorsal and plantar, and then the needle is inserted dorsally perpendicular to intermetatarsal space (Fig. 27.9). Until the tip of the needle is felt from the plantar side, the needle is advanced, and injectate is slowly injected moving the needle dorsally.

### 27.4.6  Aftercare

After the injection the patient is observed for 15 min for development of any early symptoms. Depending on the injected agent, if no contraindication is present, icing for 15 min may be helpful. After, the patient is informed about relative

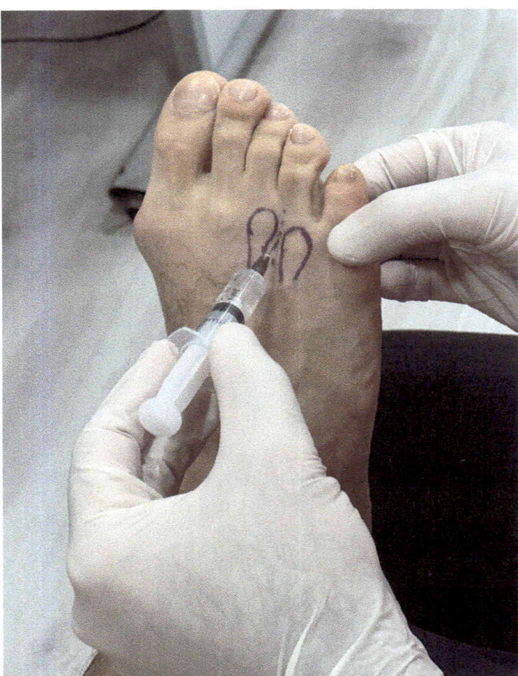

**Fig. 27.9** After metatarsal heads are marked with a sterile marker, the needle is inserted perpendicular and advance until felt underneath plantar surface

rest, low activity for the first 24–48 h, and warned about red flags for complications such as redness, swelling, or drainage from the injection zone.

## 27.5 Toe Joint: Sesamoids

### 27.5.1 Anatomy and Biomechanics

Toe joint is formed by the first metatarsal bone, two sesamoids, and proximal phalanx. It is a condyloid synovial joint supported by its synovial capsule around. Metatarsal head has two concave facets at the plantar surface, one for each sesamoid separated by a crista. The distal part of the first metatarsal is convex, and the proximal phalanx has a concave contour. There are two sesamoids in plantar surface: the medial sesamoid is in the medial head of the flexor hallucis brevis, and the lateral sesamoid is in the lateral head of the flexor hallucis brevis. Sesamoids are

joined by intersesamoid ligament. The intersesamoid ligament is the roof of a canal in which flexor hallucis longus tendon runs.

### 27.5.2 Pathologies and Indications for Injections

Pathologies in the toe joint mainly consists of impingement syndromes with dorsal osteophytes, intraarticular cartilage pathologies, and rheumatoid conditions such as rheumatoid arthritis and crystalline arthropathies and osteoarthritis. Sesamoid pathologies include sesamoiditis, fractures of sesamoid bones, and avascular necrosis. Injections can both be used to diagnose and treat the pathology. For toe joint pathologies like osteoarthritis, first-line treatment consists of the use of rocker bottom shoes, nonsteroidal anti-inflammatories, and activity modification [17]. In resistant cases, intraarticular injections can be preferred. For sesamoid pathologies insoles, shoe-wear modifications, and nonsteroidal anti-inflammatories can be used, and if these methods fail, injections to the sesamoidal area can be preferred.

### 27.5.3 Appliances

Since the toe joint and sesamoids are relatively superficial joints and easy to penetrate, usually a 25- to 27-G needle with 25–38 mm length can be used. Depending on the preference of the treating physician, thinner needles should be preferred for elite athletes to avoid injection site any discomfort or pain. Usually, a 1–3 ml of injectate is enough to avid overdistension of the joint.

### 27.5.4 Agents

Injectable agents include local anesthetics, steroid of different potencies, or a combination of these two and viscosupplementation and orthobiologic agents.

## 27.5.5 Injection Technique

### 27.5.5.1 MTP Joint Injection

The patient is placed supine with a supportive pillow underneath the hip joint to avoid external rotation of the lower extremity and better orientation. Injection area is prepared and made sterile, and anatomical landmarks is marked with a sterile marking pen. Extensor hallucis longus tendon is marked, and by gentle plantar-dorsal flexion, the head of the first metatarsal and base of the proximal phalanx is identified and marked (Fig. 27.10). By applying dorsiflexion, joint capsule on the dorsal aspect is loosened, and the needle is inserted from the medial side in an oblique fashion (Fig. 27.11). After penetrating the joint capsule, the first syringe containing local anesthetic is infused, and low resistance of joint to fluid is confirmed; then without removing the needle, the second syringe containing corticosteroid is infused after confirming the position of the needle in the joint.

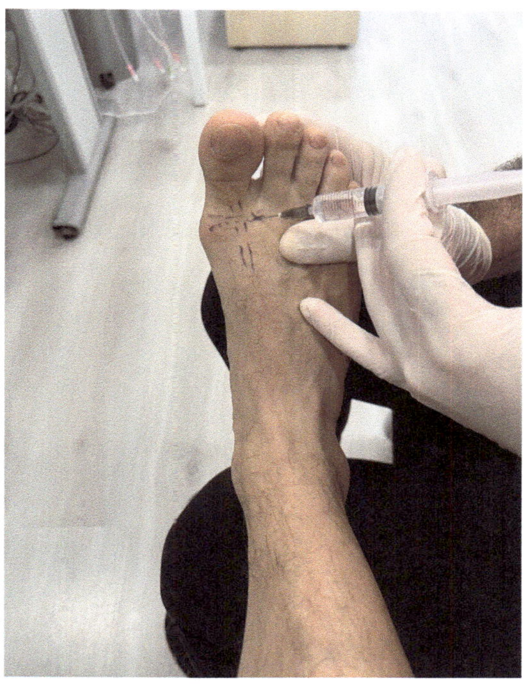

**Fig. 27.11** Medial-sided joint penetration showing the entry point

### 27.5.5.2 Sesamoid Injection

The patient is placed supine, and the injection site is prepared sterile with anatomical landmarks drawn by sterile marking pen. Medial and lateral sesamoids, the plantar surface of the metatarsal head, and the tendon are identified (Fig. 27.12). The (1) MTP joint is plantarflexed, and the space between metatarsal and sesamoids is widened. From the proximal site of medial sesamoid, a needle is penetrated in an oblique fashion and advanced distally under the metatarsal area (Fig. 27.13). After penetrating the joint capsule, the first syringe containing local anesthetic is infused, and low resistance of joint to fluid is confirmed; then without removing the needle, the second syringe containing corticosteroid is infused after confirming the position of the needle in the joint.

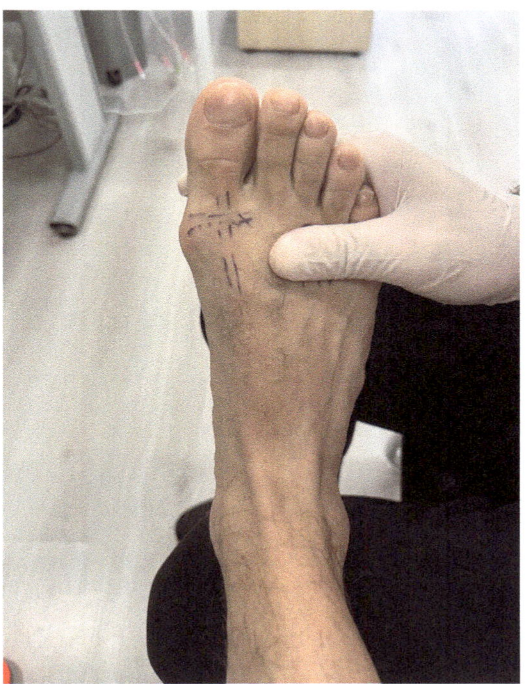

**Fig. 27.10** Skin marking of the EHL tendon and metatarsal and phalangeal surfaces is easier with gentle motion of the MTP joint

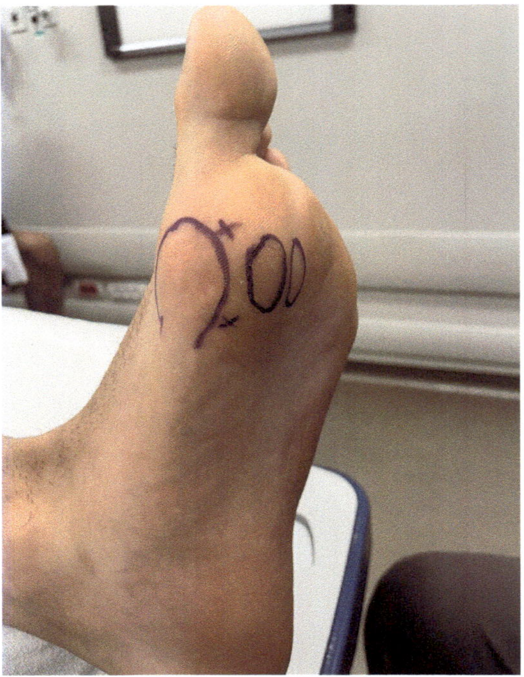

**Fig. 27.12** Metatarsal head and sesamoids are marked after palpating the structures

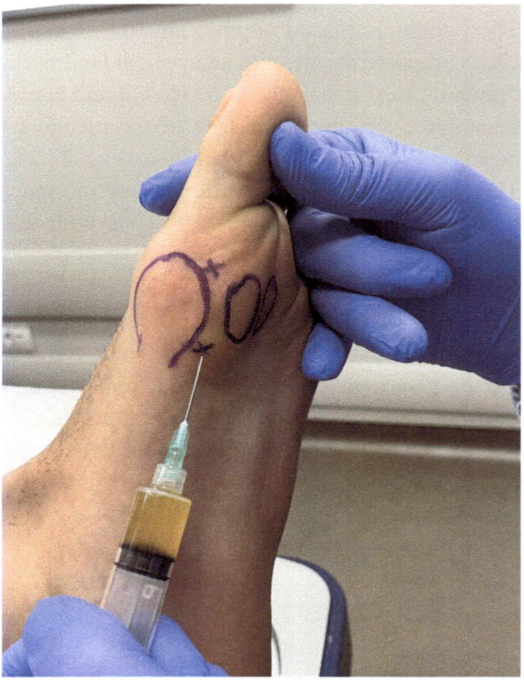

**Fig. 27.13** With gentle plantarflexion the needle is inserted in an oblique fashion penetrating the metatarsal-sesamoid joint

## 27.5.6 Aftercare

After the injection the patient is observed for 15 min for development of any early symptoms. Depending on the injected agent, if no contraindication is present, icing for 15 min may be helpful. After, the patient is informed about relative rest, low activity for the first 24–48 h, and warned about red flags for complications such as redness, swelling, or drainage from the injection zone.

## References


1. Kudaş S, Dönmez G, Işık Ç, Çelebi M, Çay N, Bozkurt M. Posterior ankle impingement syndrome in football players: case series of 26 elite athletes. Acta Orthop Traumatol Turc. 2016;50:649–54.
2. Mouhsine E, Crevoisier X, Leyvraz P, Akiki A, Dutoit M, Garofalo R. Post-traumatic overload or acute syndrome of the os trigonum: a possible cause of posterior ankle impingement. Knee Surg Surg Traumatol Arthrosc. 2004;12:250–3.
3. Nepple JJ, Matava MJ. Soft tissue injections in the athlete. Sports Health. 2009;1:396–404.
4. Moore KL, Agur AMR, Dalley AF. Essential clinical anatomy. 4th ed. Baltimore: Lippincott Williams & Wilkins; 2011.
5. Wong M, Kiel J. Anatomy, bony pelvis and lower limb, Achilles tendon. Treasure Island: StatPearls Publishing; 2020.
6. Williams SK, Brage M. Heel pain-plantar fasciitis and Achilles enthesopathy. Clin Sports Med. 2004;23(1):123.
7. Abate M, Salini V. Mid-portion Achilles tendinopathy in runners with metabolic disorders. Eur J Orthop Surg Traumatol. 2019;29(3):697–703.
8. Young JS, Maffulli N. Etiology and epidemiology of Achilles tendon problems. Achilles Tendon. 2007;15:39–49.
9. van Dijk CN, van Sterkenburg MN, Wiegerinck JI, Karlsson J, Maffulli N. Terminology for Achilles tendon related disorders. Knee Surg Sports Traumatol Arthrosc. 2011;19(5):835–41.
10. Lemont H, Ammirati KM, Usen N. Plantar fasciitis: a degenerative process (fasciosis) without inflammation. J Am Podiatr Med Assoc. 2003;93(3):234–7.
11. Gefen A, Megido-Ravid M, Itzchak Y. In vivo biomechanical behavior of the human heel pad during the stance phase of gait. J Biomech. 2001;34(12):1661–5.
12. Chen DW, Li B, Aubeeluck A, Yang YF, Huang YG, Zhou JQ, Yu GR. Anatomy and biomechanical properties of the plantar aponeurosis: a cadaveric study. PLoS One. 2014;9(1):e84347.
13. Riddle DL, Schappert SM. Volume of ambulatory care visits and patterns of care for patients diagnosed with plantar fasciitis: a national study of medical doctors. Foot Ankle Int. 2004;25:303–10.

14. Rhim HC, Kwon J, Park J, Borg-Stein J, Tenforde AS. A systematic review of systematic reviews on the epidemiology, evaluation, and treatment of plantar fasciitis. Life. 2021;11(12):1287.
15. Mak MS, Chowdhury R, Johnson R. Morton's neuroma: review of anatomy, pathomechanism, and imaging. Clin Radiol. 2021;76(3):235.e15–23.
16. Hassouna H, Singh D. Morton's metatarsalgia: pathogenesis, aetiology and current management. Acta Orthop Belg. 2005;71(6):646–55.
17. Andrews NA, Ray J, Dib A, Harrelson WM, Khurana A, Singh MS, Shah A. Diagnosis and conservative management of great toe pathologies: a review. Postgrad Med. 2021;133(4):409–20. https://doi.org/10.1080/00325481.2021.1895587. Epub 2021 Apr 4.